Springer Aerospace Technology

The series explores the technology and the science related to the aircraft and spacecraft including concept, design, assembly, control and maintenance. The topics cover aircraft, missiles, space vehicles, aircraft engines and propulsion units. The volumes of the series present the fundamentals, the applications and the advances in all the fields related to aerospace engineering, including:

- structural analysis,
- aerodynamics,
- aeroelasticity,
- aeroacoustics,
- flight mechanics and dynamics
- orbital maneuvers,
- avionics,
- systems design,
- materials technology,
- launch technology,
- payload and satellite technology,
- space industry, medicine and biology.

The series' scope includes monographs, professional books, advanced textbooks, as well as selected contributions from specialized conferences and workshops.

The volumes of the series are single-blind peer-reviewed.

To submit a proposal or request further information, please contact: Mr. Pierpaolo Riva at pierpaolo.riva@springer.com (Europe and Americas) Mr. Mengchu Huang at mengchu.huang@springer.com (China)

The series is indexed in Scopus and Compendex

Fritz Boden
Editor

Numerical Methods and Experimental Means for the Integration of Future Propulsion Systems

A Resumé of the Achievements Made within the CleanSky 2 XDC Demonstrator

DLR

Springer

Editor
Fritz Boden
Institute of Aerodynamics and Flow Technology
German Aerospace Center
Göttingen, Niedersachsen, Germany

ISSN 1869-1730 ISSN 1869-1749 (electronic)
Springer Aerospace Technology
ISBN 978-3-031-97724-4 ISBN 978-3-031-97725-1 (eBook)
https://doi.org/10.1007/978-3-031-97725-1

This Springer imprint is published by the registered company Springer Nature Switzerland AG
The registered company address is: Gewerbestrasse 11, 6330 Cham, Switzerland

Preface

It can be very dangerous to see things from somebody else's point of view without the proper training.

—Douglas Adams, Mostly Harmless

This quote provides a fitting introduction to this book, which serves as a guide for non-experts seeking to apply advanced numerical methods and experimental techniques to address complex questions, such as the integration of novel propulsion systems into aircraft design. The book offers an overview of the methods developed and demonstrated within the Cross Capability Demonstrator XDC, a key component of the Clean Sky 2 Innovative Aircraft Demonstration Platform "Large Passenger Aircraft" (LPA).

In this Preface, I would like to extend my gratitude to the scientists, engineers, and technicians who contributed to the XDC's success and made this book possible. I also want to thank the LPA coordination team and the Clean Sky 2 Joint Undertaking for their support, which enabled us to conduct our research and publish this book.

Although the manuscript took a considerable amount of time to complete, and the methods are still evolving, this book provides a comprehensive summary of the possibilities offered by the XDC methods and serves as a handy guide for those who are not yet familiar with their potential.

I hope you enjoy reading this book and find valuable solutions to the challenges you face in your design and development work. For those who were part of the XDC, I hope this book will serve as a reminder of the work we accomplished together.

Göttingen, Germany
April 2025

Fritz Boden
DLR Research Scientist

Contents

Experimental Means

Introduction

The XDC – An Overview on the CleanSky 2 Demonstrator on Common Numerical Methods and Experimental Means

Fritz Boden, Ulrich Herrmann, and Antje Stamm

Abstract Within this introductory chapter the frame work of CleanSky 2 and especially the Cross Capability Demonstrator (XDC) will be presented in order to give an overview of the research program and provide a classification of the work described below.

Keywords CleanSky 2 · XDC · Cross capability demonstrator · Numerical methods · Experimental means

1 Introduction

Reducing deleterious gaseous emission and noise is one of the most important goals being pursued in all sectors of transport. Road transport, shipping and aviation are major contributors to climate change [11]. Although aviation in absolute emission figures lies behind that of road transport, it has been much in focus as it is likely to undergo a significant future growth, it is quite difficult to decarbonize and shows a uniquely high visibility compared to the other sectors. Therefore the Advisory Council for Aviation Research and Innovation in Europe (ACARE) has set as its fundamental objective the achievement of fully climate neutral air mobility by 2050, which means that after this date emissions from this sector will no longer contribute to climate change [12, 13].

F. Boden (✉)
German Aerospace Center (DLR), Institute of Aerodynamics and Flow Technology, Göttingen, Germany
e-mail: Fritz.Boden@dlr.de

U. Herrmann
German Aerospace Center (DLR), Program Directorate Aeronautics, Braunschweig, Germany

A. Stamm
German Aerospace Center (DLR), Project Administration and Controlling, Hamburg, Germany

F. Boden (ed.), *Numerical Methods and Experimental Means for the Integration of Future Propulsion Systems*, Springer Aerospace Technology,
https://doi.org/10.1007/978-3-031-97725-1_1

1.1 Clean Sky

The development towards climate-neutrality is a long term action, requiring well-coordinated long term research activity from the aviation industry, European research centers and universities. For this reason, a public-private partnership – the Clean Sky Joint Undertaking (CSJU) [1] – was founded in 2008 in Europe to coordinate and fund research activities for the implementation of innovative and environmentally friendly technologies for all segments of civil aviation; namely, large commercial aircraft, regional aircraft and rotorcraft, as well as the associated technologies for engines, systems and the life cycle of materials. The CSJU was formed by the European Commission (EC) acting as public representative of industrial leaders and private members, like for instance the DLR. The initial Clean Sky program (CS) was set up for a period of 10 years up to end of December 2017.

1.2 Clean Sky 2

Based on the positive result of an interim evaluation of the CS program, its successor Clean Sky 2 (CS2) was launched in 2014 [2] in order to proceed with the developments initiated in CS and to contribute to the more ambitious ACARE Flightpath 2050 [12] goals. Apart from the goals of developing sustainable alternative fuels and of designing air vehicles to be recyclable, a further target is to reduce emissions of CO_2 by 75%, NOx by 90% and aircraft noise by 65%, when compared with values for typical in-service aircraft in 2000.

To be able to achieve these goals, CS2 was set up with a larger community, a higher funding of 1.7 Billion Euro and an expanded range of research areas. As shown in Fig. 1, CS2 was structured into three Innovative Aircraft Demonstration Platforms (IADP), integrating and demonstrating technologies and system innovations at full aircraft level, three Integrated Technology Demonstrators (ITD), developing and integrating new technologies, and three Transverse Activities (TA), focussing on technology challenges across the IADPs and ITDs, including one Technology Evaluator to assess the technological achievements after their implementation into a fleet of operated aircraft.

The IADPs and the ITDs were further structured. The Large Passenger Aircraft (LPA) IADP was divided into three 'Platforms':

- **Platform 1 "Advanced Engine and Aircraft Configurations"** – focussing on the development, assessment and integration of innovative propulsion systems, like Open Rotor (OR), ultra-high bypass ratio (UHBR) [6] engines and Boundary Layer Ingestion (BLI) [24],
- **Platform 2 "Innovative Physical Integration Cabin – System – Structure"** – targeting advanced fuselage structural concepts, the next generation of large passenger cabins and cargo, as well as component demonstration,

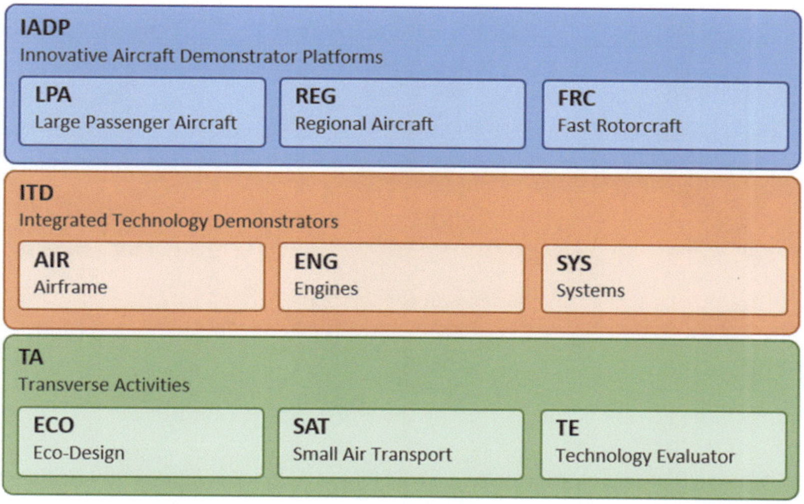

Fig. 1 Clean Sky 2 overall program structure – IADPs, ITDs and TAs

- **Platform 3 "Next Generation Aircraft Systems, Cockpit and Avionics"** – related to the development and demonstration of next generation cockpits and navigation.

A major part of Platform 1 was dedicated to the development of more efficient, advanced propulsion systems and their integration into the airframe of a passenger aircraft. Within the CS2 project ADEC (Advanced Engine and Aircraft Configurations), the focus had initially been on the concept of a Counter Rotating Open Rotor (CROR) [7], whose development had already commenced in CS. Later, after a project internal CROR eco-viability evaluation gate in July 2017, it was decided to perform a re-orientation of the project and to focus on UHBR. Examples of OR, UHBR and BLI engine types are shown in the box in the center of Fig. 2.

1.3 The Cross Capability Demonstrator XDC

One needs special tools in order to be able to develop the propulsion concepts mentioned above and to optimize their integration into an airframe. On the one hand, powerful numerical methods to predict the performance of the novel propulsion systems and to optimize their benefits are needed. This would imply at least similar simulation accuracies for the disciplines aerodynamics, acoustics and structural (e.g. aeroelastic) effects (see the left box in Fig. 2). On the other hand, experimental methods for these disciplines are required for the validation and assessment of the realized technology bricks as well as of the whole propulsion systems (see the right box in Fig. 2).

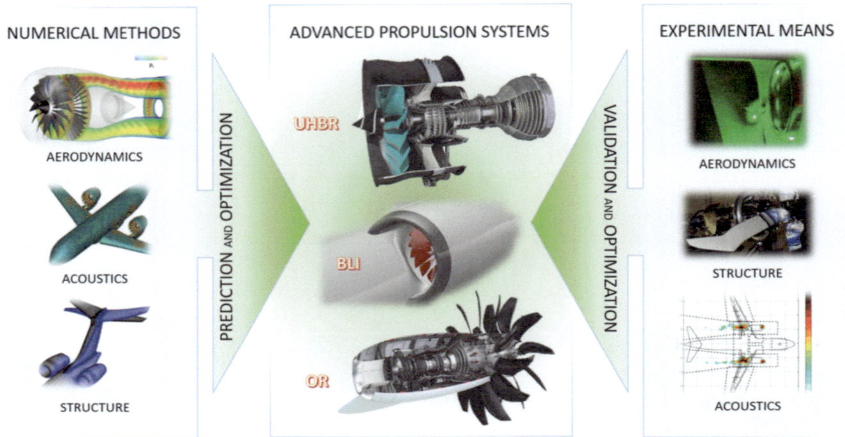

Fig. 2 Disciplines of numerical methods (left box) for prediction and optimization and disciplines of experimental means (right box) for validation and optimization of the benefits of propulsion technologies proposed to reach the Flightpath 2050 goals (box in the middle – example UHBR engine (top [7]), example BLI engine (middle [22]) and example open rotor engine (bottom [7]))

Since most of the required methodologies at the beginning of CS2 had been applied by research institutes and had not been made available for universal industrial application, the so-called Cross Capability Demonstrator (XDC) was established in 2019 as part of the LPA re-orientation. The program was led and mainly driven by DLR, although Airbus, ONERA and NLR had already contributed to this special CS2 demonstration activity.

The XDC was divided into two streams: Common Numerical Methods and Common Experimental Means; its scope was the development and demonstration of these means, enhancing a faster integration of advanced propulsion systems and assessing and verifying their respective benefits (e.g. with respect to fuel burn and noise).

The major XDC objectives were thus to establish design processes and tools applicable to all types of propulsion systems. XDC ultimately provided high-fidelity CFD tools, further developed prediction tools for noise and aeroelastics as well as advanced testing tools for all mentioned disciplines. The following Sects. 2 and 3 will give a short overview on the topics addressed.

2 Common Numerical Methods

Numerical methods play a crucial role in the development process because they predict and simulate the behavior of the aircraft design at an early stage whilst conceptual ideas are still being carried over to design details. Furthermore, they help to optimize design without the need for real and expensive hardware. DLR has long-standing experience in the development and application of numerical methods to different fields of application. In the XDC DLR developed and demonstrated the performance of numerical methods in various disciplines (aerodynamics, aeroacoustics and aeroelastic), as well as overall configuration design tools. Furthermore topics regarding material science and shielding of the fuselage against impact of engine debris have been addressed by DLR. The individual tasks are briefly outlined below.

2.1 *Numerical Simulations for Configuration Studies*

To perform an assessment of the impact and benefit of novel propulsion systems on aircraft performance, it is important to understand the underlying physical phenomena involved by performing numerical simulations, both CFD and CAA, with the focus placed on their integration into the configuration. For such investigations, DLR researchers have substantially further developed their inhouse CFD Code TAU [15], a numerical code to solve the unsteady Reynolds Averaged Navier–Stokes (uRANS) equations. Computational AeroAcoustics (CAA) simulations were conducted using the DLR CAA code PIANO [10].

One field of application was the generic ORAS engine, being an OR engine with a stator made of swirl recovery vanes (SRV) placed behind the rotor stage [25]. First studies of the isolated engine showed that changes in angles of attack were found to affect notably the aerodynamic performance of the SRVs, this being due to local separations. A detailed aerodynamic analysis highlighted these adverse impacts on performance. It was shown that SRV blade setting angle adjustments have to be made to mitigate these unfavorable flow conditions, such as flow separation. Furthermore, the occurrence of so-called 1P loads produced by the blades due to the variation of blade loading across the azimuth were found to achieve non-negligible magnitudes at high incidence angles or high flight speeds, and thus need to be considered during structural integration. Furthermore, these in-plane forces can have a strong impact on the handling qualities and controllability of the aircraft. The performed analysis led to the suggestion of installation of OR engines with opposing senses of rotation, similar to the Airbus A400M propeller aircraft on the right and left wings yielding an aerodynamically symmetric aircraft with ensured good overall handling qualities. Finally, as a result of the high-fidelity uRANS simulation of the propulsor slipstream and its interaction with the wing and flap (see Fig. 3 from [25]), it was found that the SRV tip vortices play an important role in potentially improving low-speed, high-lift performance of the wing and flap, similar to the case of a nacelle strake vortex on a turbofan powered aircraft.

Other examples using numerical simulations considered the integration of a BLI propulsion system to an airframe. Parametric studies on engine sizing, engine mass flow/fan pressure ratios as well as design refinements of an aircraft configuration with a tailcone-mounted BLI configuration were performed, with the focus on cruise flight conditions. Studies of the rear-engine aircraft concept indicated a performance

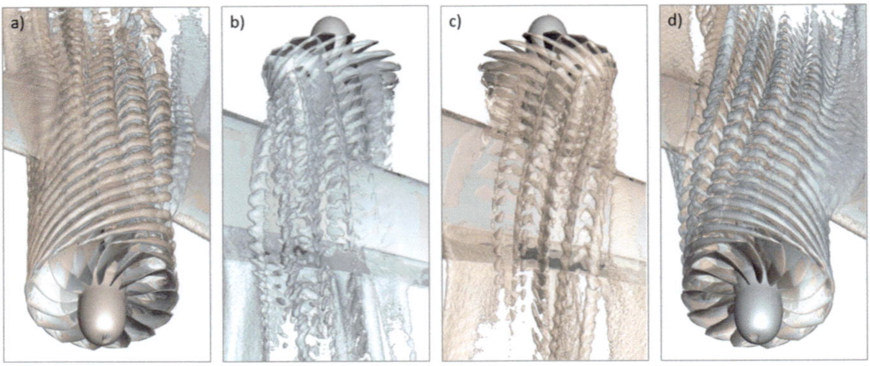

Fig. 3 Rotor and SRV tip vortex visualization using the x-component of the vorticity (**a** – Rotor and SRV Tip vortex interactions (IBUP case), **b** – SRV vortices over upper wing surface (IBUP case), **c** – SRV vortices over upper wing surface (IBDN case), **d** – Rotor and SRV Tip vortex interactions (IBDN case), positive sense of rotation = shaded pink-orange, negative sense of rotation = shaded grey-blue (figures from [25]))

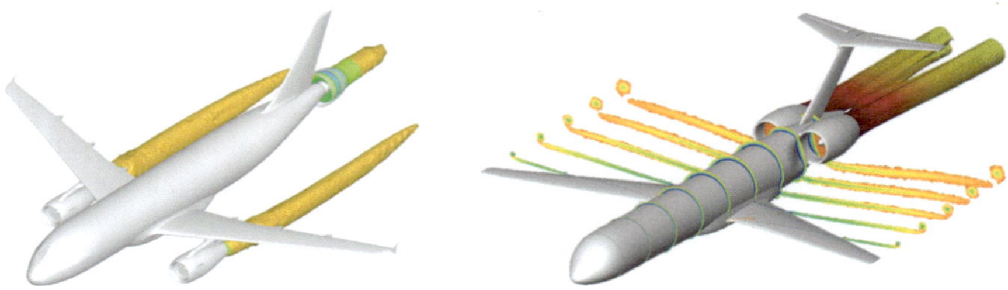

Fig. 4 Two examples of BLI configurations investigated by DLR [27]

benefit of BLI against classic under-wing mounted engines of about ∼5% [27, 29]; two example geometries are shown in Fig. 4.

Last, but not least, the integration of an UHBR engine to the airframe was examined using numerical methods. The resulting size of the nacelle for an UHBR engine raises challenges due to increasing engine-airframe interactions; hence DLR investigated a generic UHBR engine design, where the focus was placed on the interaction between UHBR and a high-lift system [26].

Aeroacoustics have been investigated for different UHBR configurations applying the CAA Code PIANO. One XDC activity was to predict the fan mode shielded by the aircraft airframe for an UHBR installed above the wing. For that purpose an acoustic model for the installation noise effect of the disturbed fan was developed and implemented in PIANO, with the consideration of installation noise, non-uniform flow effects and sound propagation. The model was applied to the SFB 880 aircraft [23] as shown in Fig. 5, which shows the acoustic pressure signature of the UHBR on wing and fuselage.

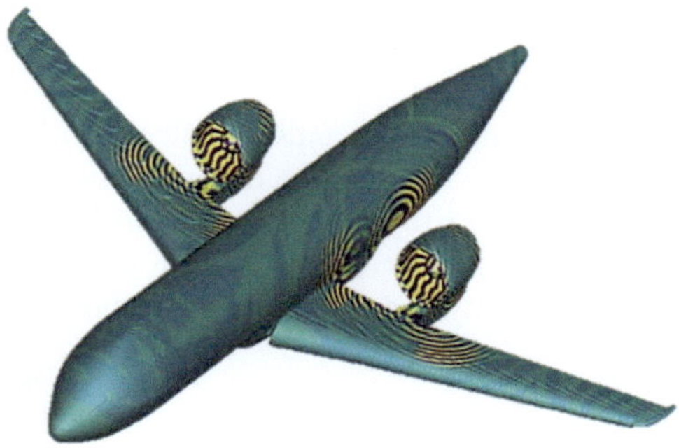

Fig. 5 CAA simulated acoustic pressure signature of UHBR on wing and fuselage [4]

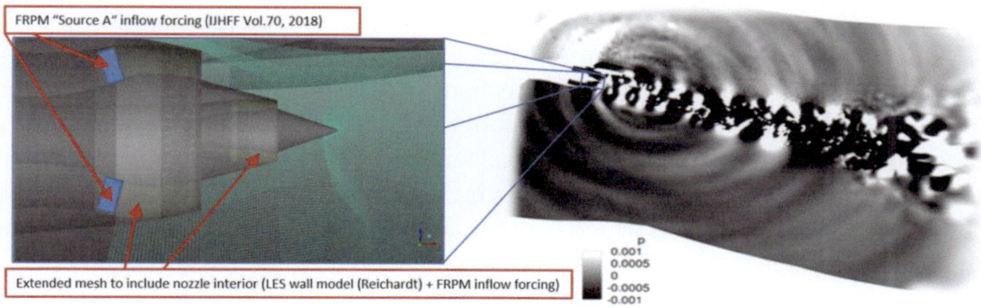

Fig. 6 View through the UHBR grid with indicated inflow-forcing locations (left) and resulting acoustic pressure fluctuations due to jet-flat interaction (right) [4]

Another important issue for the integration of large UHBR engines, especially for a retro-fit, is the impact of jet-flap interference noise. Figure 6 (on the left) shows a detail of the UHBR numerical grid, including the location of the applied inflow forcing; the resulting simulated noise field is shown on the right.

2.2 *Prediction of Fan Noise*

Complex numerical simulations are required not only for analysis, but also for fast prediction tools in the initial design phase. In order to improve the customary manner of predicting broadband noise (presently relying on empirical correlations from the 1970's), DLR has developed a physics-based aerodynamic model suited for transonic rotor blades; details of this model are given in this book in the chapter "Fast Prediction of Fan Broadband Noise with an Analytical Aerodynamic Model Suited for Transonic Rotor Blades".

2.3 Stability Analysis

A further study explored the impact of rotational effects on boundary-layer instabilities on the surface of an open rotor blade. In a rotating reference frame, terms corresponding to Coriolis and centrifugal forces appear in the linearized disturbance equations. However, the effect of these rotational terms on the stability of boundary layers has so far been only partially understood. The questions to be addressed at the DLR study were:

- Do rotational terms have to be considered in an eN-transition-prediction framework?
- Does rotation add new types of instabilities, or does it only alter transition mechanisms already existing in non-rotating cases?

Using as an example the Z49 rotor blade geometry [28], the impact of these occurring rotational terms on the transition mechanisms was then investigated by means of linear stability computations. The effect of rotation was studied by comparing sets of instability computations where rotational terms had been neglected and accounted for. The transition potential was estimated via the empirical logarithmic amplification factor, the so-called N-factor, which is meant to indicate the occurrence of transition if it is above a value of $N = 7.8$ [16].

It was concluded that there is no new type of instability linked to the rotational terms and that the rotation only changes the instability mechanisms that are already present on the rotor blade. Tollmien-Schlichting (T-S) waves seem to be the dominant instability mechanism on the investigated rotor. If the rotational terms are taken into account, their effect on the T-S growth rates is negligible (see Fig. 7 left). Cross-flow instabilities (CFI) are found to be destabilized (see Fig. 7 right), even though they do not directly trigger transition on the blade. The conclusion: rotation terms can be neglected if T-S waves dominate, whereas, if CFI are the dominant transition mechanism, the terms should be included in order to predict transition at its correct location.

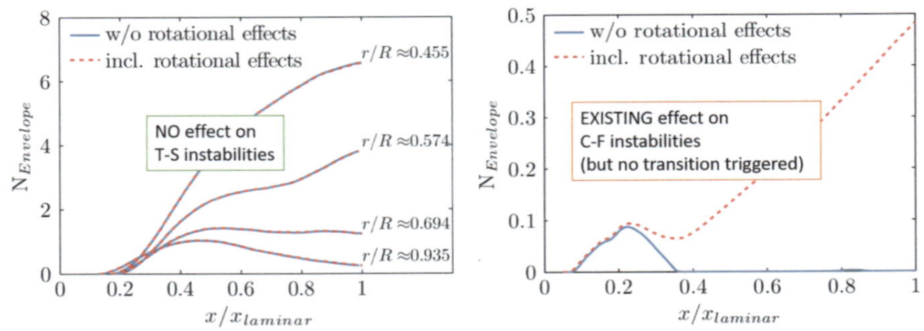

Fig. 7 Impact of the rotational effects on the stream-wise evolution of the N-factor envelope curve for Tollmien-Schlichting (T-S) instabilities (left) and Cross-Flow (C-F) instabilities (right) [4]

2.4 Aeroelastics

DLR extended the capabilities of its inhouse tool MONA [19] and generated a parametric aircraft model to realistically represent the global structural and aeroelastic dynamics of an aircraft. Within the XDC, the focus was on the extension of the modelling capabilities of MONA in order to generate fuselage models of sufficient detail for aeroelastic and vibroacoustic analyses. Those models were then passed on to project partners for their respective analyses. The chapter "Parametric Structural Modelling for Aeroelastic and Aeroacoustic Analyses of an Aircraft Fuselage" of this book describes such a model and provides a demonstration of such load analyses.

One important task for such a parametric aeroelastic model is the flutter assessment of a passenger aircraft configuration featuring UHBR engines in comparision with a common configuration. Results of such an assessment performed in the framework of the XDC are presented in the chapter "Flutter Assessment of a Passenger Aircraft with Standard and UHBR Engines" of this book.

2.5 Fuselage Shielding

In order to reduce the environmental impact of aviation, both from fuel burning and consumption of material resources, the trend has gone to advanced light weight composite materials. For safety reasons such optimized materials still have to withstand the same loads as with previously adopted materials. One important property in this case is the shielding capability that becomes all the more important if propulsion concepts such as the open rotor are implemented. For this reason, this topic was also addressed by DLR within CS2 by investigating shielding of aircraft rear ends against debris from, for example, Open Rotor propulsion. The task included the characterization of potential shielding materials and the development of simulation methodologies for the assessment of their impact behaviour.

2.6 Multi-disciplinary Design and Optimization

With the available powerful numerical simulation and prediction tools, aircraft design considering all disciplines can be carried out. By way of example, the design and integration of an UHBR propulsion system was conducted; results can be found in the chapter "Design and Integration of UHBR Propulsion Systems—An Overview of DLR Activities Within the Clean Sky 2 ASPIRE and SA^{2}FIR Projects".

Part of the design of the UHBR engine was the structural design of the fan. The overall UHBR fan design process is described in more detail in the chapter "Structural Design of a Generic High Bypass Ratio Fan".

Last but not least the multi-disciplinary design of a low noise wing-body aircraft is presented in the chapter "Hybrid Wing Body as a Low-Noise Aircraft from a Multidisciplinary Point of View".

3 Common Experimental Means

In addition to the development and demonstration of numerical methods, advanced measurement technologies were also improved as part of the XDC. These experimental means are important in order to evaluate the benefits of innovative multidisciplinary technologies. The intension is to provide common experimental means that can be easily applied to measurements as soon as first demonstration prototypes either of parts of the propulsion technologies or the whole demonstrator are available. Their application hence extends from laboratory right up to in-flight validation tests.

3.1 Acoustic Measurements

One important topic addressed by the XDC was the improvement of the sound source localization using far-field microphone measurements and the SODIX (SOurce DIrectivity modeling in the cross-spectral matriX) tool [14]. This advanced source localization technique developed by DLR is able to determine the individual contributions of aircraft engine components to the overall radiated noise and can thus help in the evaluation of noise reduction technologies applied to the single components of an engine or to the overall aircraft. The chapter "Improving Source Localization with SODIX for Far-Field Measurements in Static Engine Noise Tests" gives an example demonstration of this tool applied to far-field microphone data from a static noise test of a short-cowl turbofan engine.

Apart from an improvement in the acoustic data post-processing methods, the DLR has also started to implement MEMS microphones in aeroacoustic measurements. MEMS microphones are small and cheap. Nevertheless, in order to use them for advanced acoustic measurements, they have to be treated differently compared to classical microphones. The chapter "MEMS Microphones for Wind Tunnel and Flight Tests" provides a good overview of the preparation of MEMS sensors for in-flight aeroacoustic measurements.

3.2 Measurement Techniques for Structures

Measuring structural parameters of existing and future aircraft configurations have also been carried out within the XDC. One of these was related to the vibroacoustic characterization of airframe structures. This was done on an A320 wing

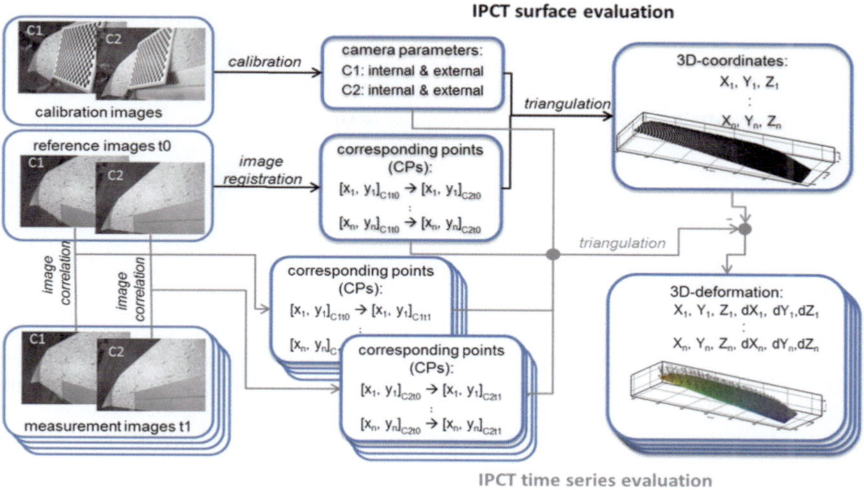

Fig. 8 Schematic diagram of the IPCT surface and time series evaluation [18]

by introducing vibrations with the help of a multi channel excitation system (see the chapter "Experimental Vibroacoustic Characterization of an A320 Wing Utilizing a Controlled Shaker System for Artificial Engine-Vibration Excitation").

The other topic was related to the further development of the Image Pattern Correlation Technique (IPCT) for propeller blade deformation measurements [5] and to its in-flight application [18]. Stereo IPCT [3], which was brought to maturity within the XDC, is a stereo photogrammetry method based on digital image correlation (DIC) methods. It enables engineers to measure the shape and deformation of optically accessible structures in a non-intrusive way. The investigated surface has a superimposed geometrical pattern to be used for image correlation. The IPCT evaluation can be separated into the main modules: camera calibration, surface evaluation and a time series evaluation. Figure 8 gives a schematic overview of these modules and their interactions; these modules have been significantly improved within the XDC in order to enable accurate 3D deformation measurements.

For an improvement in the robustness of IPCT measurements a recalibration method for corrections due to camera movements was implemented – this is considered crucial for inflight testing. Furthermore, the influence of cabin windows, which are usually made of Plexiglas, was investigated. It was found that structures within the Plexiglas material can lead to the appearance of artefact structures in the measurement results. The number of Plexiglas panes in the optical path of the cameras should therefore be minimised, for example, by either removing inner panes or even completely replacing the cabin windows with glass windows of good optical quality. If this is not possible and Plexiglas panes cannot be avoided in a measurement setup, the quality of the windows for the final alignment of the cameras should be checked, not only by a visual inspection, but also by an IPCT surface evaluation. It

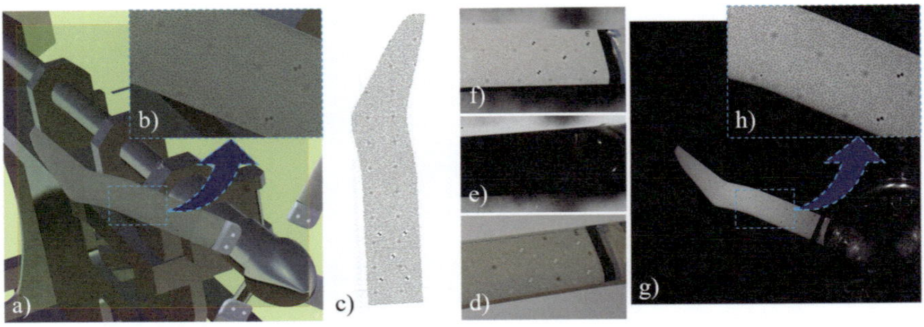

Fig. 9 Digital Mock-Up (DMU) and pattern application (**a**—example DMU image (green shaded rectangle = camera field of view), **b**—zoomed detail of the simulated pattern on the blade, **c**—final resulting pattern, **d**—paint mask on white primed blade, **e**—applying black paint, **f**—resulting patterned blade after paint mask removal, **g**—later camera image from the measurements (size equals the green shaded rectangle from **a**), **h**—zoomed in detail of realized pattern)

is recommended to perform the evaluation of the reference surface as part of the calibration procedure. This allows the camera position and orientation to be adjusted as accurately as possible, depending on window size, and if local patterns in the reprojection error distribution show local defects within the pane, followed by a repeat of the calibration images and a further re-projection error distribution check.

In addition to the advancements of the IPCT measurement and the image post-processing, the preparation process for such measurements has also been improved within the XDC. One topic was the maturation of the Digital Mock-Up (DMU) toolbox – a CATIA based toolbox used for the design of camera-based measurement systems. For flight tests and also for applications in a wind tunnel, the setup can be optimized in the DMU in terms of optical accessibility and spatial resolution for the specific measurement task without needing to access the flight test aircraft or the wind tunnel. One example – the creation of the setup with its required dot pattern – is shown in Fig. 9. The geometry of the test object is included in the DMU and cameras are positioned. Simulated camera views are created using the parameters of the cameras and the applied optics (see Fig. 9a). The pattern design which was created using a novel pattern generator, developed within the XDC, is applied to the measured surface (detail see Fig. 9b). As soon as the pattern demonstrates the correct dot sizes and arrangement, a paint mask is created from the final resulting pattern (Fig. 9c). Next, the paint mask is glued onto the white primed blade and the whole blade is then painted black (Fig. 9d, e). After the paint has dried, the mask is removed and the finished patterned blade is obtained (Fig. 9f). An example measurement image is shown in Fig. 9g (with pattern detail in subfigure h), showing a good agreement with the previously simulated DMU view. Another possible use of the DMU is to carry out virtual measurements, this enabling testing of the evaluation chain on a virtual measurement object. The IPCT provides as a direct result a 3D surface or a 3D deformation vector field in a camera-fixed coordinate system. In most cases, further processing of this data with special post-processing routines adapted to

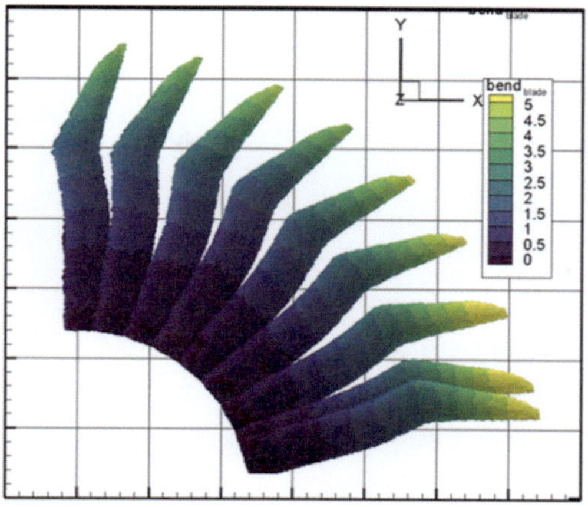

Fig. 10 Example results of the blade deformation measurements using IPCT (blade measured at different azimuthal positions during one revolution – harmonic tip oscillation/changing bending is clearly visible)

the measurement task is necessary in order to derive the actual desired measurement variables: angle of attack, relative movements of flaps, etc. Creating and testing these post-processing routines before the actual test leads to a faster provision of the results.

Using all toolbox ingredients a measurement campaign determining the deformation of the blade of a research rotor with double-swept tip [20] on the Rotor Test Rig in Göttingen (RTG) had been performed. The recorded images have been processed with the matured IPCT software including the time series correlation. The measured rotor blade is depicted in Fig. 10 in different azimuthal positions. Using the colour coded averaged blade deformation, a bending oscillation of the blade during its revolution is evident.

3.3 Surface Measurements

Additionally to the IPCT geometry measurements on the double-swept rotor blade (see Sect. 3.2), an optimized pressure-sensitive paint (PSP) measurement system was applied to investigate unsteady surface pressures on the same blade surface [30] and to examine a correlation between pressure and deformation. The PSP method uses the property of special fluorescent dyes excited by a laser or LED, and is capable of capturing transient flow phenomena (iPSP) on the surface. The unique experimental setup allowed scanning of surface pressures across the entire pitch cycle at a phase resolution of 0.225 deg azimuth for different collective and cyclic-pitch settings. The findings give a detailed and as yet unseen insight into the pressure footprints of double-swept rotor blades undergoing dynamic stall, and provide details of "stall maps", where confined areas of stalled flow on the blade can be seen as a function of the pitch phase. Experimental results were further compared to numerical simulations

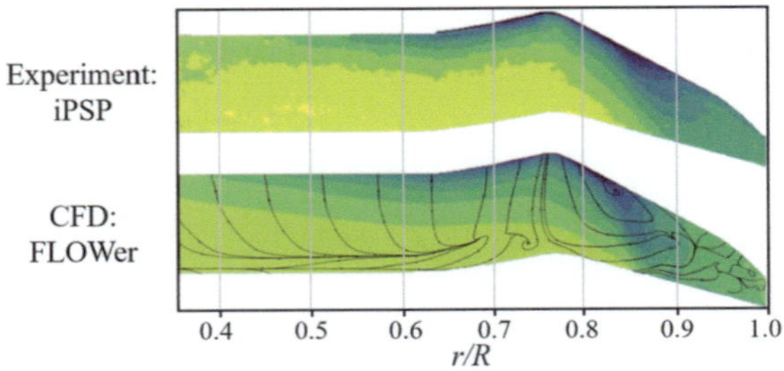

Fig. 11 Surface pressure distribution on the rotor blade measured with iPSP and computed with FLOWer [30]

Fig. 12 Model integrated TSP measurement system (left—view of the rear side of the model showing positions of cameras and LEDs, right—measured surface (below) illuminated by model integrated LEDs (top))

using the DLR flow solver FLOWer as shown in Fig. 11, where good agreement can be seen. To summarize, the study, performed within the framework of the XDC presents the first comprehensive iPSP data set on a rotor under cyclic-pitch conditions at all pitch phases and its findings are supported by the results of numerical computations. Thus it is now possible to investigate also dynamic phenomena, for example for innovative rotor designs with partially separated surface flow.

Similar to PSP, the Temperature Sensitive Paint (TSP) method exploits the dependence of the fluorescence of special colours on the ambient temperature; Here also, the dye is excited using a special light source (LEDs) and the fluorescent light is recorded by cameras. In addition to a temperature measurement, heat transfer on the surface with different flow topologies can also be visualised using an artificially generated temperature jump between the surface and the surrounding flow (e.g. by heating up the model) and thus transition locations can be measured with high precision. Within the XDC a miniaturized TSP setup was integrated directly into a wind tunnel model in order to investigate boundary layer transition induced by stationary crossflow instabilities. A photo of the camera and LED installation is shown in Fig. 12. More details of the measurement campaign and the results obtained can be found in this book in the chapter "Temperature-Sensitive Paint Method for the Investigation of Stationary Crossflow Instability Induced Boundary Layer Transition".

3.4 Measurement Techniques for Flow Field Investigations

Finally the maturation and demonstration of modern non-intrusive flow field measurement techniques such as the Particle Image Velocimetry (PIV) [21] was also part of the XDC. To demonstrate the capabilities of this method for large measurement objects, a full-scale engine ground test using the DLR research aircraft ISTAR, a Dassault Falcon 2000LX, has been performed in a noise protection hangar. An overview of this measurement campaign is presented in the chapter "Full-Scale Particle Image Velocimetry in Aircraft Related Flows Under Open-Air Conditions" of this book, as well as other full-scale and in-flight applications.

Another powerful volumetric flow field measurement technique is the Shake-The-Box (STB) method. A demonstration of this novel method applying high-resolution Lagrangian particle tracking for large and small scale aerodynamics was also part of the XDC. Details are given in the chapter "3D Lagrangian Particle Tracking with Shake-the-Box Applied to Large-Volume Measurements".

4 Conclusion and Summary

As shown so far, this "Cross Capability Demonstrator" XDC, unlike other Clean Sky 2 demonstrators, does not represent a specific product or method, but several essential numerical and experimental method improvements for aerodynamic, aero-acoustic and structural investigations of future propulsion concepts.

A brief overview of the XDC has been given. Starting with its classification within the European Framework Research Programmes and continuing with a general overview of the XDC structure, the work content in the established two main streams – Common Numerical Methods and Common Experimental Means – was briefly outlined. The investigated and further developed numerical methods and experimental procedures, which are suitable for and applicable to the prediction, optimisation and validation of the integration of modern propulsion systems, have been discussed briefly. A more detailed insight into the XDC methods is provided in the following chapters of this book.

Some of the methods established in the XDC are already used in the ongoing Clean Aviation projects COMPANION [8] and OFELIA [9] to measure special aspects of an open-rotor engine in flight tests. Within COMPANION the developed source localization methods are further developed and prepared for their application on the OpenFan Flight Test Demonstrator (FTD) [17] shown in Fig. 13. Furthermore, the developed acoustic MEMS sensor system is being adapted to the requirements of the OpenFan FTD in order to enable in-flight source localization with fuselage mounted microphone arrays. This requires a significant upscaling of the measurement system. Within OFELIA, the lessons learnt from the development of the Multi Axis Excitation System for vibroacoustic characterization are applied to develop a novel inflight thrust measurement method which avoids or at least minimizes the need for

Fig. 13 The OpenFan FTD [17]

modifications of critical aircraft components such as the engine itself or its mounting between the engine and the pylon. Further maturation of the inflight application of PIV will also be conducted in both projects. Within OFELIA the focus lies on the airborne artificial particle seeding, which is considered the most crucial aspect of inflight PIV. In addition, the challenges of a large inflight PIV setup arising from the dimensions and installation requirements on the OpenFan FTD are addressed in COMPANION in order to apply non-intrusive flow field measurements in front of the rotor, between rotor and stator as well as behind the stator.

Abbreviations

2D	2-Dimensional
3D	3-Dimensional
A/C	Aircraft
ACARE	Advisory Council for Aviation Research and Innovation in Europe
ADEC	Advanced Engine and Aircraft Configurations
AIR	Airframe
ASPIRE	Aerodynamic and acouStic for high-by-Pass ratIo tuRbofan intEgration
BLI	Boundary Layer Ingestion
CAA	Computational AeroAcoustics
CFD	Computational Fluid Dynamics
CFI	Cross Flow Instabilities
COMPANION	COMmon Platform and Advanced INstrumentation ReadIness for ultra efficient Propulsion DemonstratiON
CROR	Counter Rotating Open Rotor

CS	CleanSky
CS2	CleanSky 2
CSJU	CleanSky Joint Undertaking
DIC	Digital Image Correlation
DLR	Deutsches Zentrum für Luft- und Raumfahrt e.V. (German Aerospace Center)
DMU	Digital Mock-Up
EC	European Commission
ED; ECO	Eco Design
ENG	Engines
FRC	Fast Rotorcraft
FRPM	Fast Random Particle Mesh
FTD	Flight Test Demonstrator
GRA	Green Regional Aircraft
GRC	Green Rotorcraft
IADP	Innovative Aircraft Demonstrator Platform
IBDN	Inboard downward Sense of Rotation
IBUP	Inboard upwards Sense of Rotation
IPCT	Image Pattern Correlation Technique
iPSP	instationary PSP
ISTAR	In-flight Systems and Technology Airborne Research Aircraft
ITD	Integrated Technology Demonstrator
LED	Light Emitting Diode
LPA	Large Passenger Aircraft
MEMS	Micro Electro Mechanical System
MONA	DLR Modelling Process based on MOdGen and NASTRAN
NLR	Nationaal Lucht- en Ruimtevaartlaboratorium (Netherlands Aerospace Centre)
NOVA	NextGen ONERA Versatile Aircraft
OFELIA	Open Fan for Environmental Low Impact of Aviation
ONERA	Office National d'Etudes et de Recherches Aérospatiales (French Aerospace Center)
OR	Open Rotor
ORAS	Open Rotor and Stator
PIV	Particle Image Velocimetry
PSP	Pressure Sensitive Paint
REG	Regional Aircraft
RTG	Rotor Test Rig Göttingen
SAGE	Sustainable and Green Engines
SAT	Small Air Transport
SFWA	Smart Fixed Wing Aircraft
SGO	Systems for Green Operations
SODIX	SOurce DIrectivity modeling in the cross-spectral matriX
SRV	Swirl Recovery Vane
STB	Shake-the-Box

SYS	Systems
TA	Transverse Activities
TE	Technology Evaluator
T-S	Tollmien-Schlichting Instabilities
TSP	Temperature Sensitive Paint
UHBR	Ultra-High Bypass Ratio Engine
uRANS	unsteady Reynolds Averaged Navier–Stokes equations
XDC	Cross Capability Demonstrator

Acknowledgements This project has received funding from the Clean Sky 2 Joint Undertaking under the European Union's Horizon 2020 research and innovation program under grant agreement N° CS2-LPA-GAM-2020-2023-01.

The authors would also like to thank all the scientists, technicians, administrators and other helping hands for their valuable work and patience in further advancing these XDC topics and in helping to provide the fruitful and interconnected demonstration of all the methods presented here in this challenging project.

References

1. Council Regulation (EC) No 71/2007 of 20 December 2007 setting up the Clean Sky Joint Undertaking (OJ L 30, 4.2.2008), pp 1–20. https://eur-lex.europa.eu/eli/reg/2008/71/oj. Accessed 14 Jan 2025
2. Council Regulation (EU) No 558/2014 of 6 May 2014 establishing the Clean Sky 2 Joint Undertaking (OJ L 169, 7.6.2014), pp 77–107. http://data.europa.eu/eli/reg/2014/558/oj. Accessed 14 Jan 2025
3. Boden F (ed) (2013) AIM^2 advanced flight testing workshop: handbook of advanced in-flight measurement techniques. BoD–Books on Demand
4. Boden F, Barth HP, Hein S, Heitmann D, Kirmse T, Klein C, Krüger W, Lemarechal J, Norambuena M, Schäfer D et al (2023) Recent developments of common numerical methods and common experimental means within the framework of the large passenger aircraft program. CEAS Aeronaut J 14(1):281–293
5. Boden F, Stasicki B, Szypuła M, Ružička P, Tvrdik Z, Ludwikowski K (2016) In-flight measurements of propeller blade deformation on a VUT100 cobra aeroplane using a co-rotating camera system. Meas Sci Technol 27(7). https://doi.org/10.1088/0957-0233/27/7/074013
6. Borradaile J (1988) Towards the optimum ducted UHBR engine. In: 24th Joint propulsion conference, p 2954
7. Brouckaert JF, Mirville F, Phuah K, Taferner P (2018) Clean Sky research and demonstration programmes for next-generation aircraft engines. Aeronaut J 122(1254):1163–1175. https://doi.org/10.1017/aer.2018.37
8. Clean Aviation Joint Undertaking: COMPANION—common platform and advanced instrumentation readiness for ultra efficient propulsion. https://clean-aviation.eu/companion. Accessed 17 Feb 2025
9. Clean Aviation Joint Undertaking: OFELIA—open fan for environmental low impact of aviation. https://clean-aviation.eu/ofelia. Accessed 17 Feb 2025

10. Delfs JW, Bauer M, Ewert R, Grogger HA, Lummer M, Lauke TG (2008) Numerical simulation of aerodynamic noise with DLR's aeroacoustic code PIANO
11. Dessens O, Köhler MO, Rogers HL, Jones RL, Pyle JA (2014) Aviation and climate change. Transp Policy 34:14–20. https://doi.org/10.1016/j.tranpol.2014.02.014, Air Transportation and the Environment
12. European Commission and Directorate-General for Mobility and Transport and Directorate-General for Research and Innovation: Flightpath 2050—Europe's vision for aviation—Maintaining global leadership and serving society's needs. Publications Office (2011) https://doi.org/10.2777/50266
13. European Commission and Directorate-General for Research and Innovation: Fly the Green Deal—Europe's vision for sustainable aviation. Publications Office of the European Union (2022) https://doi.org/10.2777/732726
14. Funke S, Dougherty R, Michel U (2014) SODIX in comparison with various deconvolution methods. In: 5th Berlin beamforming conference
15. Gerhold T (2005) Overview of the hybrid RANS code TAU. In: Kroll N, Fassbender JK (eds) MEGAFLOW—numerical flow simulation for aircraft design. Springer, Berlin Heidelberg, Berlin, Heidelberg, pp 81–92
16. van Ingen JL (1956) A suggested semi-empirical method for the calculation of the boundary layer transition region. https://resolver.tudelft.nl/uuid:cff1fb47-883f-4cdc-ad07-07d264f3fd10
17. Joshi G Airbus & CFM team up to launch open-fan engine architecture testbed. https://simpleflying.com/airbus-cfm-open-fan-engine-architecture-testbed/. Accessed 17 Feb 2025
18. Kirmse T, Boden F, Meyer R, Philipp F (2021) Wing deformation measurements for manoeuvres of high load at the Airbus A320 DLR-ATRA by means of Image Pattern Correlation Technique. In: ETTC (ed) European test and telemetry conference proceedings online. Conference proceedings
19. Klimmek T, Schulze M, Abu-Zurayk M, Ilic C, Merle A (2019) cpacs-MONA—an independent and in high fidelity based MDO tasks integrated process for the structural and aeroelastic design for aircraft configurations. In: International forum on aeroelasticity and structural dynamics 2019, IFASD 2019. Savannah, GA, USA
20. Müller MM, Schwermer T, Mai H, Stieg C (2018) Development of an innovative double-swept rotor blade tip for the rotor test facility Goettingen. In: DLRK 2018—Deutscher Luft- und Raumfahrtkongress
21. Raffel M, Willert C, Scarano F, Kähler CJ, Wereley ST, Kompenhans J (2018) Particle image velocimetry—a practical guide, 3rd edn. Springer. https://doi.org/10.1007/978-3-319-68852-7
22. Romani G, Ye Q, Avallone F, Ragni D, Casalino D (2019) Fan noise boundary-layer ingestion installation effects for NOVA aircraft configuration. https://doi.org/10.2514/6.2019-2429
23. Savoni L, Rudnik R (2018) Pylon design for a short range transport aircraft with over-the-wing mounted UHBR engines. In: 2018 AIAA aerospace sciences meeting, p 0011
24. Smith LH (1993) Wake ingestion propulsion benefit. J Propul Power 9(1):74–82. https://doi.org/10.2514/3.11487
25. Stuermer AW (2024) Engine-airframe installation effects for open fan powered aircraft at take-off conditions. In: AIAA aviation forum and ascend, p 3518
26. Stürmer A (2019) Assessing turbofan modeling approaches in the DLR TAU-code for aircraft aerodynamics investigations. In: AIAA SciTech forum 2019. AIAA. https://doi.org/10.2514/6.2019-0277
27. Stürmer A, Keller D, Vinz A (2023) Engine-airframe integration studies for future efficient propulsion systems. In: UTIAS international workshop on aviation and climate change. https://elib.dlr.de/196994/
28. Theiss A, Hein S, Pascal L, Cliquet J (2022) Numerical investigations of rotational effects on boundary-layer instabilities on the Z49 open rotor model. In: AIAA SciTech 2022 forum, p 2541

29. Vinz A, Raichle A (2024) Investigation of the effects of boundary layer ingestion engine integration on aircraft thrust requirement. CEAS Aeronaut J 15:1235–1250. https://doi.org/10.1007/s13272-024-00722-0
30. Weiss A, Geisler R, Müller MM, Klein C, Henne U, Braukmann JN, Letzgus J (2022) Dynamic-stall measurements using time-resolved pressure-sensitive paint on double-swept rotor blades. Exp Fluids 63(1). https://doi.org/10.1007/s00348-021-03366-6

Numerical Methods

Hybrid Wing Body as a Low-Noise Aircraft from a Multidisciplinary Point of View

Michael Mößner, Lothar Bertsch, Jan W. Delfs, Yasim Hasan, Tobias Hecken, Michael Iwanizki, Dennis Keller, Antoine Moreau, Florian N. Schmidt, and Arne Seitz

Abstract One important goal when reducing the environmental impact of an aircraft is the reduction of noise. There is much potential for reducing noise when making it a major objective from the beginning of the design phase. Within this contribution a hybrid wing body concept as a mid-range aircraft will be presented together with detailed studies of different disciplines such as aerodynamics, acoustics, structure and also handling qualities. How much noise reduction can be achieved? What are the advantages, where are the difficulties and weaknesses of such an aircraft configuration?

Keywords Low-noise aircraft · Hybrid wing body · Preliminary design

1 Introduction

The main driver of research on blended wing body concepts is clearly efficiency. Making the fuselage a lifting surface leads to major advantages (see e.g. [1, 2]): Firstly, the additional effective wing span reduces induced drag. Secondly, the additional lift of the fuselage as well as its more integrated shape enables reduction in the overall wetted surface. However, in order to establish blended wing bodies as passenger aircraft several challenges must be met. These obstacles appear to be so high that currently no major aircraft manufacturer seriously considers such concepts for a passenger aircraft. One of the main difficulties is flight stability. Furthermore, highly integrated engines are often considered to enhance the drag benefit, owing to the propulsive gain arising from the boundary layer ingestion, but even this technol-

M. Mößner (✉) · J. W. Delfs · Y. Hasan · M. Iwanizki · D. Keller · F. N. Schmidt · A. Seitz
German Aerospace Center, Braunschweig, Germany
e-mail: Michael.Moessner@dlr.de

L. Bertsch · T. Hecken
German Aerospace Center, Göttingen, Germany

A. Moreau
German Aerospace Center, Berlin, Germany

F. Boden (ed.), *Numerical Methods and Experimental Means for the Integration of Future Propulsion Systems*, Springer Aerospace Technology,
https://doi.org/10.1007/978-3-031-97725-1_2

ogy is not ready for short-term usage. One further aspect is that for smaller aircraft of $\sim$180 passengers the center-body airfoils need to be rather thick to provide enough height in the cabin, which reduces the advantage of wetted surface and also increases drag due to sub-optimal airfoils.

One can ask the question: What if the efficiency benefits are given a lower priority and one concentrates just on a low-noise configuration? An answer to this question is being investigated within the DLR project SIAM (2020–2023). This given premise permits the design of a more conservative blended wing body with less uncertainties in exploiting design potentials, in particular with acoustic shielding for large noise reduction benefits. Adding a T-tail section with dedicated vertical and horizontal stabilizers enable the stabilization of the aircraft and by using elevated engines one can avoid the uncertainties of inflow distortion to the engines. This designed aircraft concept is called "hybrid wing body" and is related to a similar concept of Lockheed Martin [3]. In the following, the design of a hybrid wing body will be presented, together with several more detailed studies for a better understanding the benefits and weaknesses of such a configuration.

2 Aircraft Specifications

The top level aircraft requirements are set similar to an A320 as a typical mid-range aircraft, as given in Table 1. The aircraft underlies the approach category C [4] as well as the ICAO aerodrome reference code C [5] which limits the wing span to 36 m.

The conceptual aircraft design methodology as well as its design evaluation is described in [6]. In short, the methodology for aircraft design is based on simplified physical and semi-empirical methods, calibrated by high fidelity results. The initial design is derived from a classic blended wing body similar to the concept presented in [7]. In order to give this aircraft stable flight characteristics, a T-tail was added. Furthermore, podded engines are used in order to keep design uncertainties at a manageable level. The engine configuration has low noise properties for two reasons: First, it uses the center-body as large shielding geometry towards the ground and secondly, podded engines have low inflow disturbances which avoids excess noise. A low-noise high lift system is installed, which essentially is achieved by gapless

Table 1 Top level aircraft requirements for the hybrid wing body

Design range	2,600 nm
Cruise mach number	0.78
Passenger capacity	180
Design payload	17.1 t

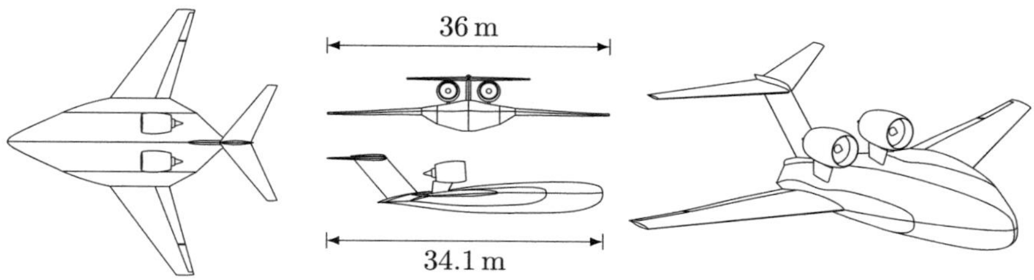

Fig. 1 Sketch of the aircraft geometry

high lift devices. A sketch of the aircraft together with its dimensions is shown in Fig. 1.

3 Stability and Control

Besides low-noise characteristics, one of the major drivers for the decision to use a hybrid wing body is stability and control. Typically, blended wing body aircraft have major drawbacks regarding their flight mechanical properties for different reasons. The vertical tailplanes and rudders have comparatively low lever arms, which may lead to lateral stability and control issues. In addition, there is no separate surface used for horizontal stabilization. This usually means that the allowable center of gravity range is rather small.

Especially for smaller blended wing bodies, it is difficult to keep the center-of-gravity travel within a small margin. The reason is that the location of the cabin compartment tends to be in the front section, leading to large variations of the center of gravity depending on the loading conditions. Indeed, for the hybrid wing body presented in this paper, the center of gravity range is comparable to an A320, while the lever arms of the stabilizers are much shorter. As a consequence, the mass configuration with the most rearward center of gravity is often associated with a negative static margin, which signifies that the aircraft is statically and dynamically unstable in longitudinal direction.

Stability augmentation systems could potentially be used to stabilize the aircraft. This has been done in military applications for several years. However, it is questionable whether an aircraft with major stability issues at certain flight points can be operated safely, even in 15 years time. This is particularly true if the naturally unstable eigenmodes are of rather high frequency such that a pilot can no longer control the aircraft, and a failure of the flight control system would then very likely result in an airplane crash. Therefore, in order to make the low-noise aircraft more realistic for the near future, preliminary stability and control considerations had to be taken into account. Here, the hybrid wing body proved to be a promising candidate, since, with its additional tail plane and its larger lever arms, it improves at least most of the typical blended wing body stability issues.

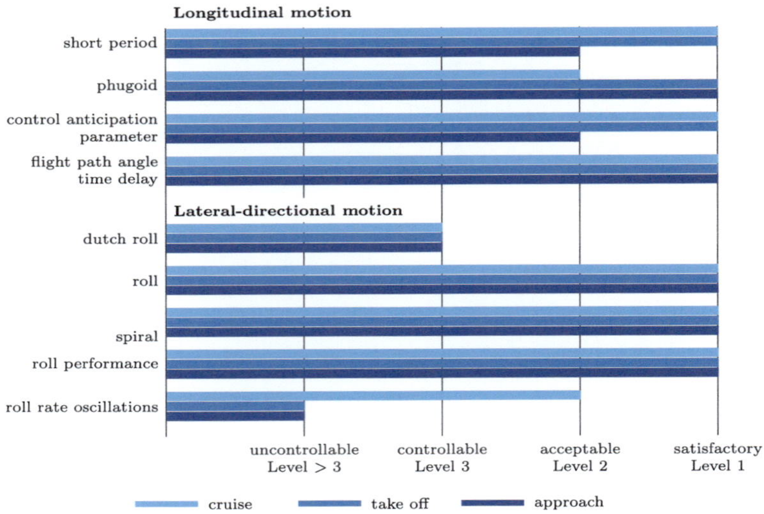

Fig. 2 Handling qualities assessment for the hybrid wing body

Figure 2 shows the assessment of some handling and flying quality criteria according to [8, 9] for the open-loop hybrid wing body obtained in the preliminary design phase. The results are given for cruise flight, takeoff and landing, taking into account different loading conditions, including the one with the most rearward center of gravity. As shown, the short period mode and the phugoid are at least acceptable or perhaps even satisfactory for all considered flight points. The control anticipation parameter, assessing the initial pitch acceleration response due to pitch inputs and thereby indicating controllability, is also satisfactory for cruise and takeoff. For the approach case, the center of gravity is more rearward, leading to a more sluggish behavior. However, even at this condition, pitch controllability is still acceptable. The flight path angle time delay evaluates the delay until a commanded change of flight path angle is established. It is also satisfactory for all flight points regarded here.

At all considered flight points the Dutch roll is the least controllable. Roll and spiral are satisfactory at all flight points and the aircraft provides sufficient roll control authority, as indicated by the roll performance criterion. For terminal flight phases, i.e. takeoff and approach, the roll rate oscillation criterion indicates uncontrollable handling. This signifies that, when a roll rate is commanded using the ailerons, the aircraft initially tends to oscillate around the commanded roll rate to a significant extent. This is distracting for pilots, as it complicates entering a turning flight. This is a remaining issue which would have to be addressed in further design iterations. Considering that the blended wing body aircraft had similar roll oscillation issues, using a hybrid wing body aircraft over a blended wing body represents a more realistic case from a flight mechanical point of view.

Besides the improvement of the aircraft's noise characteristics, noise reduction via flight procedures is possible. For instance, in order to reduce the noise during landing, approaches with steeper stable flight path angles are a good option: This is because they reduce the flight time at low altitudes and in proximity of the airport. However, this requires the use of high-lift devices, which are often accompanied by high pitch-down moments. For the case of the hybrid wing body, the separation of wing-body section and horizontal tailplane provides a higher pitch authority. Hence, the use of high-lift devices is facilitated with this configuration.

4 Planform

A hard restriction on the outer dimensions of the center-body is the 36 m limit to the wing span to fit into the limits given by the ICAO aerodrome reference code C [5]. The center-body size of the aircraft is then largely driven by the cabin and cargo layout and required space. The center-body is a lifting body, which means that it follows typical airfoil shapes. Due to the airfoil thickness distribution over the airfoil length, the passengers and cargo cannot be placed at the rear (see sketch in Fig. 3). Making the cabin and cargo space wider but shorter is difficult for several reasons. The cabin height cannot be reduced, so the relative thickness of the airfoil would have to increase, reducing aerodynamic performance. Also the travel of the center of gravity becomes larger, especially since the cabin is positioned before the center of gravity. Making the cabin longer and narrower increases the wetted area and at some point the aircraft becomes more similar to a tube and wing concept or wide-body fuselage, which are concepts which would have to be investigated on their own (see e.g. [10]). Generally, the cabin is placed in the center part where the airfoils are still thick enough, and the cargo is placed slightly outward where reduced heights are acceptable. One can see, that the passengers and cargo are located in the front

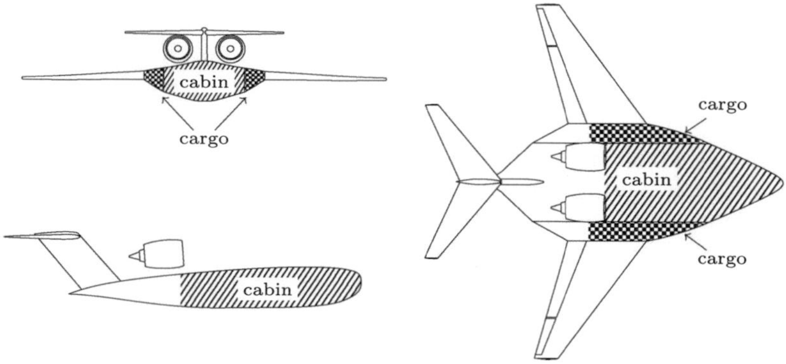

Fig. 3 Cabin and cargo layout

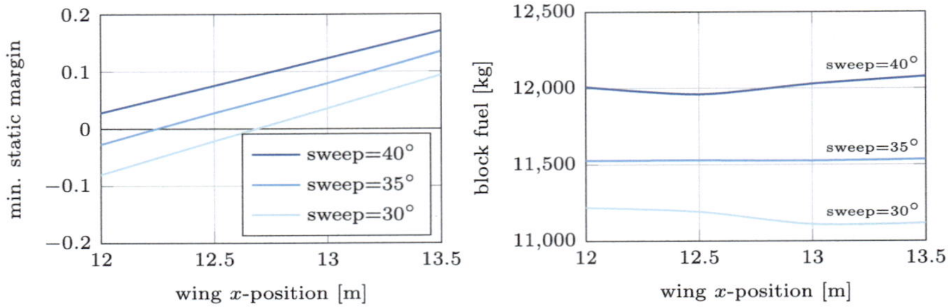

Fig. 4 Example results of the variation of the wing sweep and wing root position in x-direction showing the minimum static margin and the design mission block fuel

section, which leads to a significant range of the center of gravity position depending on the number of passengers and the amount of cargo.

The sweep, the axial position of the wings, and the wing loading are adapted by applying parametric studies. The target is the identification of a configuration with a minimum block fuel consumption subject to different constraints such as the minimum static margin or the approach speed. Examples of results related to the static margin and the block fuel for a single wing loading are shown in Fig. 4. To obtain a positive value of the minimum static margin, the wing must move backward or the sweep has to be increased, effectively moving the neutral aerodynamic center backwards. Small sweeps are favorable in terms of reduced weight, increased lift-over-drag ratio and, as a sum, reduced block fuel. For the further design process a sweep of 35° was chosen, which can still yield an acceptable static margin.

The wing area is a dominant parameter for mass and drag of an aircraft as the hybrid wing body. One of the main design drivers for the wing area (or the wing loading) is the maximum achievable lift coefficient. It can be subdivided into the maximum lift of the clean configuration and the contribution of the high-lift devices. In order to minimize the airframe noise, a high lift system consisting of gapless flaps at the trailing edge and the leading edge was chosen. As a consequence of the relatively low performance of such high lift devices, the resulting wing area of the hybrid wing body had to become comparatively large. This large wing area in turn notably increases the mass and drag of the aircraft and will diminish its overall performance.

Estimating an appropriate maximum lift is difficult with preliminary design tools. For better predictions this may be supported by high fidelity Reynolds-averaged Navier Stokes (RANS) computations. Figure 5 shows that the maximum lift coefficient is indeed overpredicted by the preliminary design methods (low fidelity). However, also the high fidelity results must be interpreted with care. There is potential of improving the behavior by design optimizations such as changing the airfoil shapes and as a result getting back closer to the low fidelity predictions which can be regarded as an optimum.

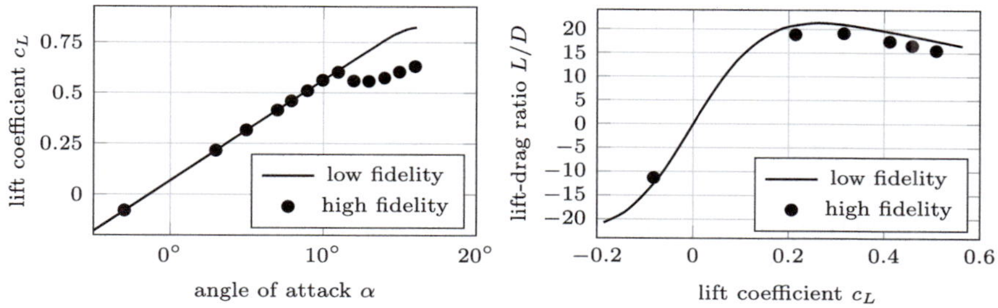

Fig. 5 Comparison of aerodynamic performance (wing and center-body only) based on preliminary design tools (low fidelity) and high fidelity CFD results for low speed conditions without high lift devices

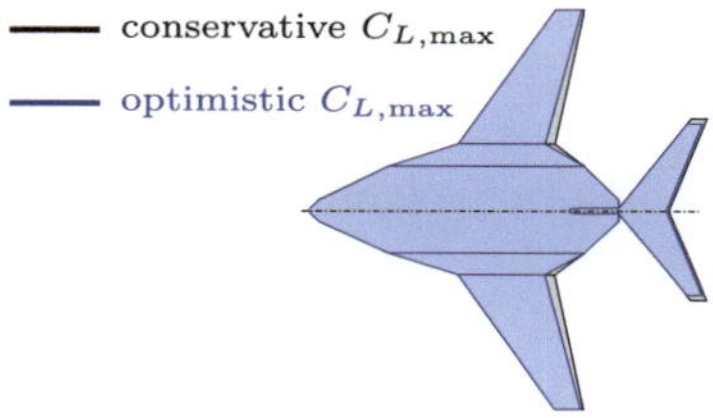

assumed increase of $C_{L,\max}$	3.5 %
block fuel improvement	4.8 %
lift over drag improvement	2.8 %
MTOM reduction	2.5 %

Fig. 6 Comparison of two different designs with conservative and optimistic values for the maximum achievable lift coefficient

The impact of the maximum lift variation can be observed by comparing two hybrid wing body designs with different assumptions for the maximum lift coefficient $C_{L,\max}$ (conservative vs. optimistic). Both configurations provide a comparable minimum static margin to ensure similar handling qualities. The concept with a higher maximum lift coefficient is designed with a higher wing loading that reduces the wing area and hence leads to less mass and drag. The results (see Fig. 6) indicate that an increase of $C_{L,\max}$ by 3.5% ($C_{L,\max}$ with full flaps) leads to a reduction in block fuel by 4.8% after a complete redesign of the aircraft. Due to the reduced mass and drag, a more efficient flight trajectory can be flown, since the fixed engines provide enough thrust to achieve higher cruise altitudes and further improve the lift over drag ratio L/D. Although this trend cannot be extrapolated linearly, compromising the low-noise high lift system by using a stronger but noisier system to achieve higher wing loading should definitely be considered in order to improve the fuel efficiency.

The tail plane is added in order to maintain acceptable handling qualities. Since the lever arms are short, its size must be quite large compared to classic tube-and-wing concepts. It is obvious that potential size reductions would immediately increase the aircraft efficiency due to decreased wetted surface and reduced weight. This could be achieved by reducing the requirements of handling qualities. Alternatively, concepts such as limiting the most rearward position of the center of gravity could be considered. This may, for example, be achieved by disallowing aircraft operation when empty (e.g. by adding extra ballast for transfer flights). One could also consider moving the vertical stabilizer to the front (canard) since this provides longer lever arms. But such a configuration brings its own challenges (e.g. high lift performance, handling qualities) and will not be further discussed here.

5 Wing Design and High Lift Devices

Starting point of the wing design was the hybrid wing body concept from the preliminary design with optimistic $C_{L,\max}$ and reduced wing area (see Fig. 6) to minimize the mass and drag of the aircraft as much as possible. The clean wing design point was defined at cruise flight conditions with a Mach number of $M = 0.78$ and a total lift coefficient of $C_{L,\text{tot}} = 0.23$ with a required local lift coefficient of approximately $c_l = 0.42$ according to the preliminary design. The shape of the clean wing is defined by two wing section profiles located at the wing's root and the wing's tip. One main goal of the wing design was the reduction of a strong suction peak, which was revealed during first aerodynamic analyses of the baseline configuration resulting in a very strong compression shock over the inboard wing. Additionally, the twist of the wing was adjusted with a good compromise between the spanwise lift distribution in cruise conditions and high-lift requirements: While the spanwise lift distribution is designed for low induced drag, high-lift requires that the outboard wing can operate at high angles of attack without separation. The resulting design does not need a leading-edge device which reduces complexity and avoids a significant noise source.

Using the inverse design capabilities of the finite volume RANS solver DLR FLOWer [11], a target pressure distribution was defined for the section profiles. The pressure distribution is given with reference to $C^*_{P,\text{crit}}$, which denotes the pressure coefficient at which the Mach number based on the velocity normal to the isobars exceeds the speed of sound (see Fig. 7). The resulting airfoil features are:

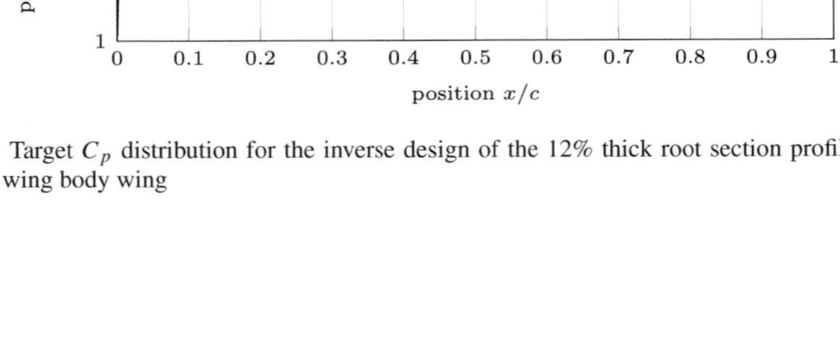

Fig. 7 Target C_p distribution for the inverse design of the 12% thick root section profile of the hybrid wing body wing

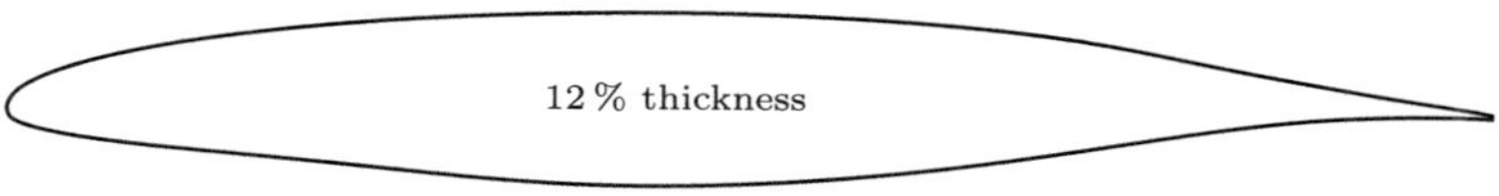

Fig. 8 Contour of the 12% thick root section profile of the hybrid wing body wing

- The pressure distribution delivers $c_l = 0.42$ at $M = 0.78$.
- The supersonic region is very limited resulting in an almost shock free recompression with a wave drag below 1 drag count.
- The pressure rise over the lower and upper rear part of the profile has a concave shape, which contributes to a low total drag, as already shown by Eppler [12].
- The pressure distribution exhibits a strong front and a low rear loading in order to keep the zero-lift pitching moment low.

The resulting airfoil shape with 12% relative thickness for the wing's root section is shown in Fig. 8. On the basis of the root section profile, the tip section profile was designed with the same features except a 13% larger nose radius for improved low speed characteristics.

Given the cruise wing design, the high-lift system of the hybrid wing body has been developed. Due to the preliminary aircraft design requirements, the high-lift configuration of the optimistic hybrid wing body needs to achieve a maximum lift coefficient $C_{L,\max}$ of 0.96 during landing at a Mach number of $M_{\text{landing}} = 0.2$. The high-lift system of the hybrid wing body was designed with the help of 2D and 3D CFD studies in such a way that the simplified wing body without engine and tailplane achieves the required lift coefficient for landing. Detailed information about the wing and high-lift system design can be found in [13]. As a low-noise configuration, the wing is equipped with a simple gapless flap and an aileron for flight control around the longitudinal axis. The flap as well as the aileron occupy 25% of the chord length. According to the CFD investigations of the 3D high-lift configuration using the DLR TAU code [14] the targeted lift coefficient of $C_{L,\text{landing}}$ of 0.96 can be achieved by the simplified hybrid wing body with the use of the designed conventional plain flap at a flap deflection of 30°, as shown in Fig. 9.

In order to decrease noise even more, an additional system of morphing flaps was investigated (Fig. 10). While a morphing flap system is difficult with respect to structural design, a form-adaptive concept can reduce flap noise significantly, as e.g. shown in [15] where noise reductions of 6 dB and more were found for mouldline technology.

It should be mentioned that the design choice to use a gapless high lift system is expected to have a significant impact on performance, as already outlined in the context of Fig. 6. Improving performance at cost of increased complexity or more noise should be considered in further design iterations.

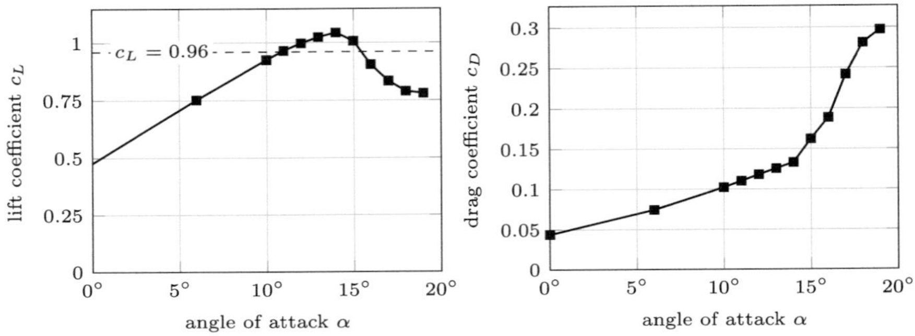

Fig. 9 Lift and drag coefficients c_L and c_D of the wing and center-body of the hybrid wing body over angle of attack α for flap deflection of 30°

Fig. 10 Wing and center-body of the hybrid wing body with deflected flaps in landing configuration. Left hand side: plain flap, right hand side: morphing flap

6 Isolated Engine Performance

When considering entry into service by 2035, it seems reasonable to assume that ultra-high-bypass-ratio (UHBR) turbofan engines will be installed. The growth in size of the engines, starting from the engine installed in the A320ceo, going to more modern geared turbofans as used in the A320neo up to ultra high bypass engines is sketched in Fig. 11. An UHBR turbofan is installed on the hybrid wing body aircraft. This has the advantage of higher efficiency. A rough estimate shows that the propulsive efficiency can be raised by $\approx 8\%$ compared to a BPR5 engine while the fan isentropic efficiency increases by $\approx 5\%$.

Besides the improved efficiency, another advantage featured by UHBR turbofans is of acoustic nature. As the fan stage produces less compression, its fan rotor may rotate at both lower tip speeds and flow Mach numbers, which cut-offs the strong fan tones. Also, jet noise is substantially decreased due to the lower jet velocity at the exhaust. As shown in [16], the reduction in fan broadband noise is not as high as might be expected, because UHBR fans operate at strongly off-design conditions at the noise certification points. This can be improved by implementing a variable-geometry system to release the fan loading at those points.

The benefits provided by the UHBR fan technology compared to the reference A320ceo engine are summarized in Fig. 12, based on the detailed study of [16]. The additional noise reduction potential provided by the variable-geometry systems is

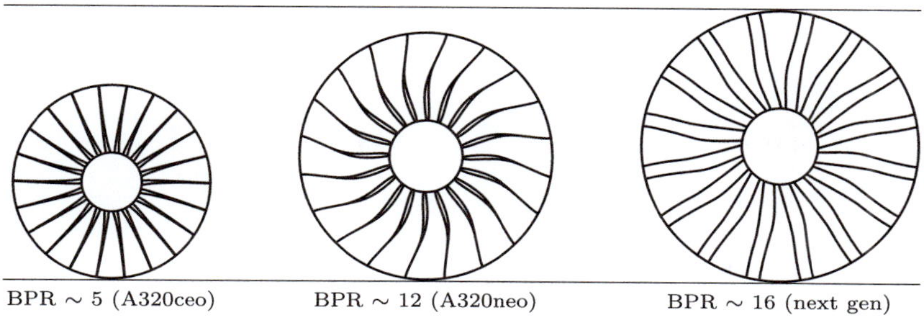

Fig. 11 Fan sizes for different generations of turbo fans (increasing bypass ratio, BPR)

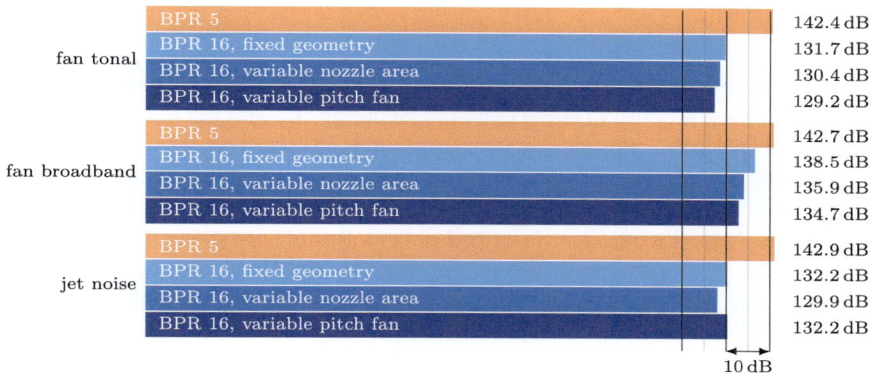

Fig. 12 Noise radiation reduction (given as overall sound power level) of ultra high bypass engine compared to the reference (as used by A320ceo) at sideline condition. Shown are different variants of fixed nozzle and fan, as well as variable nozzle area and variable pitch fan

also highlighted. As can be concluded from Fig. 12, variable-geometry systems for UHBR engines show the best reduction potential for fan tonal and jet noise, which should both lie 10 dB below the reference A320ceo engine. For a single operating point at sideline condition, the benefits even reach 13 dB reduction. However, the overall engine noise reduction should be limited by the relatively moderate decrease in fan broadband noise, which precisely becomes the dominant source in UHBR engines. The poor reduction by only 4 dB of the UHBR engine with fixed nozzle and fan shows clearly the need for a variable-geometry system, as reductions of up to 8 dB are to be expected.

7 Engine Noise Shielding

Shielding engine noise towards the ground is one of the major aspects for a low-noise aircraft design. The consequence is that the engines must be placed on top of the aircraft. Especially for large short-cowl engines fan-rearward noise becomes

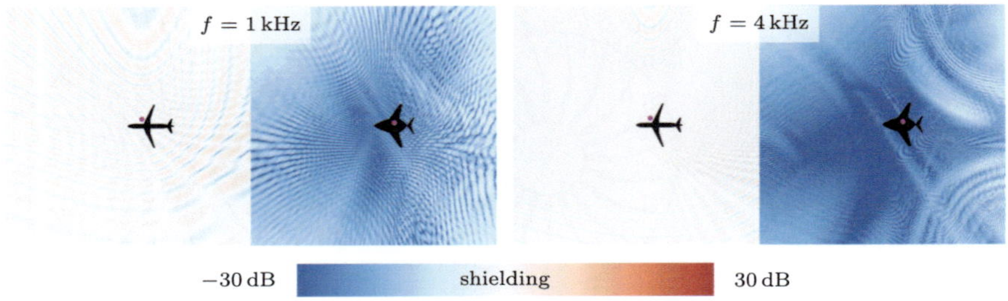

Fig. 13 Engine shielding of airframe based on a point source, 51 m below the aircraft for two different frequencies. Positive numbers mean a amplification of noise due to the airframe geometry, while negative numbers mean less noise at the ground

more and more relevant, next to the fan-forward noise. This means, that the engine should be moved forward to allow rearward shielding. The engine noise shielding is evaluated by replacing the engine with a point source. While a point source simplifies the problem in comparison to a real engine which is a much more directed source it gives a good estimate of the potential of shielding. The shielding is computed with a boundary element method as described in [17]. Results are shown in Fig. 13. For a conventional under-the-wing aircraft, engine noise is not attenuated, but slightly amplified, since parts of the wing reflect upward noise back to the ground. In contrast, a hybrid wing body with top-mounted engines shows attenuation of noise of 20 dB and more. It must be noted that compared to shielding of fan noise, jet noise shielding is hardly possible due to its source position well behind the engine. For jet noise reduction one must mainly rely on larger engines with slower jets as described in Sect. 6.

8 Aerodynamic Engine Integration

Many blended or hybrid wing body designs use highly integrated engines (see e.g. [1, 2, 18–21]). This design choice has several advantages:

- Strong reduction of wetted area for less viscous drag.
- Easier integration of large engines (less interference drag).
- Better aero-propulsive efficiency due to exploitation of boundary layer ingestion.
- Bringing the thrust vector closer to the center of gravity, reducing the need for momentum compensation.
- More noise shielding towards the ground.

Still, for the current design, podded engines have been chosen for the following reasons: Highly integrated engines represent a substantial risk for engine efficiency and most importantly for engine source noise. The interaction of the fan with the distorted inflow not only triggers new tonal and broadband sources, it also affects existing sources (e.g. wake modulation modifying the classical rotor-stator interaction) or propagation mechanisms (e.g. scattering in non-uniform flow triggering new duct modes). For these reasons, if the technology of boundary layer ingestion has not been fully understood, quantified, and implemented correctly, excess engine noise at the source by more than 5 dB may be expected and may therefore cancel the benefit from increased shielding. Further, passenger comfort can be detrimentally affected due to the extra noise and vibration.

Without any special shape treatments, the extra drag due to engine installation is extremely high. Numerical simulations of the aircraft design with the initial engine integration indicate an increase of $\Delta C_D = 0.0111$ (83%) in total drag compared to the combination of the isolated wing-body and isolated nacelle. The substantial drag increase is caused by the channel flow evolving between the center-body and the nacelle that eventually leads to a shock. Numerical simulations show that even though the shock results in a sudden pressure increase, the surface pressure downstream of the shock and the stagnation pressure at the trailing edge do not reach the values found for the case without engine. Position variations can slightly improve the situation, mainly by altering the maximum flow velocity between the center-body and the nacelle:

Moving engines up: This will indeed reduce drag, but will also increase the thrust-induced moments due to the larger distance to the center of gravity. Also pylons become longer, increasing structural weight and viscous drag. Noise shielding towards the ground will suffer. Moreover, the effect of increasing the engines' vertical position on the total drag is rather small for reasonable positions as found by a sensitivity study (see Fig. 14 on the left).

Moving engines backwards: Moving the engines backwards means moving it away from the locally accelerated flow. The drag increase due to the engine integration may thereby be notably reduced as shown in Fig. 14 on the right. If the tail is not considered, the drag coefficient can be reduced up to $\Delta C_D = 0.0094$ when the engines are moved downstream beyond the center-body's trailing edge. Considering the tail in the simulations, the maximum drag reduction is limited to $\Delta C_D = 0.0078$ (compared to the baseline position). While moving the engines towards the rear is obviously favorable regarding airframe drag, it will also move the center of gravity towards the back. As a result, the static stability margin is reduced. Moreover, the attenuation of fan rearward noise will suffer.

Besides changing the engine position, modifying the center-body shape can have a substantial effect on airframe drag. Engine integration design studies show that the installation drag in cruise flight can be brought down to a minimum by shape modifications. Optimizing the center-body section at the position of the engine's center by means of 2D-RANS computations leads to sectional shapes (see Fig. 15) that reduce the maximum flow velocities within the channel between the center-body and the nacelle and prevent the occurrence of shocks. In combination with subsequent 3D adjustments the shape modifications yield a substantial improvement in aerodynamic efficiency where the drag coefficient C_D is only 6 drag counts higher compared to the one with the isolated case (combined drag of isolated wing-body and

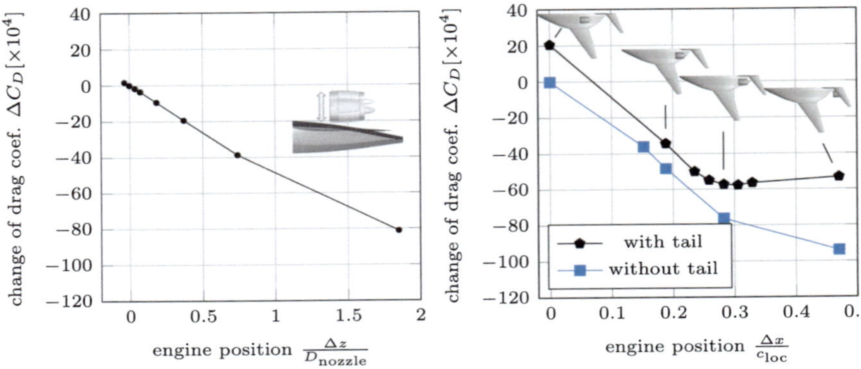

Fig. 14 Change in drag coefficient depending on the engine position (left hand side: vertical positioning, right hand side: streamwise positioning)

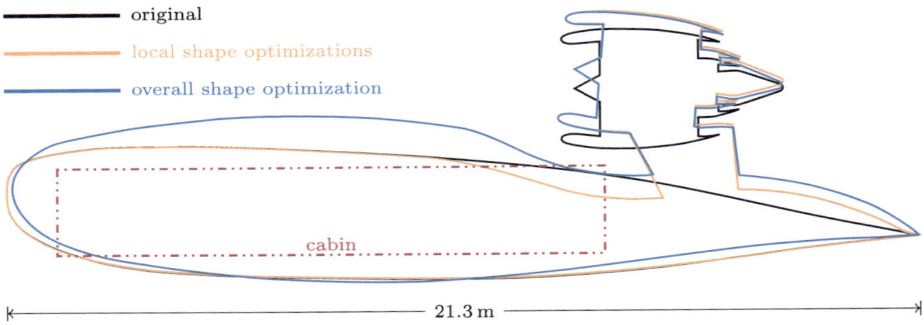

Fig. 15 Geometric shape at engine's center section before and after optimization

isolated nacelle) while also considering the engine pylons. The improvement is even higher, when the additional benefit in net thrust is considered. The total installation drag is therefore reduced by $\Delta C_D = 0.0107$ (44%) compared to the initial shape. Figure 16 compares the surface pressure distribution on the upper side of the initial and the modified shape at cruise conditions.

One critical aspect to mention at this point is that shaping the fuselage for drag reduction leads to locally thinner airfoil thicknesses. As a result, the optimal shape will affect the cabin size as shown by the locally optimized shape in Fig. 15. This is especially a problem of small-size blended wing bodies. A way to circumvent the decrease in cabin size is to increase the thickness of the airfoil ahead of the nacelle (overall shape optimization). This however leads to a shock ahead of the nacelle that eventually reduces the achievable drag reduction ending up in a improvement of only $\Delta C_D = 0.0099$. For a more detailed investigation of the engine integration, see [22].

The previous investigations lead one to ask whether the use of highly integrated engines should be reevaluated. Besides the classical advantage of better efficiency, they also provide a new advantage here, where the problem of installation-induced

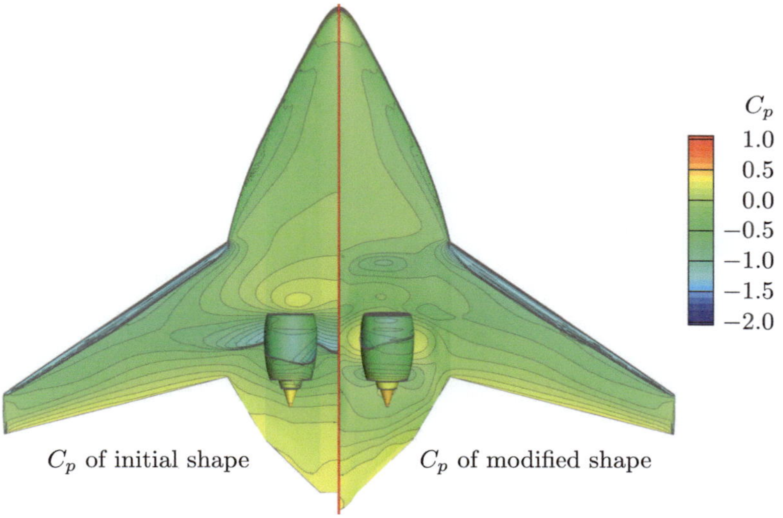

Fig. 16 Comparison of the initial shape against the modified shape for surface pressure distribution on the upper side at cruise conditions, see also [13, 22]

drag becomes much less prominent. Alternatively, an engine size reduction could be discussed. While UHBR engines seem a reasonable choice, it should be questioned whether there is a point where the higher installation drag of a large engine together with its excessive mass does not counteract the increased propulsion efficiency.

9 Structural Aspects

The preliminary structural design for the hybrid wing body is challenging because this aircraft deviates strongly from classic tube-wing concepts. Hence, the use of finite element (FE) model tools for classical aircraft—such as cpacs-MONA [23]—is complex and has its limits. In spite of its limitations, cpacs-MONA is used for structural design, where the center-body is modelled as a wing.

A FE model is created for all components of the hybrid wing body: vertical and horizontal tail plane, outer wing, center-body, control surfaces (aileron, elevator, rudder, trailing edge flaps). Apart from shell elements for the skin, spars and ribs, also beam elements for the stringers as well as ribs and spars stiffeners are modeled. Z-shaped, T-shaped and rectangle cross sections are used for the beams. The buckling fields—shell elements—are kept to a reasonable size by the distributed stiffener elements. In Fig. 17 the FE models of all components are shown. The structural design of all components except the inner wing has a common topology, i.e. a wing box with two or more spars or one spar for control surface structure. The distribution of the ribs

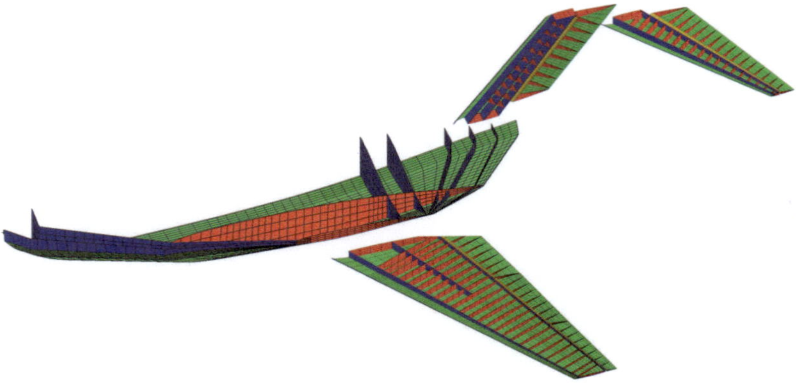

Fig. 17 Finite element models of center-body and outer wing

is similar to conventional aircraft. For the center-body, the topology of the structure is slightly different due to the required cabin volume. Here, stringers with a C-shaped cross section are used. These stringers represent the fuselage frames. Additionally, Z-shaped beams along the x-direction of the shells are defined as fuselage longerons. The rib inside the inner wing separates the cabin from the cargo area. The outer wing is attached at the outer rib of the inner wing. The spars separate the cabin from the rear fuselage and serve as fuselage main frames to transfer the loads from engine pylons and T-tail to the aircraft structure. These main frames increase the stiffness and moments of inertia in the rear fuselage.

In order to perform a static aeroelastic trim and Pratt analysis, an aerodynamic model is also created for each component. Here, a panel model based on the double lattice method is used. With the initial FE and aerodynamic model being set up, the next step will be the load calculation and the structural sizing. According to the CS25 [24] maneuver and gust loads for several flight conditions will be defined and considered in the load calculation. Based on these results, the initial estimated mass breakdown as given in Fig. 18, will be improved.

One uncertainty which remains is the load cases for pressurization of the cabin, which have not yet been considered. The FE model of the inner wing has sharp edges at the intersection between rib and spars, which is disadvantageous to pressure loads. The pressure loads are transferred via bending moments and not via in-plane loading, as is usual for circular pressure vessels. This would result in mass penalties due to heavier stiffeners and shell elements. One solution would be a combination of an integrated skin-shell concept with a segregated pressure shell concept [25–27]. Here, additional structural elements have to be added to the rear and sides of the cabin, while the upper and lower skin continue to be used for the pressurized cabin. However, there might always be a mass penalty for a hybrid wing body compared to a tube-wing concept.

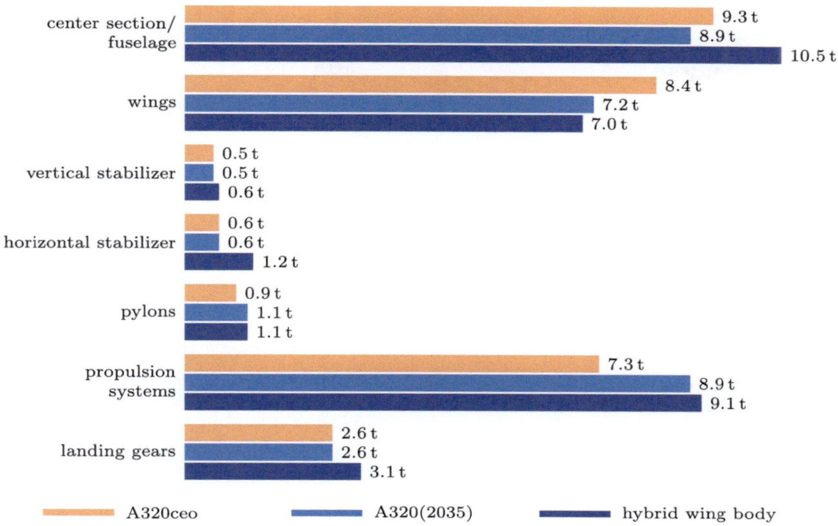

Fig. 18 Mass comparisons of the hybrid wing body against the A320ceo and an extrapolated A320 with 2035 technology. The hybrid wing body masses are still only estimates from the preliminary design

10 Landing Gears

The landing gear design of the nose and main landing gears is based on ground and landing load cases defined by FAR25 [28]. For the sizing of structural elements static loads are assumed. The landing gear consists of the following structural elements: main fitting, piston, axle, side strut, drag strut, actuator strut, tires. According to the defined load cases, the loads acting on the landing gears, namely forces and moments, are calculated considering minimal information about the aircraft and the landing gear; for example mass cases (maximum take-off mass, maximum landing mass) or aircraft geometry. T- and I-shaped cross sections are used for the beams. The maximum bending stresses and beam buckling according to Euler or Johnson [29], are considered in the design. ASTM300 steel and a standard aviation safety factor of 1.5 are applied in the landing gear sizing. In Fig. 19 the resulting nose and main landing gear designs are shown.

Landing gears are a main noise source during approach. Noise reduction techniques will be investigated in the future by conducting acoustic wind tunnel experiments. Since the landing gear structure is similar to an A320 landing gear, technologies such as those being investigated in the LNATRA project [30] are promising candidates.

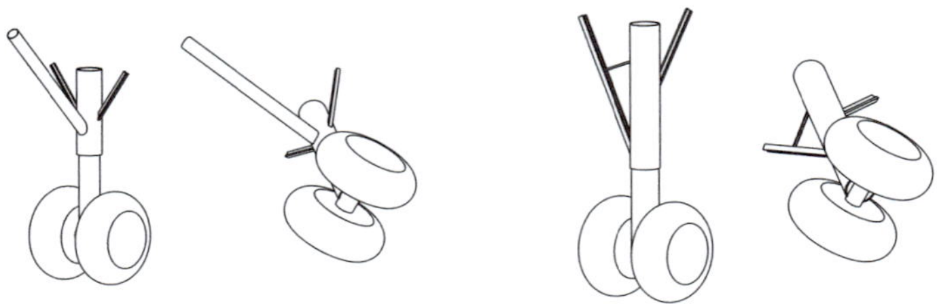

Fig. 19 Nose (left) and main (right) landing gear

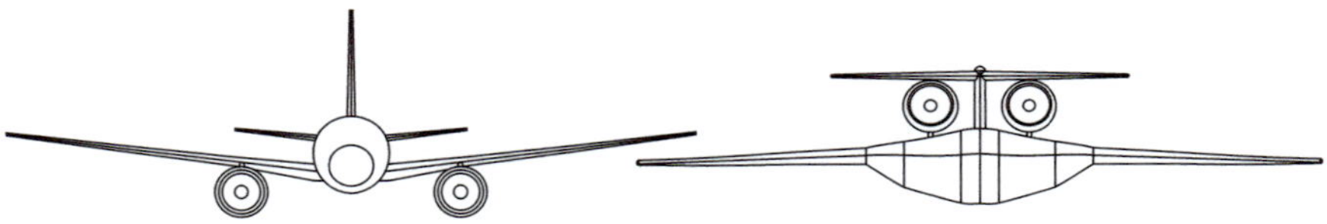

Fig. 20 Front view sketches of reference aircraft (left) and hybrid wing body (right)

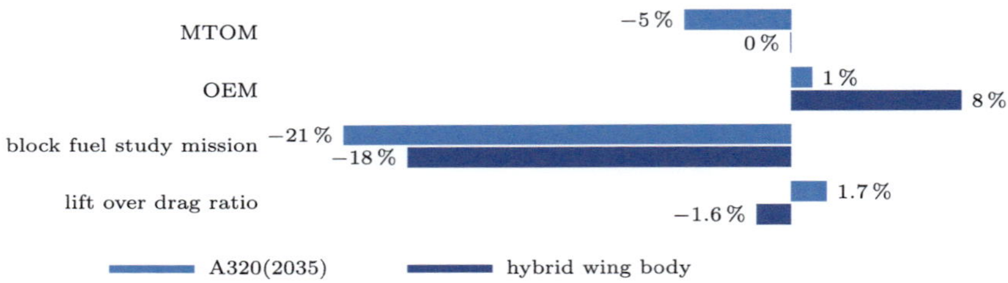

Fig. 21 Aircraft performance compared to an A320ceo

11 Entire Aircraft Performance

Although the hybrid wing body is designed for low noise as its primary goal, efficiency cannot be ignored. While high fidelity investigations are still ongoing, the preliminary design gives a good indication of how the aircraft performs. The aircraft is compared to a A320-like reference aircraft with assumed modern technology such as lighter wings and ultra high bypass engines. Figure 20 sketches the front view of the two aircraft. Figure 21 shows a performance comparison. Obviously, the block fuel is roughly 3.4% more for the study mission (800 nm), which is mainly due to the higher total weight.

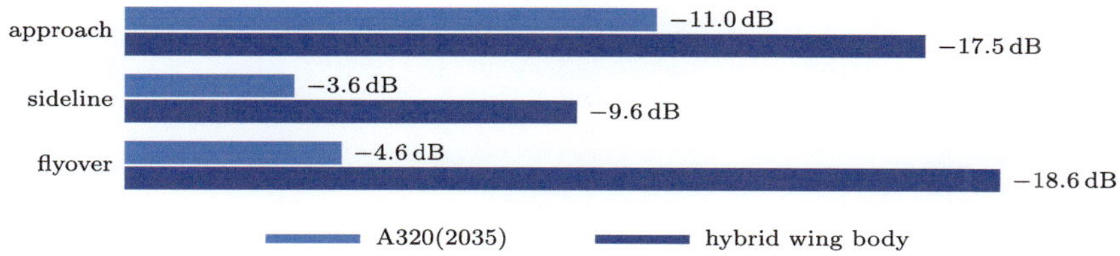

Fig. 22 Noise reduction in terms of EPNLdB for certification points, compared to A320ceo

12 Acoustic Properties of Entire Aircraft

For the noise impact at ground fast predictions are conducted by the use of semi-empirical component noise models. The simulation framework for the design and noise assessment of existing and novel aircraft concepts is described in [31]. In short, the process is comprised of engine and aircraft design synthesis codes, flight simulation, noise shielding assessment, and overall system noise prediction codes. Codes of different fidelity level are available for certain tasks, i.e. fully empirical models up to high-fidelity numerical simulation methods can be integrated. The setup of the process allows one to select reduction of aircraft noise as an objective early within the conceptual design phase. This allows an optimization of the vehicle and its individual flight performance for the lowest achievable noise impact. This process has been updated in order to capture novel vehicle concepts such as the hybrid wing body.

The simplified noise certification yields the EPNL at different noise certification locations, flyover, sideline and approach. The results are shown in Fig. 22 in terms of reductions compared to a A320ceo. Huge improvements can be seen at the certification points located directly below the aircraft (centerline). The sideline certification shows less improvements, but still almost 10 dB. The difference is explained by the fact that sideward engine shielding is not as effective as for other points (recall Fig. 13).

It is interesting to consider airframe and engine noise separately. Figure 23 shows the breakdown for the maximum A-weighted noise level for the three different aircraft. Clearly, engine noise dominates for A320-like aircraft for all certification positions. However, for the hybrid wing body, the engine is so well shielded that at least for the approach condition airframe noise becomes dominant. This makes it clear that with quiet engines and good noise shielding, treatments such as low noise high-lift and landing gears become more and more relevant.

In order to classify the noise improvements better, they are compared to the ACARE noise reduction goals [32]. Here, a reduction of perceived noise by 65% compared to a reference aircraft of year 2000 (e.g. an A320ceo) is required. This translates to a 15 dB reduction at all certification locations. The hybrid wing body design as a low noise design from ground up can be seen as being close to the maximum of what can be achieved in a mid-term perspective. It requires revolutionary aircraft designs such as the presented hybrid wing body to at least come even close

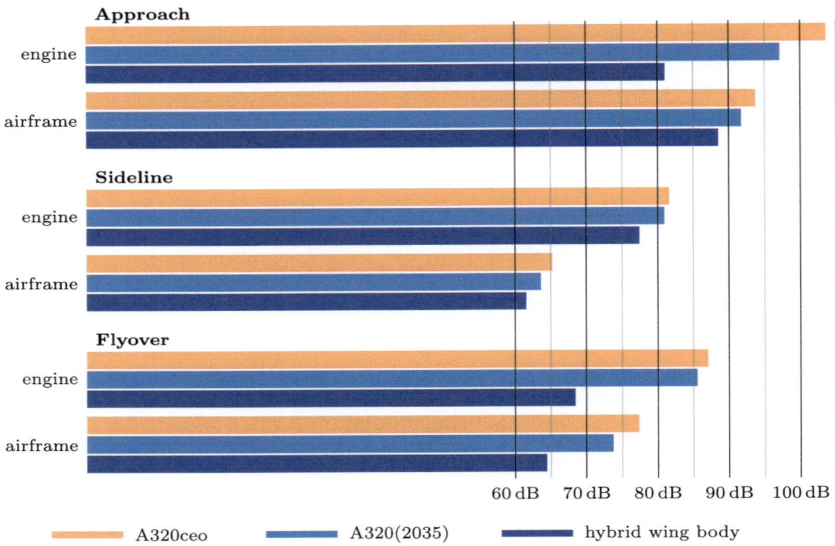

Fig. 23 Noise reduction for in terms of $L_{A\max}$ separately for airframe and engine, at different certification points

to these goals. But even then, the ACARE goals are still to be regarded as overly ambitious.

13 Other Aspects

Apart from the many studies so far in different disciplines for a better understanding of a hybrid wing body design, many more critical points must still be considered. A few aspects will be named in the following:

Aircraft family: For classical tube-and-wing aircraft it is comparatively easy to establish a whole aircraft family; for example such as in shortening or extending the fuselage length. This is much harder for blended and hybrid wing bodies due to the tight integration of the components.

Cargo: It is not easily possible to use LC3 containers for storing cargo within the aircraft. This means that loading the aircraft will be more time-consuming.

Evacuation: Finding a working evacuation concept is more difficult for blended and hybrid wing bodies. This has not been investigated for the aircraft discussed here.

Passenger comfort: Since the cabin is wider than classical tube-and-wing concepts, the outer passengers will experience stronger roll movements. For the given concept it is expected that this will not be a big problem, since the cabin is relatively

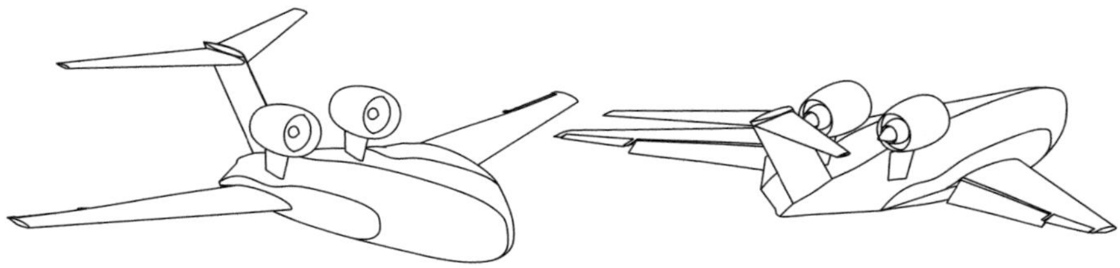

Fig. 24 Drawing of complete aircraft with flaps, ailerons, engines and tailplane

narrow anyway. Other comfort aspects are that a blended wing body will have less windows and that the engine-induced cabin noise is potentially higher.

At current understanding, all issues are not considered ultimately critical, but they rather would require compromises between community noise and cost or passenger comfort. It will require further investigations to make more definite statements.

14 Conclusion

The paper presents a hybrid wing body as a truly low-noise aircraft. A sketch of the resulting geometry is shown in Fig. 24. It is shown, that the potential for noise reduction is very high. Indeed, such an aircraft may be considered as an answer to the ACARE goals, coming close to the demanded reduction of perceived noise by 65% compared to a reference aircraft of year 2000. Due to the tail plane design, it is possible to avoid many uncertainties regarding the handling qualities and controllability compared to classical blended wing bodies.

On the other hand, there are still many shortcomings for the presented concept. First, there are obvious conflicts, such as the engine integration which needs very intrusive changes on the center-body or remaining shortcomings regarding handling qualities. Then, there is clearly the efficiency aspect: The aircraft is a low noise design with conservative design aspects such as flight safety as well as avoiding deeply integrated engines. These decisions lead to cuts regarding the efficiency. However, these aspects must be considered from the perspective that the presented aircraft is the result of only a first design iteration.

There is a lot of potential for improving the aircraft in future iterations. Besides shape adaptations, several design choices should be reconsidered. Most will come at a price of more noise or of increasing uncertainties where the latter most likely would move the expected entry into service further into the future. Potential improvements, which increase uncertainties are, for example, deeply integrated engines (boundary layer ingestion) or allowing worse handling qualities and compensate this with artificially enhanced stability. Other changes, which likely affect noise are stronger high lift systems or moving engines further to the rear. Also, it should be reconsidered whether smaller engines could be implemented. While it seems common sense, that the higher propulsion efficiency for ultra high bypass ratios is beneficial, it is not as clear for installation on top of the aircraft where integration is more difficult, and where also the increased weight might have negative secondary effects such as requiring a larger tailplane.

This work on the hybrid wing body tries to bring together multiple disciplines from the beginning in order to understand better the benefits and problems with such a configuration. Once more, this shows how difficult it is to establish revolutionary aircraft designs, but also demonstrates the potential for less aircraft noise pollution. Moreover, as a lesson learned, high fidelity analysis is essential to reveal difficulties arising from complex flow physics which are inaccessible to classical aircraft preliminary design tools, especially for new aircraft configurations.

Abbreviations

ACARE	Advisory Council for Aviation Research and Innovation in Europe
CFD	Computational fluid dynamics
c	Chord length
C_D	Drag coefficient
c_l	Local lift coefficient (in 2D section)
C_L	Lift coefficient
c_{loc}	Local chord width
$C_{L,\max}$	Maximum lift coefficient
$C_{L,\text{tot}}$	Total lift coefficient
$C_{L,\text{landing}}$	Lift coefficient during landing
C_p	Pressure coefficient
$C^*_{p,\text{crit}}$	Pressure coefficient, where Mach number based on the velocity normal to isobars exceeds speed of sound
BPR	Bypass ratio
D	Drag
D_{nozzle}	Nozzle diameter
DLR	Deutsches Zentrum für Luft- und Raumfahrt, German Aerospace Center
EPNL	Effective perceived noise level
f	Frequency
FAR	Federal Aviation Regulations
FE	Finite element
ICAO	International Civil Aviation Organization
L	Lift
$L_{A\max}$	Maximum A-weighted sound pressure level
LNATRA	Low noise ATRA (Advanced Technology Research Aircraft)
M	Mach number

M_{landing}	Mach number during landing
MTOM	Maximum take-off mass
NM	Nautical miles
RANS	Reynolds-averaged Navier Stokes
SIAM	SchallImmissionsArmes Mittelstreckenflugzeug; low-noise midrange aircraft
UHBR	Ultra high bypass ratio
x	x-Coordinate
α	Angle of attack

Acknowledgements This publication was funded by the Clean Sky 2 Joint Undertaking under the European Union's Horizon 2020 research and innovation program under grant agreement N° CS2-LPA-GAM-2020-2023-01.

References

1. Liebeck RH (2004) Design of the blended wing body subsonic transport. J Aircraft 41(1):10–25. https://doi.org/10.2514/1.9084
2. Okonkwo P, Smith H (2016) Review of evolving trends in blended wing body aircraft design. Prog Aerosp Sci 82:1–23. https://doi.org/10.1016/j.paerosci.2015.12.002
3. Chan DT, Hooker JR, Wick AT, Plumley R, Zeune C, Ol MV, DeMoss J (2017) Transonic Semispan aerodynamic testing of the hybrid wing body with over wing nacelles in the National Transonic Facility. In: 55th AIAA aerospace sciences meeting. No. AIAA 2017-0098, AIAA, Grapevine, Texas. https://doi.org/10.2514/6.2017-0098
4. International Civil Aviation Organization (2018) Aircraft operations: vol 1—flight procedures, 6th edn
5. International Civil Aviation Organization (2018) Aerodromes—Annex 14 to the convention on international civil aviation, 8th edn
6. Iwanizki M, Keller D, Schmidt F, Hasan YJ, Kurz J, Wienke F, Bertsch L, Mößner M, Balack P, Häßy J, Fröhler B (2022) Conceptual aircraft design activities related to blended wing body configurations in the scope of the DLR project SIAM. In: 9th European conference for aeronautics and space sciences. https://doi.org/10.13009/EUCASS2022-6104
7. Fröhler B, Iwanizki M, Zill T (2021) Conceptual design of a blended-wing-body for a short/medium range mission enhanced by high-fidelity aerodynamics. In: Deutscher Luft- und Raumfahrtkongress. https://doi.org/10.25967/550015
8. Military Specification, Flying Qualities of Piloted Airplanes (1980) MIL-F-8785C
9. Department of Defense, USA: Military Standard, Flying Qualities of Piloted Aircraft (2001) MIL-STD-1797A
10. Krengel MD, Bertsch L, Dahlmann K, Günther Y, Gerlinger B, Linke F, Wolters F, Vieweg M, Krüger WR, Blinstrub J (2020) Entwurf und Bewertung einer lärmarmen Kurzstreckenkonfiguration im Vorentwurf – Erkenntnisse aus dem DLR-Projekt KonTeKst. In: Deutscher Luft- und Raumfahrtkongress. https://doi.org/10.25967/530053

11. Streit T, Hoffrogge C (2017) DLR transonic inverse design code, extensions and modifications to increase versatility and robustness. Aeronaut J 121(1245):1733–1757. https://doi.org/10.1017/aer.2017.101
12. Eppler R (1990) Airfoil design and data. Springer Berlin Heidelberg.https://doi.org/10.1007/978-3-662-02646-5
13. Schmidt FN, Keller D, Iwanizki M, Seitz A (2023) High-lift-system design for a low-noise medium-range aircraft. In: Deutscher Luft- und Raumfahrtkongress. Stuttgart, Germany
14. Schwamborn D, Gerhold T, Kessler R (1999) DLR TAU-code—an overview. In: 1st ONERA/DLR aerospace symposium. Paris, France
15. Storms B, Hayes J, Jaeger SM, Soderman PT (2000) Aeroacoustic study of flap-tip noise reduction using continuous moldline technology. In: 6th Aeroacoustics conference and exhibit. no. AIAA-2000-1976, American Institute of Aeronautics and Astronautics, Lahaina, USA. https://doi.org/10.2514/6.2000-1976
16. Moreau A, Mennicken M (2023) Acoustic preliminary design of a low-noise fan stage considering a variable-area nozzle and variable-pitch rotor blades. CEAS Aeronaut J 14:325–341. https://doi.org/10.1007/s13272-023-00658-x
17. Lummer M (2019) Aircraft noise generation and assessment—installation: numerical investigation. CEAS Aeronaut J 10:159–178. https://doi.org/10.1007/s13272-019-00382-5
18. JetZero (2023). https://www.jetzero.aero. Accessed 10 Sep 2023
19. Gauvrit-Ledogar J, Tremolet A, Defoort S, Morel F, Liabocuf R, Méheut M (2022) Multidisciplinary design and optimization of the blended wing body configuration SMILE. In: 9th European conference for aeronautics and space sciences. Lille, France
20. Hooker JR, Wick AT, Hardin CJ (2017) Commercial cargo derivative study of the advanced hybrid wing body configuration with over-wing engine nacelles. Tech. Rep. NASA/CR-2017-219653, NASA
21. Leifsson L, Ko A, Mason WH, Schetz JA, Grossman B, Haftka RT (2013) Multidisciplinary design optimization of blended-wing-body transport aircraft with distributed propulsion. Aerosp Sci Technol 25(1):16–28. https://doi.org/10.1016/j.ast.2011.12.004
22. Keller D (2023) Aerodynamic investigation of propulsion integration for a low noise hybrid wing-body with podded UHBR turbo fan engines. In: 13th EASN international conference. Salerno, Italy, paper under submission
23. Klimmek T (2014) Parametric set-up of a structural model for FERMAT configuration for aeroelastic and loads analysis. J Aeroelast Struct Dyn 3(2):31–49. https://doi.org/10.3293/asdj.2014.27
24. European Aviation Safety Agency: Certification Specifications and Acceptable Means of Compliance for Large Aeroplanes—CS-25 (2011) Amendment 11
25. Hoogreef MFM (2012) The oval fuselage—a new structural design concept for blended wing body cabins. MA thesis, Delft University of Technology
26. Mukhopadhyay V (2005) Blended Wing Body (BWB) fuselage structural design for weight reduction. In: 46th AIAA/ASME/ASCE/AHS/ASC structures, structural dynamics and materials conference. AIAA, Austin, Texas. https://doi.org/10.2514/6.2005-2349
27. Vos R, Geuskens FJJMM, Hoogreef MFM (2012) A new structural design concept for blended wing body cabins. In: 53rd AIAA/ASME/ASCE/AHS/ASC structures, structural dynamics and materials conference. AIAA, American Institute of Aeronautics and Astronautics, Honolulu, Hawaii. https://doi.org/10.2514/6.2012-1998
28. Aviation Regulations (2023) Part 25—airworthiness standards: transport category airplanes. https://aviation-regulations.com/part/25. Accessed 05 Oct 2023
29. MechaniCalc (2023) Column buckling. https://mechanicalc.com/reference/column-buckling. Accessed 05 Oct 2023
30. Pott-Pollenske M, Almoneit D, Buchholz H (2021) Low noise ATRA—an aircraft noise reduction study based on retro-fit technologies. In: AIAA AVIATION 2021 forum. No. AIAA 2021-2117, AIAA. https://doi.org/10.2514/6.2021-2117
31. Wienke F, Bertsch L, Iwanizki M, Balack P, Häßy J (2023) System noise assessment of conceptual tube-and-wing and blended-wing-body aircraft designs. In: AIAA aviation forum. No. AIAA 2023-4170, AIAA. https://doi.org/10.2514/6.2023-4170

32. European Commission (2012) Flightpath 2050—Europe's vision for aviation report of the high-level group on aviation research. Tech. rep. https://doi.org/10.2777/50266

Fast Prediction of Fan Broadband Noise with an Analytical Aerodynamic Model Suited for Transonic Rotor Blades

Antoine Moreau and Robert Meier zu Ummeln

Abstract The usual and affordable way to predict broadband noise generated by aero-engine fans relies on empirical correlations established by NASA since the 70's. The present contribution depicts an alternative approach, based on analytical models for the noise sources and their aerodynamic key driving parameters. A simple new physical loss model suited for slender fan blades improves the prediction of aerodynamic stage performance and the rotor wake size, as well as providing a much more realistic sensitivity of fan broadband noise to incidence and working line variations.

Keywords Fan broadband noise · Meanline performance model · Transonic fan stage

1 Motivation and Background

1.1 Importance of Fan Broadband Noise and its Relation to Aerodynamic Loading

With the advent of ultra-high-bypass-ratio (UHBR) engines in the coming years, the ranking of engine noise sources is expected to change significantly. As shown by the authors in a previous study (see [14]) and depicted schematically in Fig. 1, old-generation engines were dominated at take-off conditions by jet noise and the fan tones, which were respectively the result of high exhaust flow velocity and supersonic rotor tip speed. With the next generation of UHBR engines, the substantial decrease in jet velocity and rotor tip Mach number is expected to reduce very efficiently those sources. A reduction in fan broadband noise is also expected, only to a lesser extent, owing to the shift of off-design operating points toward higher fan loading (details are found in [14]).

A. Moreau (✉) · R. Meier zu Ummeln
Department of Engine Acoustics, DLR, Institute of Propulsion Technology, Berlin, Germany
e-mail: Antoine.Moreau@dlr.de

F. Boden (ed.), *Numerical Methods and Experimental Means for the Integration of Future Propulsion Systems*, Springer Aerospace Technology,
https://doi.org/10.1007/978-3-031-97725-1_3

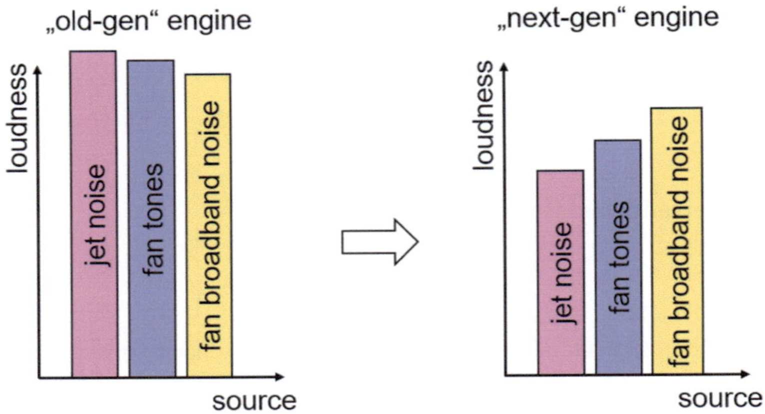

Fig. 1 Modification of the noise source ranking at take-off conditions accompanying the evolution of aero-engines

As a result, the relative importance of fan broadband noise has increased significantly in engine noise research. This source is known to be particularly sensitive to the aerodynamic loading of the fan, with sound power increasing by as much as 1 to 2 decibels for each change in degree of incidence experienced by the blades of the fan rotor ([3–6]). Moreover, UHBR engines tend to operate at strongly off-design conditions during take-off and approach ([14]) with substantial positive blade incidence. This can be mitigated by implementing a variable-area nozzle or a variable-pitch fan, which both affect the blade incidence and thus the generation of fan broadband noise, as shown by [14].

Therefore, for the researchers and engineers to develop a physical understanding of the UHBR-engine noise design problem, fast prediction methods for fan broadband noise prediction are needed, which should capture correctly the impact of aerodynamic loading and rotor blade incidence on noise generation based on a physical modelling approach rather than ad-hoc (single-case) empirical models or data-based approaches, which are currently becoming widespread with the growth of machine learning but do not easily provide insight. It is the purpose of the present contribution to introduce a method responding to these needs at the conceptual design level.

1.2 Methods to Predict Fan Broadband Noise at the Conceptual Design Level

The accurate prediction of fan broadband noise usually requires numerical simulations which resolve turbulent scales to some extent, such as large-eddy simulations (LES), or at least hybrid numerical approaches that are much more demanding than Reynolds-averaged Navier-Stokes simulations (RANS). At the conceptual level, this

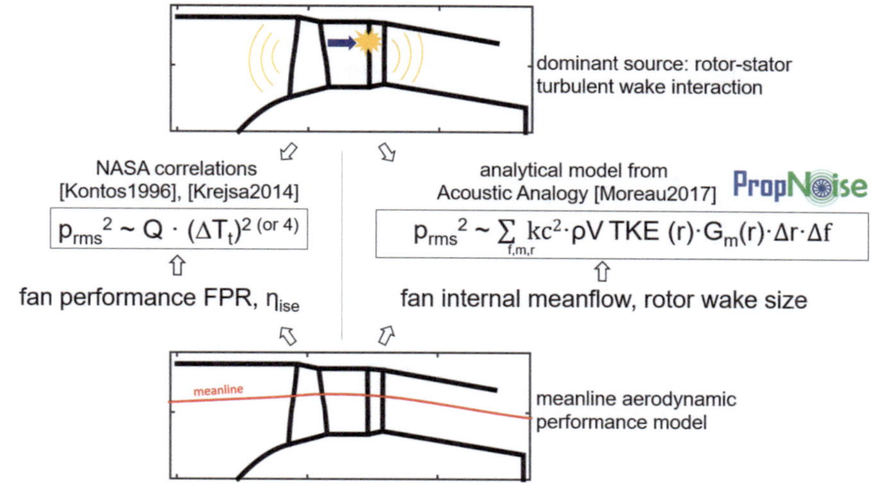

Fig. 2 Alternative methods to predict fan broadband noise at the conceptual design level

approach is not well suited. Hence, more simple, faster methods are usually geared to predict fan broadband noise in a realisable way.

Figure 2 depicts such methods, which may be classified in two categories. The first category represented on the left side of the figure consists of semi-empirical correlations derived from experiments. The common methods of this category are the correlations by NASA originally developed by Heidmann and later adapted by Kontos [9] and Krejsa [11] to more modern engines. These correlations basically express the average sound pressure in terms of global engine parameters. The second category is based on analytical models derived from Lighthill's Acoustic Analogy: the formulation for the sound pressure is based on the individual physical parameters of the flow inside the fan stage and thus describes in a physically more accurate way the generation of noise than the correlations of the first category ([13], fan noise prediction code PropNoise).

It should be noted at this point that the empirical correlations in the left part of Fig. 2 do not need to assume a priori which sound source mechanisms within the fan stage are responsible for the emission of broadband noise, whereas the analytical method of the right-hand side assumes that the fan broadband noise is dominated by the unsteady-lift response of the stator vanes to incoming rotor wake turbulence. This latter source is of a dipole-type. There may exist other mechanisms, such as quadrupole-type sources (interaction of incoming turbulence with the potential field of a blade row), and these may be also fairly sensitive to overall aerodynamic loading of the fan. However, owing to the difficulty in modelling these sources and the relatively moderate Mach numbers at the stator entry ($M < 0.65$), we will not consider these sources in the present study and will restrict the modelling work to the well-known dipole-like acoustic rotor-stator interaction.

The correlation by Heidmann and Kontos proposes the following scaling law for the squared sound pressure p^2_{rms}, where Q is the mass flow and T_t the total-temperature rise across the stage. This scaling law can be reformulated in terms of rotor tip speed U and the usual non-dimensional flow and work coefficients ϕ and ψ (see Abbreviations):

$$p^2_{rms} \sim Q \cdot (\Delta T_t)^2 \rightarrow p^2_{rms} \sim U^5 \cdot (\varphi \cdot \psi^2) \tag{1}$$

The effect of flow speed is given by the term U^5, with the expected 5th power dependency. The effect of loading is given by the product $(\phi \cdot \psi^2)$, which, however, predicts a too weak sensitivity to incidence variations, as will be seen later. Some years after the work of Heidmann and Kontos, an alternative correlation was proposed by Krejsa, which scales the squared sound pressure with the 4th power of the temperature rise:

$$p^2_{rms} \sim Q \cdot (\Delta T_t)^4 \rightarrow p^2_{rms} \sim U^9 \cdot (\varphi \cdot \psi^4) \tag{2}$$

This change leads to a modified loading term $(\phi \cdot \psi^4)$, which is slightly more sensitive to incidence variations, as shown later. However, the scaling with the 9th power of tip speed, instead of with the 5th, is not applicable for a dipole-type source, as would be the case for the rotor-stator turbulent wake interaction. The scaling laws by Heidmann/Kontos and Krejsa have been derived from static fan rig tests where the operating conditions were only varied along a given working line, so that significant variations of the aerodynamic loading were not part of the dataset used to derive those scaling laws. As will be seen in Sect. 3, this is of importance regarding the fan broadband noise prediction.

The alternative analytical approach presented in Fig. 2 relies more strongly on a physical modelling of flow conditions inside the fan stage; as a result, a markedly different form of scaling law has been obtained from the analytical model detailed in [13]:

$$p^2_{rms} \sim \sum_{f,m,r} kc^2 \cdot \rho \cdot TKE(r) \cdot G_m(r) \cdot \Delta r \cdot \Delta f \rightarrow p^2_{rms} \sim U^5 \cdot (\varphi^2 + \psi^2)^{\frac{3}{2}} \cdot A^2 \tag{3}$$

In Eq. 3, kc represents the stator chord-based non-dimensional wavenumber, $G_m(r)$ the in-duct Green's function for an acoustic mode of order m, and V is the flow velocity at the entry plane of the stator. The root-mean square pressure is obtained by summing up over all frequencies f, radial positions r, and modes m. The turbulent kinetic energy TKE is related to the rotor wake size A and an empirical proportionality factor K discussed in [16]:

$$TKE = K \cdot V^2 \cdot A^2 \tag{4}$$

The loading term is formed by $(\phi^2 + \psi^2)^{\frac{3}{2}} \cdot A^2$, where a new parameter appears which is related to the physics of the noise generation inside the fan stage: the size of the rotor blade wakes, mathematically described by the non-dimensional wake area A.

1.3 New Aerodynamic Performance Model

Both methods presented in the previous section, the empirical correlation and the analytical approach, need to be accompanied by a meanline performance model in order to be applicable at the conceptual level. This section first describes the new aerodynamic loss model implemented in the meanline approach, and how, in a second step, the rotor wake area can be derived from this model.

1.3.1 Original Semi-empirical Model

The model for the prediction of aerodynamic loss originally implemented in PropNoise [13] is based on widely used semi-empirical correlations that consider the off-design loss being proportional to the on-design loss by adopting an empirical correction factor based on the off-design incidence angle. The on-design loss is that for an incidence angle equal to zero or equal to the design incidence for which the blades shall operate most efficiently. The off-design loss is the loss at incidence values well away from the design incidence. The off-design loss coefficient is therefore usually modelled as $\omega = (1 + K \cdot (i - i_{des})^2) \cdot \omega_{des}$, where K is empirical and calibrated for a specific fan rotor and Mach range.

Despite the widespread use of this approach, there is no theoretical or physical evidence that legitimates it. The constant K is usually single-case empirical and only suited for a particular blade row and Mach number range. The main advantage of the approach is its simplicity and the empirically accepted and verified observation that the off-design loss curve scales with incidence angle and has a minimum around zero values. The new model presented in the next section respects that feature, while being theoretically founded and still remaining simple to derive and to understand.

1.3.2 New Aerodynamic Loss Model

Blades of fan rotors installed in aero-engines are usually very thin and weakly cambered over a large part of their span; this is particularly the case for supersonic rotors of engines of the previous generations but it is also true for lower-tip speed UHBR-engine fans. The reason for this is that a rotor with near-sonic or supersonic tip speed needs slender blades to operate efficiently during its main flight phase, which is cruise. However, this type of blade is prone to leading-edge flow separation at subsonic flow conditions if it is operating away from its optimal incidence. Even if separation can be avoided, it tends to produce relatively thick boundary layers at off-design incidence, which directly affects the size of the rotor blade wakes being convected downstream. An analytical model suited for those blades was derived in two independent past studies by Kramer and Stanitz [10] and by Freeman and Cumpsty [2]. As shown in Fig. 3, the model assumes that the forward part of the blades can be represented by thin flat plates placed at an angle of incidence.

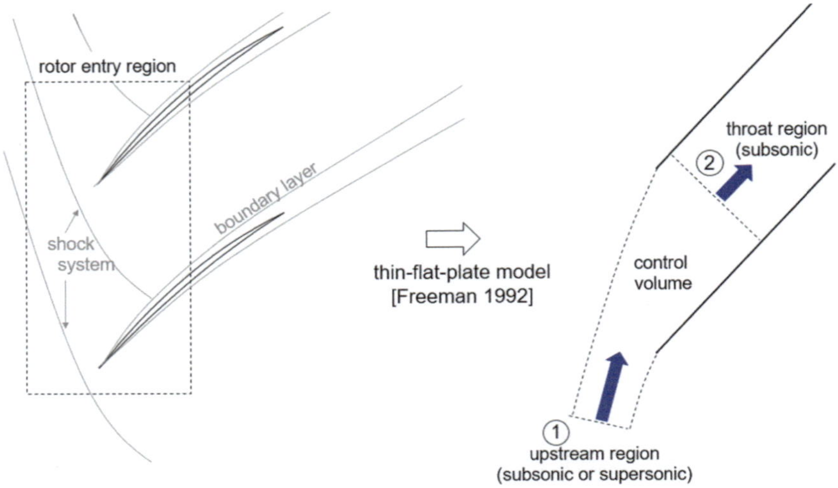

Fig. 3 Simplification of the rotor blade geometry according to the thin-flat-plate model and control volume considered for derivation of the model

A control volume is defined between the upstream region (1) and the throat region (2) inside the blade passage. By applying the conservation of mass flow and of enthalpy, and by assuming that the flat plate cannot exert any force along its own orientation direction, the subsonic flow conditions can be calculated analytically. This, in turn, provides an expression for the aerodynamic loss coefficient induced by the shock system and the boundary layer between positions (1) and (2).

$$\begin{aligned}
&(1)\text{ mass flow: } Q_2 = Q_1\\
&(2)\text{ enthalpy: } T_{t2} = T_{t1}\\
&(3)\text{ momentum along blade direction:}\\
&P_2+\rho\cdot V_2^2+F_{blade} = P_1+\rho\cdot V_1^2\cdot\frac{\cos\beta_1}{\cos\chi}\cdot\cos(\beta_1-\chi),\text{ with } F_{blade}=0
\end{aligned} \tag{5}$$

From the system of Eqs. 5, the loss coefficient defined according to the usual compressor conventions can be expressed as a function of the inflow Mach number and angle in the relative reference frame, where it also depends on the rotor meanline stagger angle:

$$\omega = \frac{P_{t1}-P_{t2}}{P_{t1}-P_1} = f(M_1,\beta_1,\chi),\text{ with } \beta_1 = \chi + i \tag{6}$$

As this new loss model inherently includes the effect of Mach number, rotor blade stagger angle and incidence, it does not need to be calibrated for a specific fan

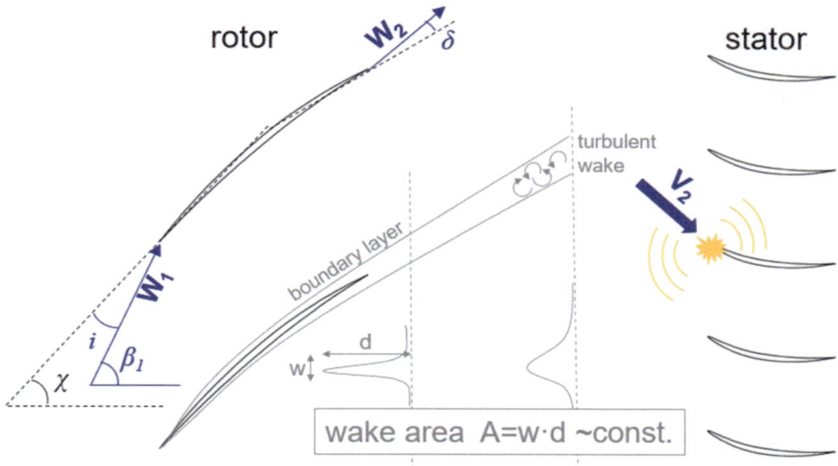

Fig. 4 Simplified model for the interaction of the rotor turbulent wake with the stator vanes based on the non-dimensional wake area

application, unlike the original semi-empirical model mentioned previously, and can be applied as long as the basic assumption of slender blades is fulfilled.

1.3.3 Wake Area Model

The size of the rotor blade wakes is represented by a non-dimensional quantity, the wake area, denoted A, which is the product of the also non-dimensional wake depth d and the wake width w, see Fig. 4 and [13] for details of the definitions. As a wake decays downstream of a rotor, its velocity deficit decreases but at the same time its spatial extent increases such that the product of both quantities is nearly constant in the streamwise direction. For that reason, the wake area is a convenient quantity to model the problem. As presented in the previous section, the aerodynamic loss coefficient calculation results from a control volume analysis where the share of the viscous and shock loss contribution is a priori not known. In the present model, it is assumed that the viscous loss contribution to the total loss is the same as that obtained in an incompressible flow, for which the analytical solution assumes a very simple form:

$$\omega_{visc} = \omega_{incomp} = \omega(M \rightarrow 0) = \frac{i^2}{\cos^2(\chi)} \tag{7}$$

The rotor blade incidence is given by i, and the angle of the rotor blades (representative angle of the forward-part blade geometry, at meanline radial position) is denoted χ. In addition to the viscous loss contribution given in Eq. 7, and which results from flow separation at the leading edge of the blades, another term should be added that is related to the friction of the flow along the blade surfaces:

$$\omega_{fric} = C_f \cdot \frac{2c}{h} \cdot \frac{\cos^2(\beta_1)}{\cos^2(\chi)} \quad (8)$$

In Eq. 8, C_f represents the average skin-friction coefficient of the boundary layer and the ratio 2c/h represents the ratio of blade-wetted surface to flow passage area. The sum of the viscous and friction loss coefficients can be related to the boundary layer momentum thickness according to the approach proposed in [12]. It is also assumed that the boundary layer momentum thickness θ divided by the flow passage height h is equal to the wake area in the far wake.

$$\omega_{visc} + \omega_{fric} = \omega_{BL} = 2 \cdot \frac{\theta}{h} \cdot \frac{\cos^2(\beta_1)}{\cos^2(\chi)}, \text{ with } \frac{\theta}{h} \sim A \quad (9)$$

Finally, we obtain a simple expression for the non-dimensional wake area:

$$A = C_f \cdot \frac{c}{h} + \frac{1}{2} \cdot \frac{i^2}{\cos^2(\beta_1)} \quad (10)$$

The first term of Eq. 10 scales with the overall wetted surface in the flow passage and does not depend on loading. This is the dominant term in blade rows where the boundary layers remain well attached and no flow separation occurs. The second term starts to increase rapidly as soon as the incidence angle on the rotor blades becomes significant (only a few degrees). This is the term mostly responsible for the sensitivity of fan broadband noise to aerodynamic loading in transonic fans with thin blades.

2 Validation of the Performance Model

2.1 Validation Test Cases

Two realistic fan applications are considered for validation of the model. The first test case is the NASA Source Diagnostic Test (SDT) fan stage, which is a well-documented fan stage dedicated to aero-acoustic research. The test case was made available by Ed Envia from NASA as part of a fan broadband noise benchmark held at the AIAA Aeroacoustics Conference in Atlanta, 2014. A photograph and a sketch of the fan are shown in Fig. 5. The NASA SDT fan has been thoroughly characterized aerodynamically and acoustically in [8, 17], respectively. The fan pressure ratio FPR and relative tip Mach number M_{tr} at the design point are detailed in Table 1; the NASA SDT is designed to operate at supersonic speeds over a wide range of blade spans. Two validation exercises are performed based on the data from that fan:

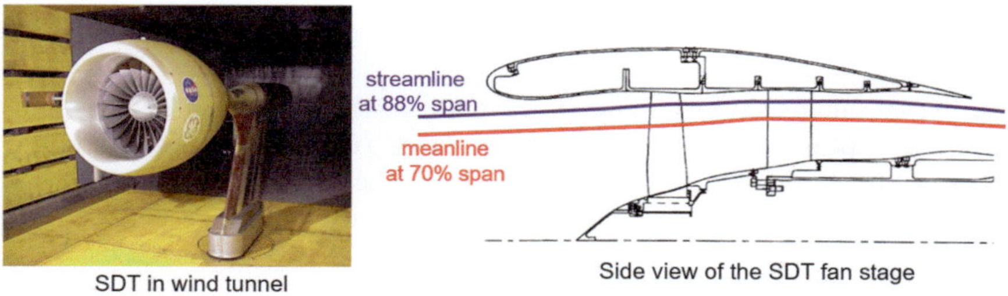

Fig. 5 Sketch of the NASA SDT fan stage with R4 rotor and baseline radial stators, from [8, 17]

Table 1 Performance of the test fan stages at design point

Fan stage	FPR	M_{tr}
NASA SDT	1.5	1.27
ACAT1	1.4	1.14

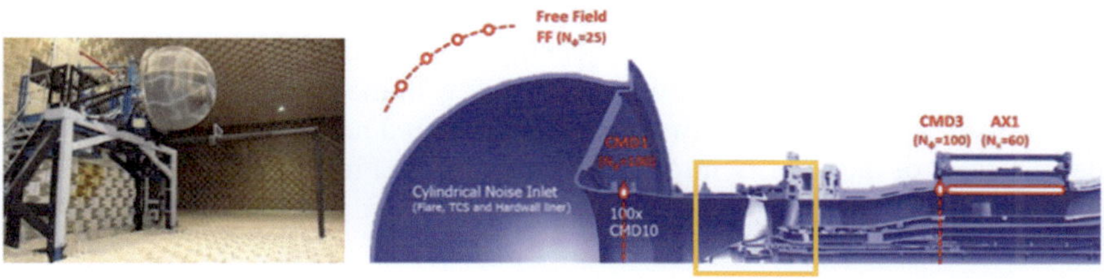

Fig. 6 Sketch of the UFFA fan rig at AneCom AeroTest and installed ACAT1 fan stage with instrumentation, from [1, 7]

- the 2D performance of the rotor can be validated by comparison with CFD simulations originally carried out and presented in [15] to validate a model for rotor shock noise. In that case, the rotor was simulated along a stream-line at 88% span (see Fig. 5), sufficiently far away from the endwall boundary layer, but still sufficiently far in the outer blade part so that the assumption of slender blades is still applicable.
- the full-stage and full-span aerodynamic performance is validated by comparison with the experimental fan map published in [8]. In that case, the representative meanline is taken at 70% of the blade span (see Fig. 5).

The other fan application originates from the EU-funded project TurboNoiseBB [1], in which a fan stage was measured at the AneCom AeroTest (ACAT) aero-acoustic UFFA fan rig, in Wildau, Germany, see Fig. 6 and [7]. This fan (designated ACAT1) has a slightly lower pressure and tip speed than the SDT fan, but still operates supersonically at design conditions and has slender rotor blades.

2.2 Validation of Performance

The first validation exercise is limited to a two-dimensional test case of the rotor-alone performance of the NASA SDT fan at 88% blade span. A wide range of operating conditions has been simulated, as shown in Fig. 7. The different lines represent lines of constant rotational speed along which the aerodynamic loading of the rotor is varied by increasing the back-pressure behind the rotor, resulting in lower mass flow, lower Mach number and an increase in pressure ratio. The new model is depicted by the blue line, the semi-empirical model originally implemented is in yellow, and the reference data obtained from 2D CFD simulations shown as black asterisks. This same color code will be retained throughout the paper. It should be mentioned at this point that the CFD simulations are considered correct as the numerical mesh was carefully designed around the blades in order to capture properly the decay of shocks, the blade boundary layer was fully resolved with an y^+ around 1, and finally, the absence of endwalls and three-dimensional effects mitigates the risk of modelling uncertainties [15]. As can be seen from the upper part of Fig. 7, a very good match with CFD results is achieved with the new model in terms of rotor pressure ratio Π_{rotor}. The rise of that quantity is less pronounced when decreasing the Mach number along a speed line, compared with the original model. However, the most striking difference between both original and new model is observed with the variations of the rotor isentropic efficiency η_{rotor}. First, the decay of efficiency

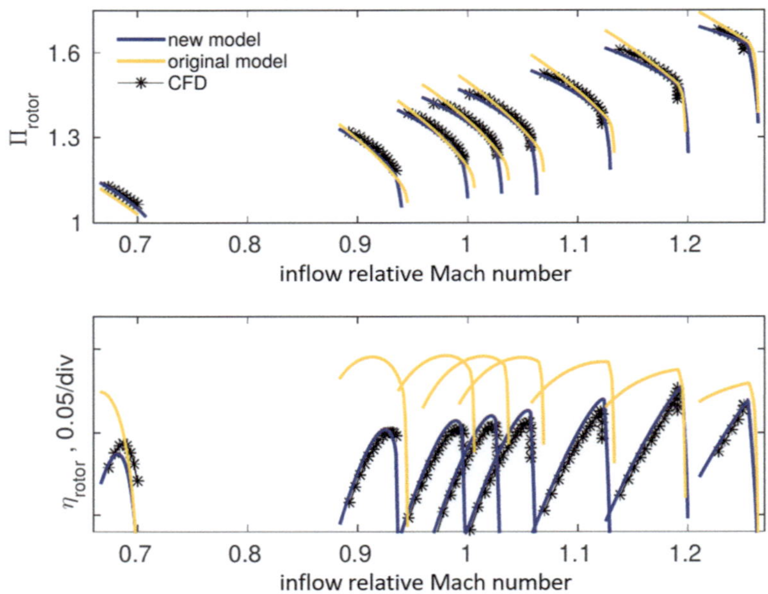

Fig. 7 Validation of the 2D performance model for the rotor-alone NASA SDT test case at 88% blade span

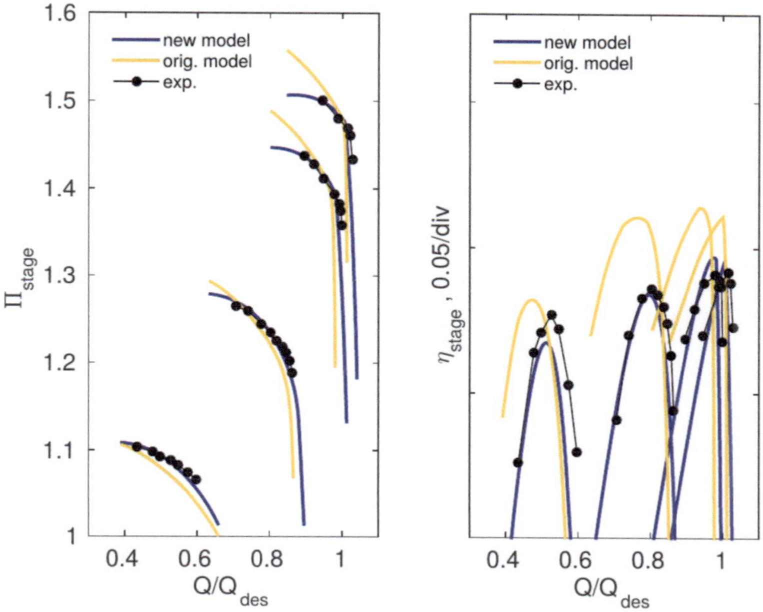

Fig. 8 Validation of the full-span, full-stage performance of the NASA SDT fan, predicted at 70% blade span with the meanline models

while loading the rotor at constant speed is much higher and more realistic with the new model. Secondly, the values of efficiency are more accurately predicted with the new model. The absolute values of efficiency are not disclosed for intellectual-property reasons, only the variation of 5 points of efficiency per division (0.05/div) is given; the peak efficiency lies in the range of 90%, which is typical of well-designed transonic rotor blades. As a summary, the present validation test case demonstrates the validity of the thin-flat-plate assumption and the relevance of the analytical model presented in the previous section, at least whenever the blades have a slender shape and for the case of 2D flow.

The next validation test case addresses the question of whether such a simplified 2D model at a representative meanline radius can be applied to represent properly the full-span performance of the rotor, and also to represent the performance of the full stage including the stator vanes located behind the rotor. The result of this validation is shown in Fig. 8. Here, the reference data are experimental, obtained from integration of the full-span full-stage performance with pressure rakes. On the horizontal axis, the flow Mach number has been replaced by the ratio of mass flow to the design mass flow Q/Q_{des}.

The conclusions drawn from Fig. 8 are similar to those discussed in the 2D test case. The new model provides significantly better predictions. However, the match is here less accurate than for the 2D case because in this full-span, full-stage case the stator vanes and some of the rotor sections, especially those in inner region toward the

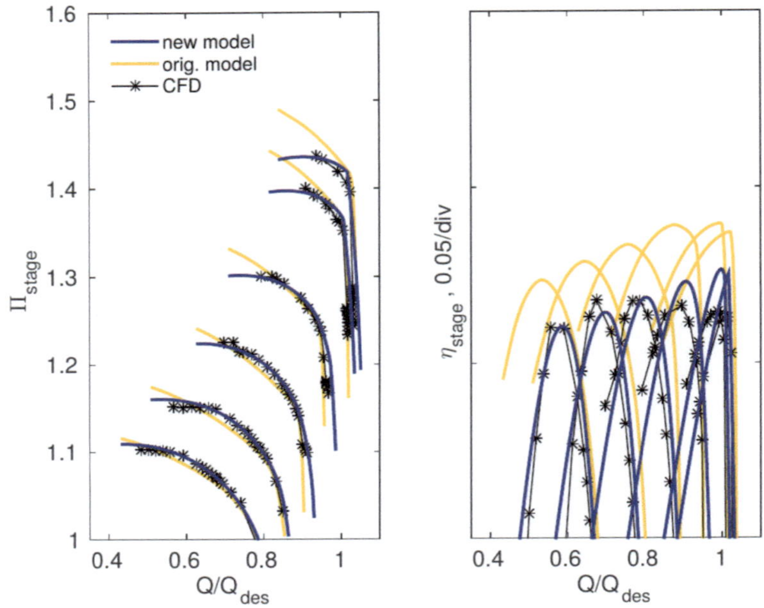

Fig. 9 Validation of the full-span, full-stage performance of the ACAT1 fan, predicted at 70% blade span with the meanline models

hub, do not fully fulfill the slender-blade assumption as well as in the outer sections. However, these inner sections operate at lower Mach numbers so their contribution to the overall performance would be limited.

The full-span, full-stage performance model has also been validated with the ACAT1 fan test case, as can be seen in Fig. 9. Here the reference data are extracted from 3D CFD RANS simulations [16]. Here again, the new model predicts some features of the fan map more realistically than the original model:

- less pronounced rise of fan pressure ratio while increasing the loading.
- steeper decay of efficiency with increasing loading.
- more accurate absolute values of pressure ratio and efficiency.

2.3 Validation of Wake Area

As the fan map of the ACAT1 stage was calculated with 3D CFD simulations, it is also possible for that fan test case to extract the rotor loss coefficient and the rotor wake area at 70% blade span and compare them with the values obtained from the meanline model presented in Sect. 1.3. The comparison is shown in Fig. 10 for a selection of four speed lines.

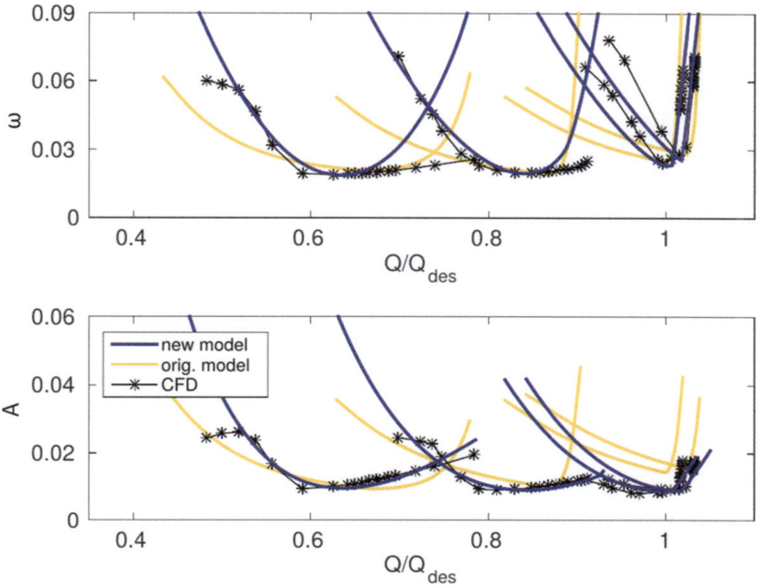

Fig. 10 Validation of the predictions for the rotor loss coefficient ω and the wake area A, in the ACAT1 fan test case, at 70% blade span

For each speed line, a minimum of loss coefficient and wake area is observed, which corresponds approximately to the design incidence condition. Near this region, both new and original models predict realistic and similar absolute values, except for the high-speed domain where the original model overestimates the wake area. In the region to the left of the design condition, which corresponds to a higher aerodynamic loading and lower mass flow, the original model underpredicts both loss coefficient and wake area (except for the wake area at high speeds), whereas the new model provides a better match with the extracted CFD data. We will see in the next section that this has an impact on the sensitivity of the predicted broadband sound power levels to changes in aerodynamic loading. It should be noted that the region right of the design condition, toward choking or negative incidences, is less of interest as the fan usually does not operate in that region at the acoustically relevant points.

3 Acoustic Validation

This section presents the validation of the predicted broadband sound power levels, according to the methods detailed in Sect. 1.2. The data are compared with the experimentally determined sound power levels from the TurboNoiseBB acoustic tests with the ACAT1 fan stage. The effect of aerodynamic loading is examined by considering

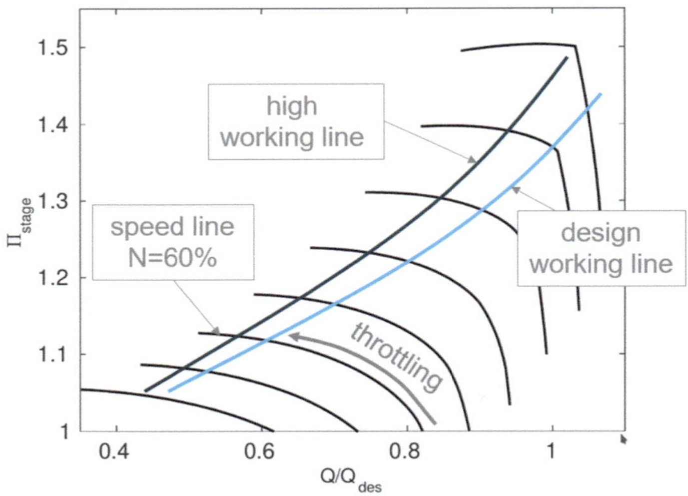

Fig. 11 Variation of operating points in the ACAT1 fan map in order to examine the effect of aerodynamic loading on broadband noise

two operating-point variations, as depicted in Fig. 11: the first one is a variation of the loading along the approach speed line at the rotational speed N=60% of the design speed; the second one is the variation between two working lines, the design working line (also called cruise or low-noise working line and crossing the design point) and a working line located at a higher aerodynamic loading (also called sea-level static working line).

3.1 Variation Along a Speed Line

The first variation is obtained by throttling the fan along the constant-speed line at N = 60% of the design speed, which roughly corresponds to approach conditions in terms of the flow Mach number range. Throttling the fan increases the aerodynamic loading by decreasing the mass flow and by increasing the incidence angle on the rotor blades. The variations of upstream and downstream sound power PWL against incidence i are shown in Fig. 12. In contrast to the previous Fig. 7, 8, 9, 10 and 11, the low-loading region is now towards the left of the horizontal axis whereas the high-loading region is to the right, toward larger incidence values.

The color and symbol code used in the figures presented so far have been retained here: the blue and yellow curves refer, respectively, to the analytical acoustic results calculated based on the new and original meanline performance models. Two sets of reference data are shown: the black points are experimentally measured data, whereas the black curve with asterisk symbols results from calculations with the

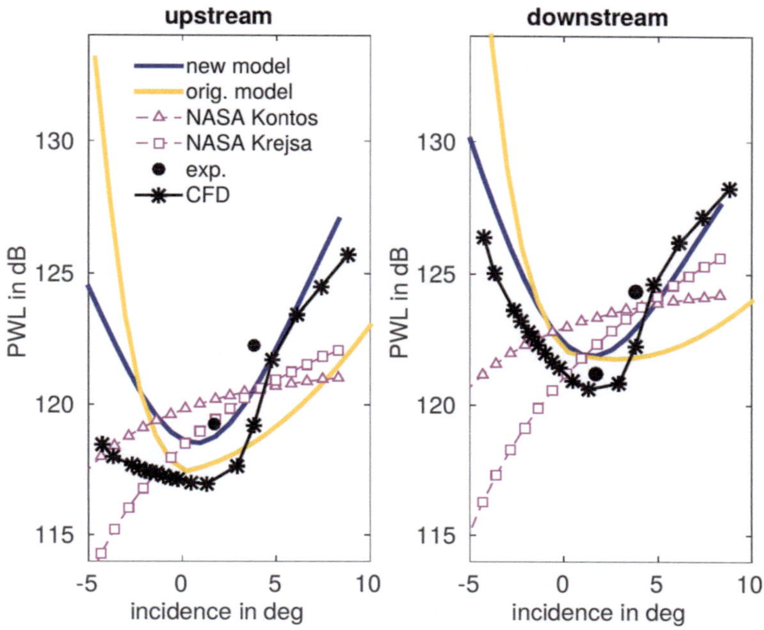

Fig. 12 Variation of upstream and downstream emitted broadband sound power PWL versus rotor incidence angle i

same analytical acoustic model which has been fed by 3D RANS CFD data. Finally, the pink curves with symbols show the correlations by Kontos and Krejsa.

Several conclusions can be drawn by comparing the predictions with the reference data:

- both meanline performance models (new and original) show a minimum of sound power near the zero-incidence angle of the rotor. This is attributable to the minimum of wake size produced by the blades for vanishing incidence. This is confirmed by the CFD-fed acoustic prediction, for which the aerodynamic input is much more detailed than with the meanline approach and can therefore serve as a good reference.
- The NASA correlations do not predict any minimum of sound generation near zero incidence, but rather predict continuously decreasing sound power as the incidence is reduced toward negative values. As detailed in Sect. 1.2, these correlations only consider global fan stage parameters and do not include any quantity related to the size of the rotor wakes, so they do not capture the minimum in the sound generation.
- In the range of positive incidence angles, the slope of sound power increase against incidence is higher and apparently more in line with the reference data in the case of the new model, compared with the original model. This is related to the slope of the wake area discussed in Sect. 2.3 and shown in Fig. 12. Moreover, the slope of

sound power is largely underestimated by the NASA correlations, with especially the Kontos value being much too low.

- Similar conclusions are valid both for the upstream and for the downstream emitted sound power, except for the slightly higher levels on the downstream side, mainly due to the convective effect of the flow on sound propagation.

3.2 *Variation Between Two Working Lines*

From the previously discussed variation shown along a speed line, it could be qualitatively observed that the sensitivity of sound power to variations of rotor incidence angle differs quite significantly depending on the models considered, and seemed to be improved with the new aerodynamic model introduced in this present study. This section aims at providing a more quantitative estimate of that noise sensitivity to incidence.

Figure 13 shows the gradient of sound power dPWL/di relative to variations of rotor incidence angle i. This quantity is calculated from the relative changes between the high and the design working line introduced in Fig. 11. The same color and symbol code as in Fig. 12 has been implemented here.

In the range of relative tip Mach number M_{tr} between 0.5 and 0.9, the reference data (purely experimental and estimated from CFD) indicate a gradient of roughly

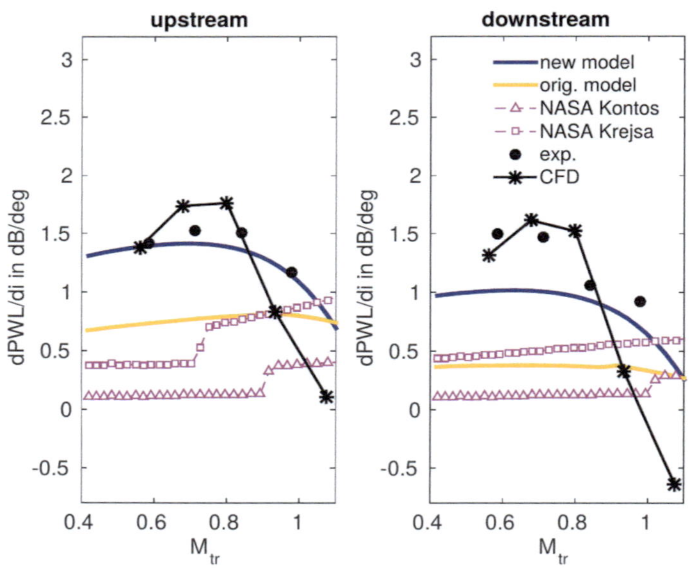

Fig. 13 Gradient of sound power relative to rotor incidence angle, for various values of the tip relative Mach number

1.5 decibels per degree of incidence. At higher Mach numbers approaching supersonic conditions, the gradient obtained from the experimental data seem to decrease slightly, and the CFD-based model data drop rapidly. In the past, experimental ([3, 5]) and theoretical ([4, 6]) studies have already documented the sensitivity of fan broadband noise to incidence angle variations: a relatively wide range of values between 1 dB/deg up to 2.5 dB/deg has been reported. That spread of values may be attributed to the different nature of broadband source mechanisms involved in each experiment/study and to the fact that quadrupole-type sources, expected to be dominant at high Mach numbers, are also more sensitive to variations of aerodynamic loading, and hence to the incidence angle. Nevertheless, the reference data obtained from the ACAT1 fan tests are in line with past studies, even if they are located at the low end of the range.

From the different models tested and compared, the best match is given by the new model (blue curve), with a noise sensitivity of around 1–1.3 dB/deg. This is substantially closer to the reference data than what is predicted by the original model. The NASA correlation by Krejsa is roughly in agreement with the original model however, as discussed in Sect. 1.2, the dependency on tip speed U follows a 9th-power scaling law, which is known to be unrealistic. The NASA correlation by Heidmann/Kontos with the correct 5th-power scaling law for U, provides, however, a much too low noise sensitivity of only 0.1–0.4 dB/deg.

4 Conclusions

An analytical model for the prediction of aerodynamic loss caused by thin flat plates under incidence has been presented and integrated into a meanline performance code. It is particularly well suited for slender rotor blades of transonic fans. Based on the prediction of the aerodynamic loss coefficient, the viscous loss related to the development of the blade boundary layer is calculated, which also provides an estimation for the size of the rotor blade wakes, an important parameter driving the generation of fan broadband noise.

With this model, the prediction of fan meanline aerodynamic performance has been significantly improved compared with the original semi-empirical model. Absolute values of fan pressure ratio and fan isentropic efficiency are well matched over a wide range of operating conditions for at least two transonic fan rotors. In particular, the drop of fan efficiency and the rise of loss, both observed when throttling the fan away from its design point, are predicted with a very good accuracy. As a result, the growth of the rotor wake size caused by incidence is higher and more realistically predicted with the new model.

This has consequences for the prediction of fan broadband noise. This noise component is dominated by the interaction of the rotor wakes–and its associated flow velocity perturbations and turbulence–with the downstream stator vanes; it is thus strongly influenced by the rotor wake size. It is shown in the present study that the sensitivity of fan broadband noise to incidence variations is higher and more realistic when calculating the rotor wake size based on the new aerodynamic flat-plate model. A sensitivity of 1.0–1.3 decibels per degree of incidence is obtained in that case, whereas the usual NASA correlations by Kontos and Krejsa provide much lower values of only 0.1–0.7 dB/deg. Experimentally measured sensitivities are in the range of 1–2 dB/deg. As the NASA correlations were derived from measured data distributed along a single working line for each fan, it is expected that they are not able to capture properly the effect of incidence. Owing to its more physical description of fan broadband noise generation, the present model captures more correctly the incidence effect.

Based on the improved prediction of the aerodynamic loading effect, further fan broadband noise sources other than the rotor-stator turbulent interaction may also be able to be predicted more accurately; this includes rotor trailing edge noise, whose generation is driven by the boundary layer thickness of the rotor blades, a quantity closely related to the rotor wake size. In the future, new sources may become relevant in strongly integrated engines, such as the inflow distortion noise. The prediction of these sources may benefit from the improved wake model to describe the rotor wake modulation caused by the distorted inflow.

Abbreviations

A	non-dimensional rotor blade wake area
FPR	fan-stage total-to-total pressure ratio
i	rotor blade incidence angle, in deg
M_{tr}	relative inflow Mach number at the rotor tip
N	rotational speed of the rotor, in percent of the design speed
PWL	sound power level, in dB
p_{rms}	root-mean-square of broadband sound pressure, in Pa
Q, Q_{des}	mass flow, design mass flow, in kg/s
U	rotor tip speed, in m/s
V_x	rotor-face axial flow velocity, in m/s
T_t	total-temperature rise across the stage, in K
Π	total-to-total pressure ratio
η	isentropic efficiency
ϕ	flow coefficient $\phi = V_x/U$
ψ	work coefficient $\psi = c_p \Delta T_t/U$
β_1	inflow angle at meanline in the rotor frame, in deg
χ	representative rotor blade stagger angle at meanline, in deg
ω	aerodynamic loss coefficient $\omega = (P_{t,rel,in} - P_{t,rel,out})/(P_{t,rel,in} - P_{in})$

Acknowledgements Parts of this work have been funded from the Clean Sky 2 Joint Undertaking under the European Union's Horizon 2020 research and innovation program under grant agreement N° CS2-LPA-GAM-2020-2023-01.

References

1. European Commission CORDIS presentation webpage of the EU research project TurboNoiseBB ("Validation of improved turbomachinery noise prediction models and development of novel design methods for fan stages with reduced broadband noise"), time period 2016–2020. https://doi.org/10.3030/690714
2. Freeman C, Cumpty N (1992) Method for the prediction of supersonic compressor blade performance. J Propuls 8:199–208. https://doi.org/10.2514/3.23461
3. Ginder R, Newby D (1977) An improved correlation for the broadband noise of high-speed fans. J Aircr 14:844–849. https://doi.org/10.2514/3.58864
4. Glegg S, Jochault C (1998) Broadband self-noise from a ducted fan. AIAA J 36
5. Gliebe P (1979) The effect of throttling on forward radiated fan noise. In: AIAA-79-0640
6. Gliebe P, Ho P, Mani R (1995) UHB engine fan broadband noise reduction study. Technical Report Final report
7. Guérin S, Kissner C, Kajasa B, Jaron R, Behn M, Hakansson S, Pardowitz B, Tapken U, Meyer R, Enghardt L (2019) Noise prediction of the ACAT1 fan with a RANS-informed analytical method: success and challenge. In: 25th AIAA/CEAS Aeroacoustics conference. No. AIAA-2019-2500, AIAA (2019). https://doi.org/10.2514/6.2019-2500
8. Hughes C (2001) Aerodynamic performance of scale-model turbofan outlet guide vanes designed for low noise. Technical Report NASA TM-2001-211352, 2001, NASA
9. Kontos K, Janardan B, Gliebe P (1996) Improved NASA-ANOPP noise prediction computer code for advanced subsonic propulsion systems, vol 1. Technical Report NASA CR-195480, NASA
10. Kramer J, Stanitz J (1955) Prediction of losses induced by angle of attack in cascade of sharp-nosed blades for incompressible and subsonic compressible flow. Technical Report NACA Technical note 3149
11. Krejsa E, Stone J (2014) Enhanced fan noise modelling for turbofan engines. Technical Report NASA contractor report 2014-218421, NASA
12. Lieblein S (1959) Loss and stall analysis of compressor cascades. J Basic Eng 81:287–400 (1959). https://doi.org/10.1115/1.400848
13. Moreau A (2017) A unified analytical approach for the acoustic conceptual design of fans of modern aero-engines. Phd thesis, Technical University of Berlin. https://doi.org/10.14279/depositonce-5935
14. Moreau A, Schnell R, Mennicken M (2023) Acoustic preliminary design of a low-noise fan stage considering a variable-area nozzle and variable-pitch rotor blades. CEAS Aeronaut J 14:325–341. https://doi.org/10.1007/s13272-023-00658-x
15. Moreau A, Staggat M, Gscheidle C (2019) A fast prediction method for rotor buzz-saw noise based on an analytical approach. In: 25th AIAA/CEAS aeroacoustics conference. No. AIAA-2019-2641, AIAA.https://doi.org/10.2514/6.2019-2641
16. Meier zu Ummeln R, Moreau A, Schnoes M (2023) Influence of different flow solvers and off-design conditions on the determination of fan-rotor wakes for broadband prediction. J Eng Gas Turbines Power 145(3). https://doi.org/10.1115/1.4055753
17. Woodward R, Hughes C, Jacky R, Miller C (2002) Fan noise source diagnostic test– far-field acoustic results. Technical Report NASA TM-2002-211591, NASA

Design and Integration of UHBR Propulsion Systems–An Overview of DLR Activities within the Clean Sky 2 ASPIRE and SA^{2}FIR Projects

Rainer Schnell and Marco Trost

Abstract A propulsion system for future aero engines with ultra-high bypass ratio (UHBR) engines was designed and extensively evaluated applying different numerical methods. The propulsion unit under consideration featured an engine representative, robust and highly efficient fan stage as it had been designed by DLR, as well as a nacelle and an intake designed by an industrial partner. The paper focuses on the design process with subsequent high-fidelity analysis, first of all of the full-scale fan stage, aiming at balancing all kinds of performance-related as well as structural and operational requirements under various operating scenarios. The application of efficient multi-disciplinary (MD) optimization and simulation techniques is addressed, for example, to better understand different phenomena such as fan operation under angle-of-attack conditions or rotor flutter sensitivity due to an acoustic feedback loop with the intake, which stem from the integration of low fan pressure ratio fans.

Keywords UHBR engine integration · Fan stability · Variable area nozzle (VAN) · Flutter · Fan/intake interaction · Cross-wind · Angle-of-attack (AoA) · Propulsor

1 Motivation and Scope of the Paper

This paper summarizes the design and assessment efforts related to UHBR propulsion systems undertaken by the DLR Institute of Propulsion Technology in the context of the Clean Sky 2 initiative. The overview given in this Chapter is carried out predominantly from a methodological perspective, aiming at highlighting what methods are needed from the fan conceptual design until the final assessment of the entire propulsion system under various angle-of-attack conditions, with which a study in detail of the tight aerodynamic and acoustic coupling between the fan and the intake is performed. All methods will be described by examples of dedicated studies, either published in detail before, or newly carried out to demonstrate a specific capability. A brief look into the fan stage conceptual design and choice of major top-level para-

R. Schnell (✉) · M. Trost
German Aerospace Center (DLR), Institute of Propulsion Technology, Cologne, Germany
e-mail: Rainer.Schnell@dlr.de

F. Boden (ed.), *Numerical Methods and Experimental Means for the Integration of Future Propulsion Systems*, Springer Aerospace Technology,
https://doi.org/10.1007/978-3-031-97725-1_4

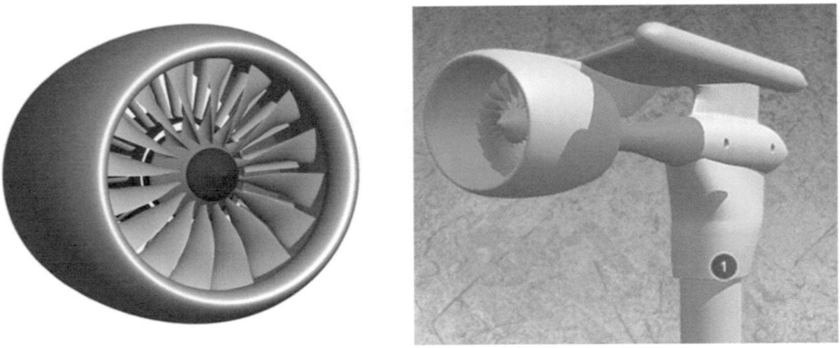

Fig. 1 Full scale baseline fan and intake from ASPIRE project (left, [9]) and SA^{2}FIR integration rig with downscaled and matured ASPIRE fan (right, ©Clean Sky JU 2011. Reproduction with kind permission https://cleansky.virtualfair.be/hangar)

meters will be given, followed by some insights into fully coupled multi-disciplinary and multi-objective, RANS- and FEM- based optimization studies. In the following section, the performance assessment will be addressed, focusing initially on the entire fan map, and thereafter looking at the isolated fan using steady state RANS analysis. The different topics in the context of fan intake interaction will be summarized, again predominantly from a methodological perspective. Classical non-axisymmetric inflow distortion effects induced by an angle-of-attack to the nacelle will be considered as well as one example for the effect of acoustic feedback and its impact on the fan rotor flutter characteristics. In the last section the bypass will be considered and a dedicated non-axisymmetric outlet guide vane (OGV) design will be described and studied. One of the main objectives, apart from providing a method overview, is to give recommendations as to what method is needed to address a particular problem most efficiently and accurately.

The baseline and full-scale propulsion system considered in this paper and shown in Fig. 1 (left) stems from the Clean Sky Initiative ASPIRE, where the fan being developed by DLR was integrated into a nacelle, intake and bypass system designed by the industrial lead Airbus. A derivative and technologically matured downscaled version of this fan was then developed in the subsequent Clean Sky ADEC initiative, contributing directly to the SA^{2}FIR fan and integration rig (Fig. 1 (right)).

2 Fan Stage Design

In this section, at first the design of the fan stage comprising of the rotor with spinner, bypass stator or outlet guide vane (OGV) and the engine section stator (ESS) will be addressed. As described in [9], the entry point for the fan design is the specification data for all mission points, all of which need to be considered in different ways during the design process. Details on the specification can also be found in [9].

2.1 Conceptual Design

The crucial role of the conceptual design phase of any turbomachinery component cannot be sufficiently stressed, since deficits due to unfavorable parameter choices made at this early stage can hardly be compensated for in later design stages. Focusing firstly on design operation, some top-level parameters such as fan rotor reduced tip speed can be estimated based on correlations taking into account the experience gained over the last decades. This may be done based on an assumption of the work coefficient ψ and the target fan pressure ratio at design conditions stemming from the overall specification. The resulting tip speed can be directly extracted from Fig. 2 (left), where various designs, all featuring a high technological maturity and collected over a long period of time, are shown by plotting tip speed over design fan pressure ratio (FPR). It can be seen that most design are clustered along a line of constant ψ between 0.45 and 0.5. In a similar way, values for the flow capacity, translating directly into a fan face Mach number and, for a given mass flow rate, rotor diameter, can be defined, as reported in [9]. A number of other parameters can be derived in a similar manner, resulting in a robust design basis to enter into the actual design phase of the fan stage.

For the conceptual blade and annulus design of the fan stage, an in-house streamline curvature method from Schnoes [11] was used. As also reported in [9], this can be used for efficient sensitivity studies, and in particular the design of annulus and initial blade based on classical blade loading parameters. All geometrical constraints to ensure compliance with adjacent components can be considered at this early stage.

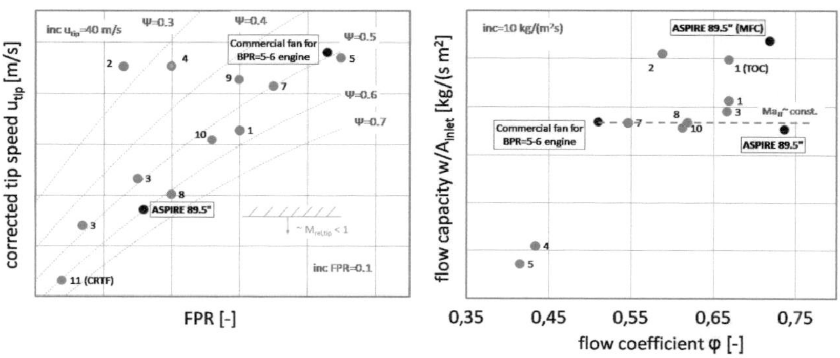

Fig. 2 Fan design top level parameters: tip speed and fan pressure ratio (FPR) together with lines of constant work coefficient ψ (left) and flow capacity over flow coefficient ϕ (right). Both figures reproduced from [9] ©DLR

As reported in [11], DLR has put much effort into a seamless release of 3D blade geometries based on pre-optimized profiles stacked along the radius, resulting in full 3D geometry ready to be evaluated with CFD. Typically, at design conditions, the resulting performance is within the limits of $+/-1\%$ in terms of major performance parameters (mass flow rate, fan pressure ratio and efficiency).

2.2 Detailed Design and Multi-disciplinary Optimization–MDO

DLR has long standing experience in complex component optimization, all the way up to a very high technology readiness level (TRL) involving complex material properties [4] [6]. In the context of the current study, one difficulty arose from different partners being in charge of the performance-related design and the detailed structural design with subsequent manufacturing (the latter was performed by NLR). Since all disciplines are inherently coupled and need to be always considered simultaneously, at least if aiming for a certain technology level, a compromise had to be found to ensure both high aerodynamic performance as well as structural and aeroelastic integrity, without having to iterate excessively between partners. The optimization procedure for the downscaled fan to be used in the SAAFIR rig will be briefly described in the following. To reduced complexity, a simplified structural model and FEM analysis procedure based on public domain FEM method CalculiX (http://www.dhondt.de/) was first established as part of the aero design of the fan on DLR side. Due to the complex nature of the composite material used for the fan rotor, this simplification required extensive verification which had to be carried out prior to the actual optimization. Isotropic material properties were calibrated to mimic the behavior of the ply layup in terms of eigen frequencies and displacements based on refined FEM results for the realistic setup provided by NLR. The optimization then carried out featured a number of constraints and requirements, the most important, performance related objectives were then to:

- maximize performance at aerodynamic design point (ADP) which was equal to cruise (CR) conditions,
- maximize flow capacity at mission point maximum flow capacity (MFC) and
- ensure sufficient stall margin at sideline (SDL) conditions

Structural constraints where also imposed aiming for:

- minimization of the individual displacement of the rotor tip at leading and trailing edge and difference thereof at different operating conditions and
- avoid crossings of first bending mode with first engine order.

A large number of geometric parameters was considered to allow for complex 3D blade shapes including sweep and lean, all within constraints imposed by the intake and test rig. Performance at multiple operating points was assessed using 3D-RANS

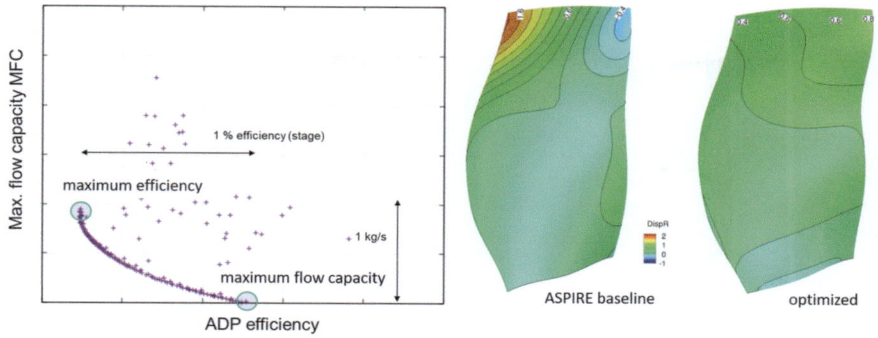

Fig. 3 Pareto front from two-point, multi-disciplinary RANS and FEM design optimization aiming at balancing flow capacity, high efficiency and stability (left)–Resulting optimized radial displacements from FEM analysis (second left) compared with baseline (right)

steady state simulations using TRACE [2], with subsequent FEM analysis, taking advantage of the aerodynamic loads from each individual CFD result per operating point. A rather coarse mesh was used with ca. 2 million cells for a single passage rotor, outlet guide vane (OGV) and engine section stator (ESS), which entailed the necessity for a refined post-optimization analysis. All in all, one optimization loop, several of which were carried out, lasted for 5–7 days in a high-performance computing (HPC) environment. Due to the complexity of the optimization setup and the large amount of data, only two main results shall be highlighted: In Fig. 3 (left) a typical pareto front plotting both objective functions is shown, It stems from an optimization carried out to quantify the trade-off one has to make typically between high flow capacity or choke margin and high levels of ADP efficiency. The resulting shock systems and flow field for two selected examples from the optimization, one yielding maximum efficiency and one maximum flow capacity, is shown in Fig. 4. At maximum efficiency, the incidence at design conditions should lead to a weak shock (left). If aiming for maximum flow capacity, the location of the design point in the performance characteristics needs to be shifted closer towards the stability limit, resulting in a stronger positive incidence and increased shock strength already at design conditions (Fig. 4, right). As one example for the structural results, Fig. 3 (right) shows the radial displacements of one of the optimized individuals in comparison with the ASPIRE baseline fan. Note that minimized radial displacement is one important operational parameter to retain small tip gaps over the entire operational regime. In the results one can see that the resulting values for the shown individual were reduced by more than a factor of 4 near the leading edge and also much better balanced in order to yield similar displacement characteristics at leading and trailing edge. This resulted in tip gap variations over the entire operating range of about 0.5 mm between cold shapes and resulting hot shapes under maximum centrifugal forces and aero loads. It was then of paramount importance to verify that the resulting displacements (and differences thereof for different operating conditions)

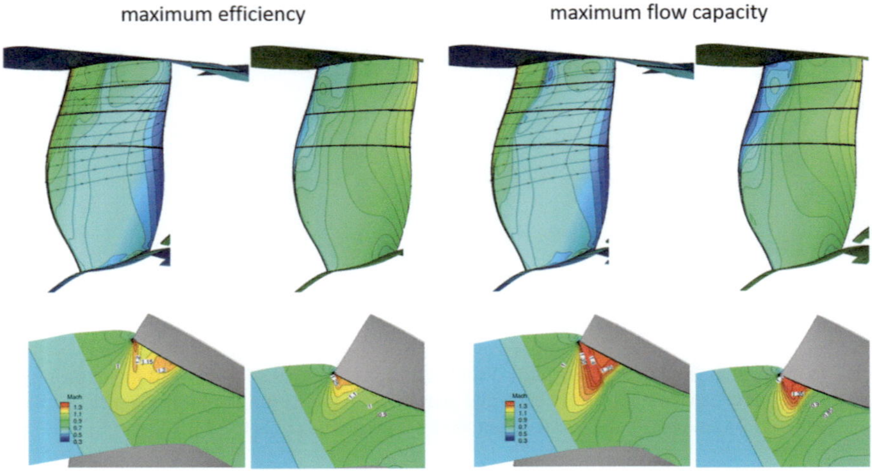

Fig. 4 Flow fields and shock structures in the fan to highlight compromise between flow capacity and ADP efficiency: design yielding maximum efficiency and compromised flow capacity (left) and maximum flow capacity with compromised efficiency (right). Contours indicating wall shear stresses, together with blade surface streak lines (top row) and relative Mach numbers in blade-to-blade plane (bottom row)

from this simplified structural analysis translated in the same amount to the refined FEM results with ply layup carried out by NLR; this comparison is shown Fig. 5. Shown are deviations from the running tip gap at ADP conditions which is indicated by the solid black line at a value of 0.5 mm. Higher rotational speeds at operating conditions top of climb (TOC) or maximum flow capacity (MFC) decrease the tip gap by up to 0.3 mm, ensuring a pre-defined minimal gap to be retained at red line conditions. Those results, as well as optimized Campbell characteristics which are not discussed here and in combination with the optimized aerodynamic performance already highlighted, were the very basis for achieving such a high TRL component design with shared responsibility.

3 Propulsion System Assessment

In this chapter several examples are given to demonstrate the application of different CFD approaches with increasing meshing and modelling complexity to address different problems that arose during and after the design process. The results will focus on both, the isolated fan stage as well as the entire propulsion system including the nacelle, intake and the bypass channel. One of the main objectives of this chapter is to demonstrate how the variety of methods and approaches that are typically involved in a such complex design task were carefully chosen in order to find the best compromise between modelling accuracy and cost (in terms of both, computational effort

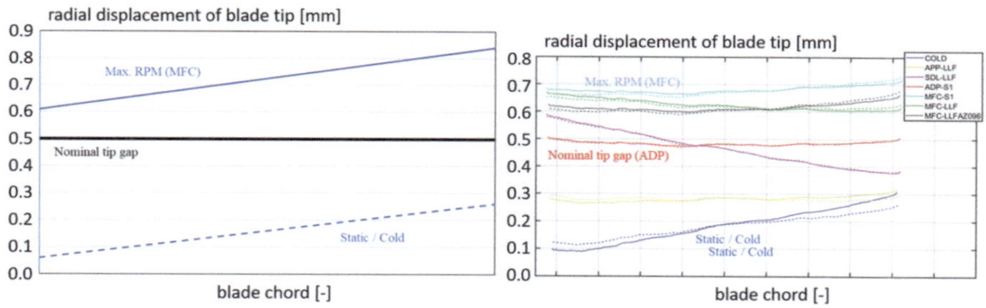

Fig. 5 Radial displacements (ordinate) along the blade chord (abscissa) for different operating and wind tunnel conditions–Results from simplified FEM analysis with CalculiX (left) in comparison with refined structural analysis using ply layup as carried out by NLR using commercial FEM software NASTRAN at different wind-tunnel conditions LLF/S1 (right, results NLR courtesy)

as well as engineering effort to setup, run and analyze the CFD). All CFD results presented in the following chapter were obtained with the flow solver TRACE [2].

3.1 Isolated Fan Performance–RANS Single Passage

First, the range of application of conventional 3D-RANS based approaches with a mixing plane in between stationary and rotating blade rows will be shown, predominantly focusing on average performance of the isolated fan stage. Most results were obtained with a Menter-SST turbulence model in combination with both wall functions and a low-Reynolds formulation for the solid walls, where mesh convergence studies and the validity of different wall law formulations were continuously assessed and reviewed for the whole design process. Here a compromise needed to be found between rapidly advancing the actual design, either within the MDO, or by manual design iterations, enabling for many design cycles with quick turnaround times using a rather coarse mesh with low fidelity modelling approaches and design robustness. The latter always required a continuous refined assessment of the design focusing on critical areas prone to pre-mature separation such as OGV end walls.

3.1.1 Fan Map

Within the MDO typically only a few operating points are included here to highlight the most challenging and competing requirements such as efficiency, stability and flow capacity. A coarse CFD mesh as shown in Fig. 6 comprising the fan rotor, the bypass channel extending only a few chords downstream of the OGV, as well as the core section with the ESS, was used during these optimizations. This allowed for rather quick optimization cycles, where one full MDO as described in the previous chapter typically ran for 2–3 days on a state-of-the-art HPC environment. One of the

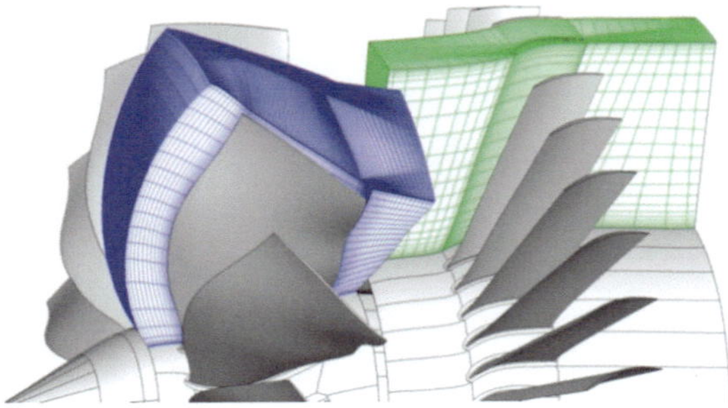

Fig. 6 Coarse mesh of the fan stage comprising of 1 million cells per passage

first tasks after each design cycle is to evaluate the fan characteristics for selected individuals to verify the design over the entire fan map up to the stability limit. Typically, this was done with the same grid as during the design cycle, yielding turnaround times of only a few hours for the entire fan map. Only for the most promising individuals was this repeated with a much finer grid up to 5 million cells to consolidate the design or identify critical regions not monitored during the automated optimization. During the design ISA conditions were often used to simplify data handling, but at some point, real conditions which pertain during flight or in a wind tunnel need to be considered as well, resulting in substantial Reynolds number variations and a high number of different ambient conditions to be considered. One important aspect was to automate all aforementioned procedures, from automatic surge detection to running a high number of speedlines with different ambient conditions in parallel for the sake of efficiency as well as minimizing user-induced errors. A fan map for the full scale ASPIRE fan is shown in Fig. 7 as an example, reflecting the main design objectives such as high levels of efficiency near cruise conditions, a sufficient stability limit for sideline (SDL) and maximum take-off (MTO) conditions as well as optimized flow capacity (Figs. 6 and 7).

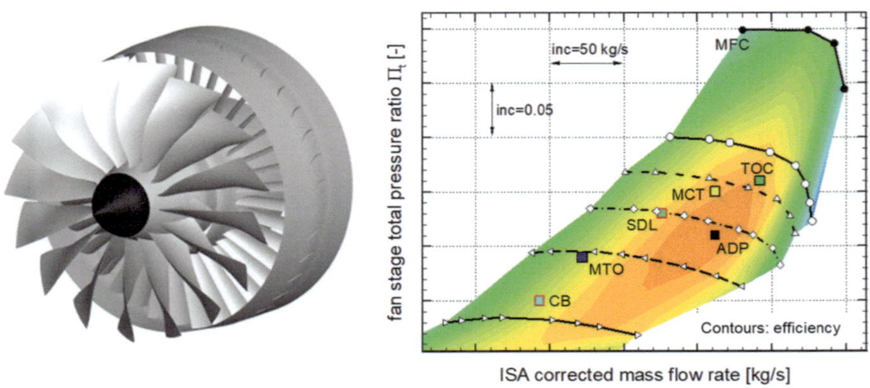

Fig. 7 Fan stage geometry (left) and performance map of the full scale ASPIRE fan from [9] (right, efficiency contour increment 2%) figures ©DLR.

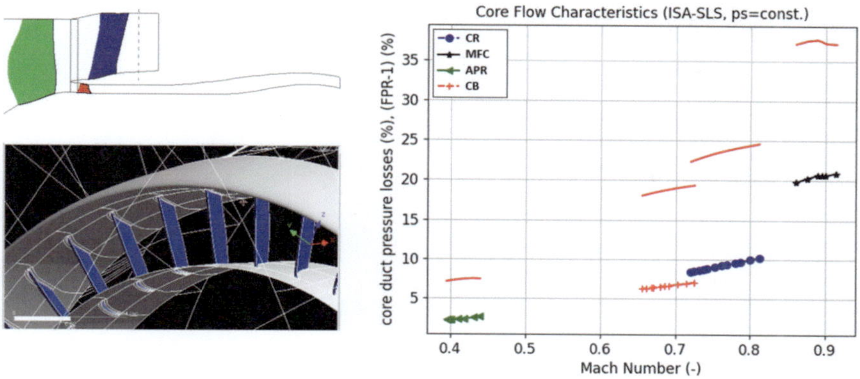

Fig. 8 Flow channels in a meridional view (top left) and engine core section stator (bottom left in blue). Resulting core flow performance plot (right): loss development along the fan charac-teristic for different mission points

3.1.2 Core Section Analysis

Since there is no booster or LPC stage in the core section, it was identified as a critical region potentially limiting the stability of the entire fan if the channel has been blocked due to excessive losses in the rather long core duct (see Fig. 8, top left). It was decided to introduce and design an engine section stator (ESS), whose function is to reduce the swirl and hence limit the losses in the core duct (see Fig. 8, bottom left). To assess this, and via that de-risk the tests, the core section domain was first of all extended up to the core duct exit in order to include the entire flow channel involved in the loss development. No mixing planes were present downstream of the ESS, so that the swirling flow in the core duct could be studied without needing to be averaged. Up top, one of the main requirements was for the pressure potential entering the core channel (this is equal to the fan total pressure ratio near the hub) to be higher than the integrated total pressure losses in the downstream duct. Figure 8 (right) compares both of the entire fan map with the pressure entering the core channel indicated as "FPR core" and the actual losses measured from ESS leading edge to core duct exit marked with the colored symbols for each individual operating point. After finalizing the core stator design, the core pressure potential was always much higher than the resulting loss levels in the duct including the ESS losses itself. Flow details around the splitter were also investigated in a comprehensive sensitivity study with varying core mass flow at different operating conditions to avoid any flow separations in that region potentially leading to a failure of the OGV due to the high loading in the OGV hub section; an example is shown in Fig. 9.

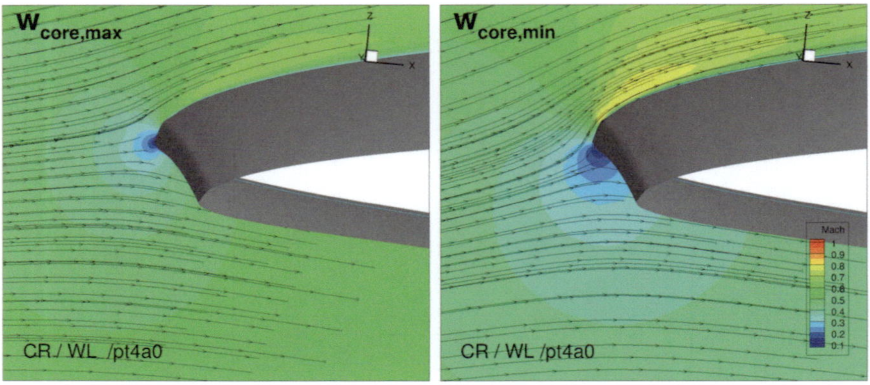

Fig. 9 Flow around the splitter for different mass flows relative to given baseline score with $w_{core,max} = 110\%\ w_{core}$ and $w_{core,min} = 90\%\ w_{core}$

3.1.3 Running Shape Impact and Tip Gap Study

Another important study using steady-state RANS simulations involved the performance assessment of the hot (=running) blade shapes in comparison with the original design intent. This was particularly important due to the many different wind tunnel conditions that the rig will be operated at, with the hence substantially varying pressure loads on the rotor. Whereas this assessment is often carried out post-test to optimize the validation effort and describe the geometry as accurately as possible, here it was done pre-test and also in a de-risking attempt to ensure mass flow capacity as well as fan stability was not compromised. In this study the hot shapes were provided by NLR and stem from a high-fidelity FEM analysis using the original ply layup of the composite rotor. The corresponding pressure fields were provided by DLR for the different wind tunnel conditions based on CFD simulation with the nominal shapes; note that no additional iterations, where the updated pressure fields were fed back again into another hot-to-cold transformation, where carried out. The results shown in Fig. 10 were then used to quantify the performance impact of the change in rotor blade shape on overall stage performance.

Another typical real geometry-related effect is a change in tip clearance during fan operation. Since off-design running gaps in the rig can only be estimated during the design and are influenced by a large number of parameters such as geometric deformations under different pressure loads present in different wind tunnels, as discussed in the previous section, a typical de-risking task during the design phase is to quantify the tip gap sensitivity on stability and overall performance. This is often done by steady state RANS and the main objective here is not only to quantify the impact of a change of tip clearance, but also to see if its increase leads to a non-linear change of the fan stage stalling characteristics (e.g. resulting from a change of the stall-critical location, for example, from rotor tip to OGV hub) and respective premature stalling compared with the intended design. Results from such a sensitivity

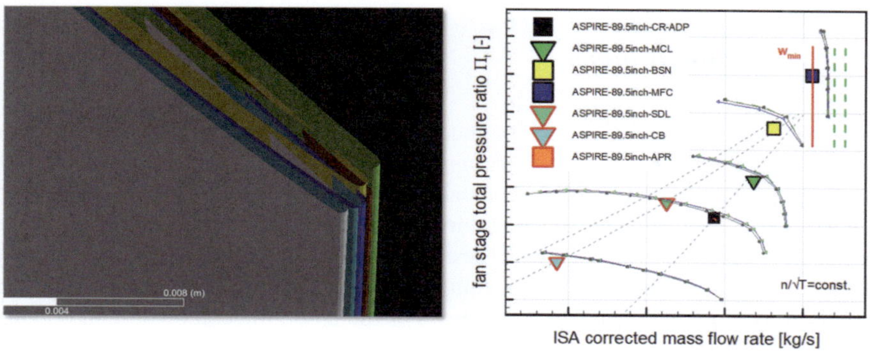

Fig. 10 Blade geometries at different operating conditions (left: hot shapes as provided by NLR). Right: resulting performance characteristics for nominal shape, as used during the design and for hot shape

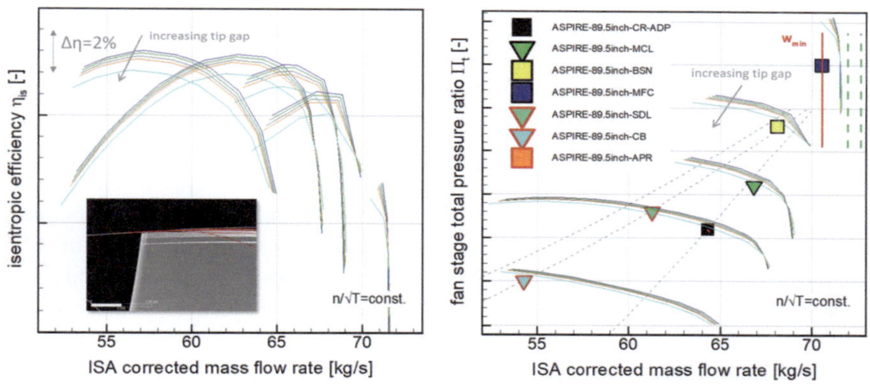

Fig. 11 Results from the tip gap sensitivity study. Simulated were six different gaps from one half (blue) up to four times the nominal gap size (light blue/cyan). Inset in left figure shows the tip section of the rotor blade for different gap sizes

study are shown in Fig. 11: here the tip gap was varied in the range between half and four times the targeted design gap at ADP. This variation showed only a small impact on the stability margin of the fan, but results in efficiency penalties of up to 1.5%. Typically, tip gaps are monitored during rig operation and their values are passed back to the post-test data analysis.

3.2 Fan and Intake Performance–RANS 360°, uRANS and Harmonic Balance

Whereas global and average performance can be evaluated with a sufficient level of fidelity over a wide range of the fan map based on steady state analysis, as shown in the previous chapter, two phenomena will be discussed in this chapter which are inherently unsteady and hence require a different choice of CFD method. These phenomena are both associated with the close coupling of the intake with the rotating blades at axisymmetric and non-axisymmetric inflow conditions which are present at cross winds or angle-of-attack conditions. For the latter, the computational domain must cover the full 360° intake at least up to the aerodynamic interface plane (AIP) just upstream of the fan within the intake and with an adequate distance between the far-field boundaries and the nacelle. The grid used for these studies is shown in Fig. 12; it comprises about 20 million cells for the inlet domain and single passage rotor, OGV and ESS domains (a justification for this will be provided later). Coupling the fan with its rotating blades to this 360° intake domain was carried out in different ways: in a classical uRANS approach, either 360° meshes in the fan domain or phase shifted boundaries requiring only a single rotor passage were used [7, 10]. Note that the computational effort for classical uRANS and hence time-accurate approaches compared with the steady state simulations with local time-stepping is approximately one order of magnitude higher on the same grid, resulting in rather high computational costs. This limits the computational effort to only a limited number of operating points to be studied in detail. A cost efficient and accurate alternative to uRANS simulations is using a non-linear, frequency domain harmonic balance method (HB), requiring only a single passage of the rotor, OGV and ESS with substantially reduced computational time being comparable to a RANS analysis [1, 3, 8] and with almost no compromise in physical accuracy.

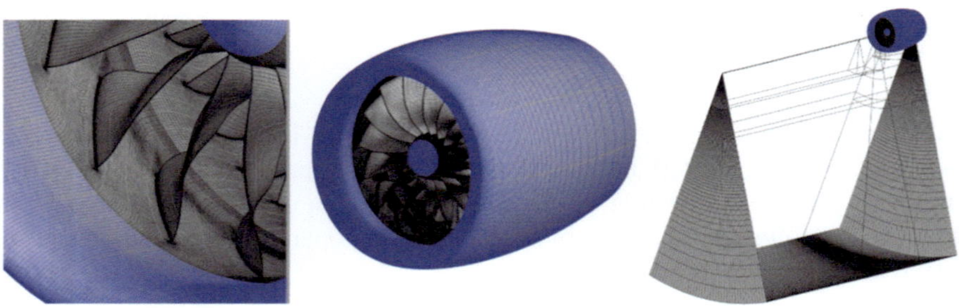

Fig. 12 Computational domain and single passage meshes to study fan intake coupling effects; note that for angle-of-attack simulations the intake mesh was expanded to cover the full annulus [8] ©DLR

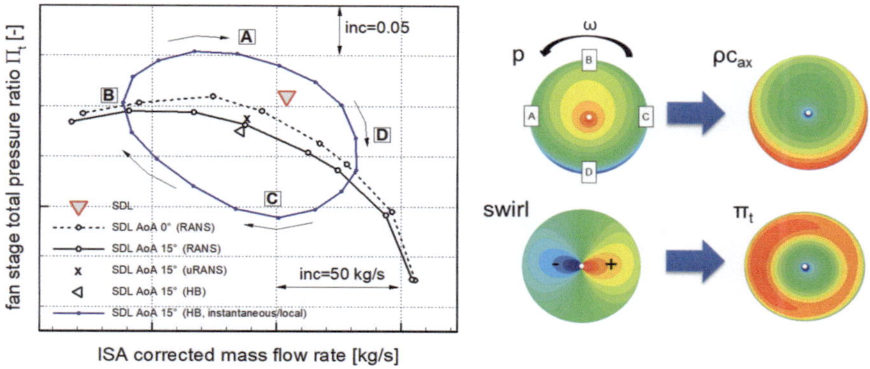

Fig. 13 Performance orbits and main regions of interest with co- and contra-swirl (A–C) and increased or decreased flow capacity (B–D) from [9] ©DLR

3.2.1 Fan Intake and Aerodynamic Performance

Since most details have already been provided in the given references, only two example results are given in the following to highlight the application of different numerical approaches: The first example, shown in Fig. 13, studied fan transient performance due to intake coupling at typical angle-of-attack conditions for a near sideline (SDL) operating point of the fan. Extensive code benchmarking studies for the very same scenario have been reported in [5], showing comparisons between different mostly uRANS-based methods. At the DLR a uRANS approach was used with phase-shifted boundary conditions as well as harmonic balance computations for the SDL point in consideration, complemented by steady-state simulations of the fan and the intake at an angle-of-attack condition along the speedline for different mass flow rates. Figure 13 shows the average impact of the presence of non-axisymmetric inflow along the characteristics (solid lines vs dashed line) for 15° angle-of-attack towards the nacelle. Averaged operating points are from uRANS and HB simulations and are denoted by the markers (triangle and cross), which almost coincide with the steady state RANS calculation where the 360° intake was connected to the rotor with a simple mixing plane. Obviously, the latter approach was not able to cover the unsteady interaction represented by the blue orbit, which is the excursion the blade passage experiences over one full revolution. From this orbit, different areas A-D can be identified; those areas are characterized by either expansion and diffusion around the intake lip (B and D) or positive and negative swirl induced by the incidence angle to the nacelle and respective up and down movement of the blade relative to this swirl (A and C). To cover fan intake interaction in a simplified manner, there are other steady state approaches such as introducing body forces in the CFD where the turbomachinery blades are located. A comparison with methods introduced here can be found in [12, 13], where those methods are typically being applied in the context of exterior aerodynamics where the fan is not of major interest.

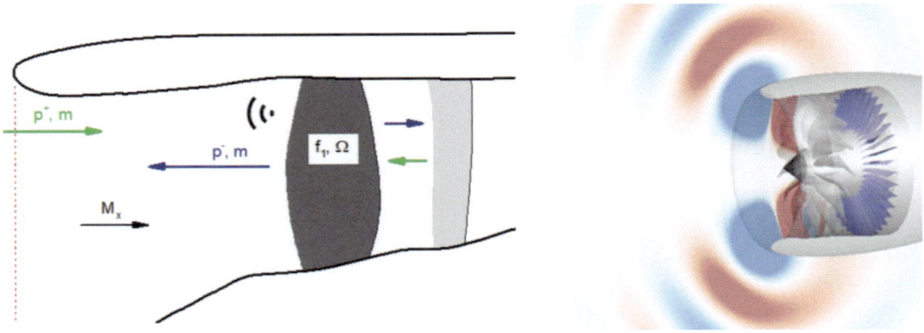

Fig. 14 Schematic of the acoustic coupling between intake and fan (left) and contours of instantaneous pressure perturbations induced by structural blade motion emitted to the far field from coupled CFD/FEM simulations with Harmonic Balance solver (right), figures ©DLR [8]

3.2.2 Acoustic Impact on Fan Flutter

Another phenomenon that was studied during the fan design was the acoustic coupling between the fan and the intake and the impact this feedback loop has on flutter characteristics of the rotor. This potentially dangerous and flutter-induced phenomenon is a complex physical process involving the propagation of acoustic modes along the duct, their respective reflection at the intake, together with blade motion which leads to the acoustic modes based on given structural eigen modes; this process is sketched in Fig. 14 (left), with an example of the acoustic field induced by the structural blade motion being emitted in the far field in Fig. 14 (right). An extensive parametric study over a wide range of parameters and operating conditions is reported in [8], taking advantage of TRACE's harmonic balance approach. Here in particular the efficiency of the numerical approach enabled the creation of a unique and large scale CFD-based data base which was then used to derive and calibrate a simple analytical model for quick flutter sensitivity estimations to be used during the design process.

3.3 Bypass Duct Analysis and Non-Axisymmetric OGV Design–RANS 360°

In addition to investigations focusing on the fan intake system, the realistic bypass flow channel was studied in detail with two major objectives: the first one being the mitigation of the negative impact of the large bifurcation on the upstream located guide vanes, and the second one being the quantification of bypass channel pressure losses at different Mach and Reynolds numbers. Two bifurcations were present in the bypass duct, an upper and a lower one with their dimensions relative to the OGV and the rotor shown in Fig. 15. In a rather complex design effort, individual OGV were re-designed in order to suppress the end wall separation that was induced in

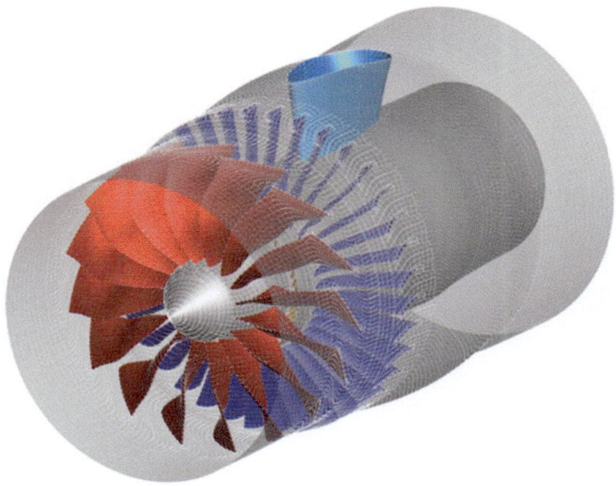

Fig. 15 Numerical setup for the non-homogeneous outlet guide vane (OGV, blue) design including bypass duct and simplified bifurcation (cyan) comprising 20 million cells [14]; the rotor is indicated in red and CFD block structure shown in white dashed lines

regions near the bifurcation due to its upstream potential effect. This required for the full annulus to be meshed since there was no longer any periodicity that could be used to reduce the computational effort; this resulted in a grid with about 20 million cells. However, since the major effect is non-transient, steady state simulations were used with the rotor being coupled with a classical mixing-plane to the OGV, whilst the OGV and the bifurcation domain being coupled with a conservative coupling approach, as first reported in [15]. The detailed results of this design exercise, including the various sensitivities that were studied, can be found in [14]. Figure 16 shows selected results from this study where the impact of the clocking between the bifurcation and the guide vanes was investigated. One main result from this study was the individually redesigned guide vanes that almost entirely suppressed separated regions by predominantly introducing a 3D-OGV with more pronounced lean in the hub region. Also, the local re-staggering of several OGV in the vicinity of the bifurcations upstream potential field led to further reducing the separated regions by taking advantage of passage wise acceleration and deceleration of the axial flow.

In another numerical study the loss development in the realistic bypass duct has been investigated over the entire operational regime of the fan, including ISA and altitude conditions at reduced Reynolds number. Loss levels are of major importance for low fan pressure ratio fans and may result in substantial performance losses of the entire propulsion system. In terms of the computational setup, compared with the previously described mesh shown in Fig. 15, the domain here included the nozzle exit and the far field, which was necessary due to the convergent nature of the bypass duct. In Fig. 17 (left) an example for a fan operating point near cruise at altitude condition with a fully choked nozzle is shown. The resulting loss contours over the entire fan map, are shown on the right side of the figure for both ISA static as well as altitude conditions at given flight Mach numbers. The latter results show substantially higher loss levels due to substantially lower Reynolds numbers. Apart from studying loss characteristics of the bypass duct, fan working lines were derived for a given nozzle area and over a wide range of flight Mach numbers (see Fig. 18, light grey

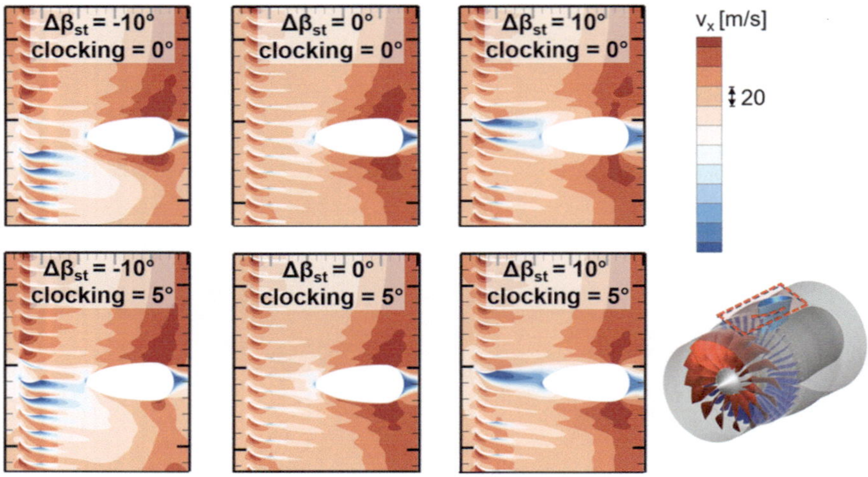

Fig. 16 The impact of OGV and bifurcation clocking–CFD results at mid-span for three values of $\Delta\beta_{st}$ of -10°, 0° and +10°, and clocking (= relative azimuthal position between OGV and bifurcation) values of 0° and 5° [14]

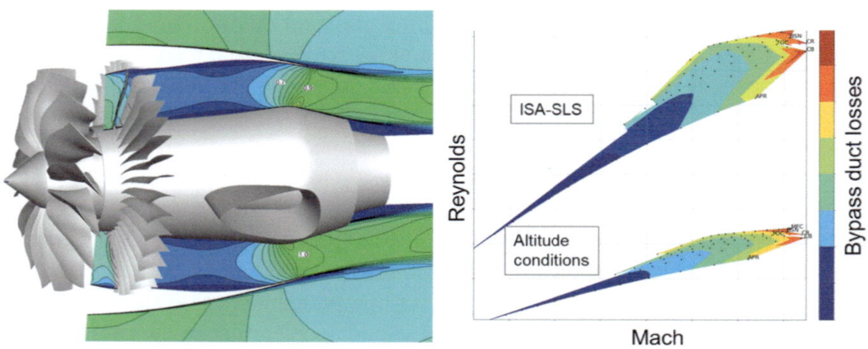

Fig. 17 Bypass channel flow field with bifurcation and choked nozzle (left, contours showing Mach number with 0.1 increment) as well as resulting loss characteristics over the entire operational regime of the fan (right, 0.25 contour increment of stagnation pressure loss coefficient, $\Delta\text{Re} = 0.5 \cdot 10^6$ and $\Delta\text{Mach} = 0.05$)

lines). These working lines may also be estimated by simple gas dynamic relations and given loss characteristics of the bypass duct together with the given fan map, but obviously full CFD including the release of the flow to the ambient allows for a much better understanding of the actual duct performance and the flow physics. This is shown in Fig. 19 by example of Cruise (ADP) and Sideline (SDL) conditions in terms of iso-Mach number contour levels. Two cross sections are shown, one intersecting both the lower and the upper bifurcation as well as one perpendicular to that and in the unaffected part of the bypass nozzle where no bifurcations are present.

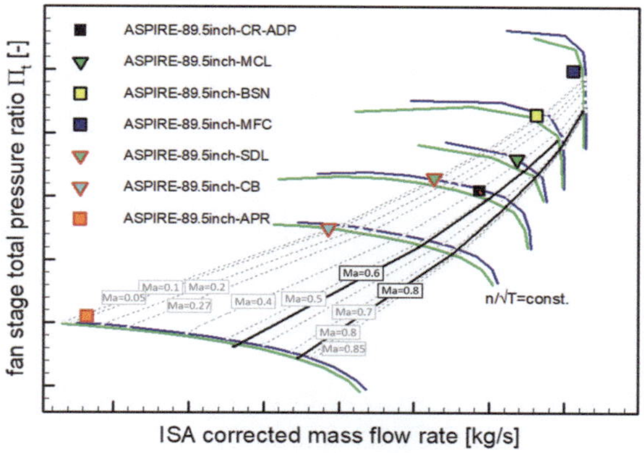

Fig. 18 Fan map and working lines from CFD for different flight Mach and Reynolds numbers (solid black lines represent simulations at altitude conditions, solid light grey lines are all at ISA sea level conditions). Solid lines in green and blue represent speed lines arising from back pressure variations at each respective rotational speed, for the design setup without the bypass channel at ISA (blue) and altitude conditions (green)

This reveals interesting choking characteristics, which confirm a choked nozzle in the latter regions, whereas for the same conditions the nozzle is unchoked in the regions of the bifurcations due to their respective wake; note that the upper bifurcation is a simplified geometry and typically would extend outside the actual nozzle in a real engine (Fig. 19).

4 Summary and Conclusions

This paper summarized the effort that was carried out by DLR Institute of Propulsion Technology dealing with the design and integration low fan pressure ratio and hence UHBR propulsion systems. The main results were summarized predominantly from a method perspective, highlighting the necessity of involving different numerical methods dependent on the physical nature of the problem and balancing the computational cost with required accuracy. Several examples were provided from the conceptual design of the fan or propulsor, to more complex design tasks based on 3D-RANS and FEM based MDO or full annulus vane designs, resulting in highly efficient and robust fan design which is mature enough to be used within the SAAFIR test bench. More complex physical phenomena such as the fan intake acoustic coupling and the impact on fan rotor flutter were investigated with highly efficient non-linear, frequency domain-based approaches, which are also available in DLRs flow solver TRACE. For performance-related fan intake interaction studies, these results have been complemented by classical uRANS simulations in either single-passage mode

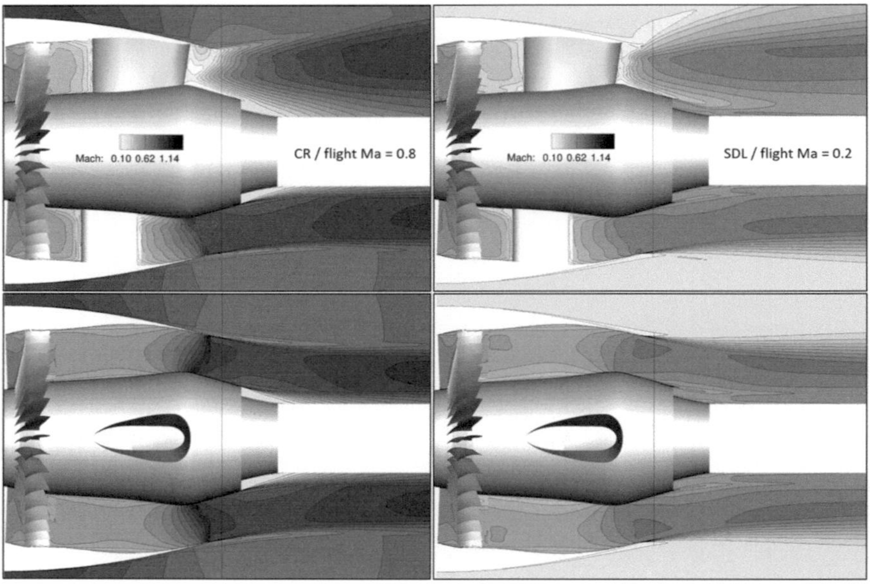

Fig. 19 Bypass channel Mach numbers for two different flight conditions: Cruise CR (left) and Sideline SDL (right). Top figures show slice intersecting both lower and upper bifurcations at 6 and 12 o'clock, respectively, bottom figure shows the corresponding slice rotated by 90°

with applied phase-shifted boundary conditions, or by meshing the full annulus. All numerical methods have been properly validated, either by extensive code benchmarks or by comparison with results from rig tests.

Abbreviations

ADP	Aerodynamic Design Point
ADEC	CleanSky 2 Project within the CS2 LPA-IADP
AIP	Aerodynamic Interface Plane
AoA	Angle-of-attack
APR	Approach
ASPIRE	CleanSky 2 Project "Aerodynamic and acouStic for high-by-Pass ratIo tuRbofan intEgration"
BPR	Bypass-Ratio
BSN	Nuzz-Saw Noise
CB	Cut-Back
CFD	Computational Fluid Dynamics
CR	Cruise
CS2	CleanSky 2

DLR	Deutsches Zentrum für Luft- und Raumfahrt (German Aerospace Center)
DLR BT	DLR Institute of Structures and Design
DNW	Deutsch-Niederländischer Windkanal (German-Dutch wind tunnel)
ESS	Engine Section Stator
FEM	Finite Element Method
FPR	Fan Pressure Ratio
HB	Harmonic Balance (frequency domain CFD method)
HPC	High-Performance Computing
IADP	Innovative Aircraft Demonstration Platform
inc	increment
ISA-SLS	ICAO Standard Atmosphere - Sea Level Static
JU	Joint Undertaking
LLF	Large Low-Speed Facility of the DNW
LPA	Large Passenger Aircraft
Ma	Mach number
MD	Multi-Disciplinary
MDO	Multi-Disciplinary Optimization
MFC	Maximum Flow Capacity
NLR	Nationaal Lucht- en Ruimtevaartlaboratorium (Aerospace Center of the Netherlands)
OGV	Outlet Guide Vanes
ONERA	Office National d'Etudes et de Recherches Aérospatiales (French Aerospace Center)
pt4a0	Internal operating point nomenclature (here: CR)
RANS	Reynolds-Averaged Navier-Stokes
Re	Reynolds number
S1MA	Transonic Wind tunnel at Onera / Modane S1MA I ONERA
SA2FIR	Simulator of Aerodynamic and Acoustic Fan Integration Test Rig
SDL	Sideline
TOC	Top Of Climb
TRACE	Turbomachinery Research Aerodynamic Computational Environment -
TRL	Technology Readiness Level
UHBR	Ultra High Bypass Ratio engine
uRANS	unsteady RANS
VAN	Variable Area Nozzle
WL	Working Line
ϕ	Flow coefficient
ψ	Work coefficient

Acknowledgements Most results summarized in this paper stem from funding that has been received from the Clean Sky 2 Joint Undertaking (JU) under grant agreement N° CS2-LPA-GAM-2020-2023-01 (ADEC) and N° 681856 (ASPIRE). The JU receives support from the European Union's Horizon 2020 research and innovation program and the Clean Sky 2 JU members other than the Union.

Furthermore, the contribution of NLR is acknowledged for providing detailed FEM results of the composite blade with ply layup which was used to calibrate the simplified FEM results used in this study. DLR colleagues from different institutes have supported individual activities, either actively or indirectly by providing guidance and support. The help from DLR-BT colleagues is gratefully acknowledged for supporting the conversion of FEM results. Finally, the discussions and support from Airbus colleagues in Toulouse and Bremen and the entire SAAFIR team is also gratefully acknowledged.

References

1. Ashcroft G, Frey C, Kersken HP (2014) On the development of a harmonic balance method for aeroelastic analysis. In: 6th European conference on computational fluid dynamics (ECFD VI). Barcelona, Spain, pp 5885–5896. (07 2014)
2. Becker K, Heitkamp K, Kuegeler E (2010) Recent progress in a hybrid-grid CFD solver for turbomachinery flows. In: V. European conference on computational fluid dynamics ECCOMAS CFD 2010. Lisbon, Portugal. (06 2010)
3. Frey C, Ashcroft G, Kersken HP, Voigt C (2014) A harmonic balance technique for multistage turbomachinery applications. In: ASME turbo expo 2014: turbine technical conference and exposition. Düsseldorf, Germany, pp 5885–5896 (06 2014). https://doi.org/10.1115/GT2014-25230
4. Görke D, Le Denmat AL, Schmidt T, Kocian F, Nicke E (2012) Aerodynamic and mechanical optimization of CF/PEEK blades of a counter rotating fan. In: ASME turbo expo 2012: turbine technical conference and exposition. Copenhagen, Denmark. (06 2012)
5. Meheut M, Sartor F, Vergez M, Laban M, Schnell R, Stuermer A, Lefevre G (2019) Assessment of fan/airframe aerodynamic performance using 360° uRANS computations: code-to-code comparison between ONERA, DLR, NLR and Airbus. In: AIAA Scitech 2019 Forum. San Diego, California, USA (01 2019). https://doi.org/10.2514/6.2019-0582
6. Schmid T, Lengyel T, Schmidt T, Nicke E (2019) Optimization of a carbon-fiber composite blade of a counter-rotating fan for aircraft engines. In: European conference on turbomachinery fluid dynamics and thermodynamics. Lausanne, Switzerland (01 2019). https://doi.org/10.29008/ETC2019-432
7. Schnell R (2004) Investigation of the acoustic nearfield of a transonic-fanstage by time-domain CFD-calculations with arbitrary blade counts. In: ASME turbo-expo 2004: power for land, sea, and air. Vienna, Austria. (06 2004)
8. Schnell R, Frey C (2022) Acoustic feedback and its impact on fan flutter in short aeroengine intakes. J Propuls Power 38(4):631–646. https://doi.org/10.2514/1.B38567. (05 2022, paper: 061016)
9. Schnell R, Goldhahn E, Julian M (2019) Design and performance of a low fan-pressure-ratio propulsion system. In: Internation society for air breathing engine-ISABE conference 2019. Canberra, Australia, p. 16. (09 2019)

10. Schnell R, Yin J, Voss C, Nicke E (2012) Assessment and optimization of the aerodynamic and acoustic characteristics of a counter rotating open rotor. J Turbomachin 134(6):1–15. https://doi.org/10.1115/1.4006285. (paper: 061016)
11. Schnoes M (2020) Eine Auslegungsmethodik für mehrstufige Axialverdichter auf Basis einer Profildatenbank. PhD thesis (08 2020). Accessed 01 Feb 2023
12. Spinner S, Keller D, Schnell R, Trost M (2020) A blade element theory based actuator disk methodology for modeling of fan engines in RANS simulations. In: AIAA aviation 2020 forum. Virtual event (06 2020).https://doi.org/10.2514/6.2020-2749
13. Stuermer AW, Spinner S, Trost M, Schnell R (2022) Aerodynamic analysis of a transport aircraft with a boundary layer ingesting aft-propulsor at cruise flight conditions. In: AIAA aviation 2022 forum. Chicago, Illinois, USA & Virtual. https://doi.org/10.2514/6.2022-3301. (06 2022)
14. Trost M, Schnell R (2025) Mitigation of the Bifurcation Impact in Low Fan Pressure Ratio Propulsion Systems by Individual OGV Designs Available to Purchase, ASME Turbo Expo Memphis. Paper No: GT2025–153933, V010T29A035. https://doi.org/10.1115/GT2025-153933
15. Yang H, Nürnberger D, Weber A (2002) A conservative zonal approach with application to unsteady turbomachinery flows. In: DGLR Jahrestagung. Stuttgart, Germany. https://elib.dlr.de/49430/. (09 2002)

Structural Design of a Generic High Bypass Ratio Fan

Paul-Benjamin Ebel, Peter Winkelmann, Rainer Schnell, and Matthias Schuff

Abstract Novel engine concepts for climate-friendly propulsion are accompanied by several deviations from today's aero engine designs. Finding feasible propulsor designs for these concepts is a highly multidisciplinary task and has to unite, at least, aerodynamic goals with structural requirements. The presented work provides an insight into DLR's approach on turbomachinery optimization and layout. The paper focuses on DLR's multidisciplinary design process that is carried out for a generic ultra-high bypass ratio (UHBR) fan within the ASPIRE project as well as the lessons learned out of this, which have been incorporated into a new "Virtual Engine" approach.

Keywords UHBR engine · Fan design · Multidisciplinary design and optimization (MDO) · Computational structural mechanics (CSM) · Virtual engine

1 Introduction

This article highlights the work carried out by several DLR institutes in the field of turbomachinery research as part of the Clean Sky 2 project ASPIRE (***A***erodynamic and acou***S***tic for high-by***P***ass rat***I***o tu***R***bofan int***E***gration), the aim of which was to design a generic but realistic fan stage for a future UHBR engine [2, 6]. The focus of the paper is on the challenges of the interdisciplinary task and the coordinated structural-mechanical design and evaluation methods of the participating institutes. Due to the very limited resources available within the framework of these Clean Sky activities, deficits became apparent in the approach taken at the time, which led to

P.-B. Ebel (✉) · P. Winkelmann
German Aerospace Center (DLR), Institute of Structures and Design, Stuttgart, Germany
e-mail: Paul.Ebel@dlr.de

R. Schnell
German Aerospace Center (DLR), Institute of Propulsion Technology, Cologne, Germany

M. Schuff
German Aerospace Center (DLR), Institute of Aeroelasticity, Göttingen, Germany

F. Boden (ed.), *Numerical Methods and Experimental Means for the Integration of Future Propulsion Systems*, Springer Aerospace Technology,
https://doi.org/10.1007/978-3-031-97725-1_5

only limited usable results on the part of the structural mechanics work packages. The methods and processes used at DLR for the development of designs and the layout of turbomachine components have since been continuously improved as part of an enhanced collaborative approach called "Virtual Engine" in order to better address such problems.

2 DLR Within the ASPIRE Project

As part of the Clean Sky program, the DLR Institute of Propulsion Technology in Cologne (DLR AT), the DLR Institute of Aeroelasticity in Göttingen (DLR AE) and the DLR Institute of Structures and Design in Stuttgart (DLR BT) were involved in the ASPIRE project to develop, among other things, a generic yet realistic design for a fan stage consisting of a fan rotor and an Outer Guide Vane (OGV) for a future UHBR engine. The ASPIRE project was the response of a consortium of several European institutions to the call JTI-CS2-2014-CFP01-AIR-01-01: "Aerodynamic and acoustic capabilities developments for close coupling, high bypass ratio turbofan Aircraft integration." A network of European research centers, namely ONERA, NLR, DLR and Russia's TsaGl (contributing with national funding) was involved. Since the efforts to further reduce emissions in air traffic arising from the propulsion units have been accompanied by a tendency to steadily increase the bypass ratio of novel aero engine concepts, this increase in size will in future result in the challenge of fitting the propulsion systems, which are significantly larger in diameter, to existing aircraft configurations.

In addition to the obvious difficulties such as the necessary ground clearance (see Fig. 1), which already leads to limitations in the feasibility of engines on single-aisle aircraft with very-high bypass ratio (VHBR) engines already flying today (an example is the Boeing 737MAX with the CFM LEAP 1B engine), aspects such as integration and interaction with regard to aerodynamics and aeroacoustics of engine, airframe and wing are to be comprehensively investigated [11]. In order to carry out these numerical investigations, it is of fundamental importance to also adapt the existing numerical tools in the field of Computational Fluid Dynamics (CFD), Computational Aeroacoustics (CAA) and Computational Structural Mechanics (CSM) to the changed boundary conditions and requirements and to validate them with experiments. One essential building block for this is the availability of an unrestricted engine design in the project. The responsibility for this design lay with the network of the three DLR institutes named beforehand and was placed in work package 1.1 "Generic Fan Design". The ASPIRE project was originally scheduled to run for 2 years, starting in January 2016, but had to be extended later. However, the duration of this elementary work package had been limited to 6 months. During this time, an aerodynamic design for fan and OGV had to be developed and subjected to a "basic" structural mechanical and aeroelastic evaluation. In the context of this paper, only the work on the fan stage will be described. The requirements for the engine design were provided by the project partner Airbus and included specifications for the corre-

Fig. 1 Numerous challenges arise from the integration of engines with ultra-high bypass ratios (Usage with kind permission of NIPSE project, ©NIPSE: Novel Integration of Powerplant System Equipment, 2017, https://web.archive.org/web/20170616211428, http://www.nipse.eu/)

sponding missions and boundary conditions. They were based on thrust requirements for a short-to-medium range single-aisle aircraft with its relevant certification points [21]. From these Top-Level requirements, a thermodynamic cycle was derived at the Institute of Propulsion Technology and a fan blade design for an UHBR engine with a bypass ratio larger than 16 was developed. Other design features included a fan stage with low fan pressure ratio as is the case for modern Geared Turbofan (GTF) engines, use of an ultra-short intake, and little cross wind and angle of attack sensitivity of the fan as well as a high flow capacity [21]. Subsequently, the blade-disk design, here implemented as a blisk (bladed disk) for simplicity reasons, was subjected to a structural-mechanical evaluation as well as aeroelastic assessments, which would normally have been followed by an iteration to optimize any anomalies and deficiencies that arose in the design. However, the key design driver for the results presented in Sect. 4 was the time constraint of 6 months for the full design process, so it was not possible to improve the design by various iterations.

3 Multidisciplinary Design of Turbomachinery Components at DLR

Component design in general is an interdisciplinary task, as a variety of often conflicting objectives usually have to be considered to get a satisfactory final result. In order to achieve discernible progress in the further development of the already highly matured turbofan engines that are used in aircraft today, close collaboration between all the disciplines involved is absolutely essential. Figure 2 gives an impression of the scope of disciplines and boundary conditions which need to be taken into account, and which can vary greatly depending on the component and its intended use.

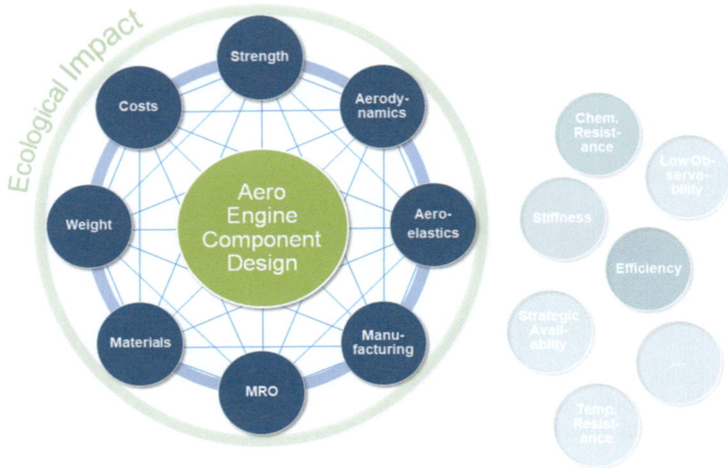

Fig. 2 Component design is a highly multidisciplinary task in which many disciplines are involved and numerous, sometimes conflicting, requirements and constraints must be reconciled to each other in order to achieve the demanding goals (MRO—Maintenance, Repair and Overhaul)

An ever-present goal in aviation is weight reduction in the form of reliable lightweight solutions. In addition, the fulfilment of efficiency requirements, strength to be achieved or a specific operating behavior, to name just a few, play an elementary role. Cost aspects and the availability of raw materials, suppliers and established production technologies have also to be considered. In face of the immense global climate change, the overarching objective nowadays is to reduce the ecological impact over the entire life cycle of a component or system. The demand for the development of a usable UHBR fan design in a period of only 6 months additionally results in the need for fast and reliable design and evaluation processes. At the time of ASPIRE project implementation, there had already been an established multidisciplinary collaboration between the DLR Institutes of Propulsion Technology, Aeroelasticity and Structures and Design in the field of design and optimization of turbomachinery components, primarily applied to blade geometries [13, 24, 25]. Figure 3 illustrates in a simplified manner the iterative procedure for the design and optimization of blade geometries for a compressor stage. In ASPIRE itself, a design point and the relevant aerodynamic parameters such as fan pressure ratio (FPR) and associated blade tip speed were derived from the Top-Level requirements before the start of the optimization, resulting in an initial concept of the fan stage as a starting point. A more comprehensive description of the methods and tools used in ASPIRE on the aerodynamic side can be found in [21]. This aerodynamic predesign for the flow channel and stage blading was the basis from which an optimization of the parameterized geometry with DLR's own algorithm "AutoOpti" [8, 26] can take place, mostly with regard to major aerodynamic target variables. In addition to these target functions, structural-mechanical and aeroelastic restrictions can be set, which must be fulfilled

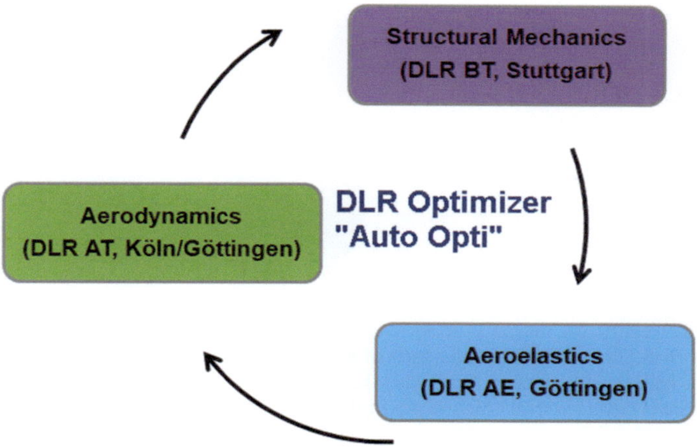

Fig. 3 Schematic sequence of the multidisciplinary iterative aero engine blade design process at DLR (as of 2016). Basic structural mechanical and aeroelastic assessment can be included within the optimization, as well as in more detail for chosen members out of the optimization afterwards

by the generated members of the optimization. At the time of the ASPIRE project, it was possible to consider basic geometric, structural-mechanical and aeroelastic restrictions within the optimization: these included among others blade thickness distribution, eigenfrequencies of a single blade, equivalent stresses and simplified flutter assessments. However, these were restrictions that affected only the blade geometries under consideration, which meant that only simplified boundary conditions for the single blade could be applied. For more comprehensive analyses, also considering extended boundary conditions such as the influence of the rotor disk and further surrounding structures or blade rows behind and in front of the designed stage, selected members out of the optimization process were subjected to a more comprehensive and more time-consuming manual evaluation at DLR AE and DLR BT after optimization.

4 ASPIRE Results Overview—Structural Assessment

At the end of the design and optimization process described above (see also [21]), a selected geometry for the fan blade became available. Extensive CFD simulations (Throughflow and RANS) and less extensive CSM calculations (stress values and natural frequency position of the single blade) had already been carried out for this blade during the design process and optimization, and an estimate of the flutter tendency had also been done. Figure 4 compares the state at the end of this design with a real engine. It is easy to see that the evaluation of structural properties on the basis of these aerodynamic shapes of the fan stage and the OGV alone would come

Fig. 4 At the end of the design process, there exists an aerodynamic shape for the optimized components (here: fan blade, OGV and parts of the flow channel, left picture). All of the "internal values" of a real engine, some of which can have a major influence on the structural behavior, have not yet been designed (right picture, © Rolls-Royce plc., https://www.n3eos.com/en/engine-overhaul/engines/rolls-royce-trent-900/)

along with a great simplification of the boundary conditions, which would inevitably lead to deviations especially in the structural behavior and subsequently also in aeroelastic and aerodynamic results. The most obvious example is the significant influence of the rotor disc on the deformation behavior and natural frequencies of the blade. These eigenfrequencies of the blade were calculated by finite element analysis (FEA) in the optimization without a disc but with a fixed clamping for all motional degrees of freedom at the blade root instead, representing an unnaturally stiff behavior and providing eigenfrequencies above the real values. The aim of all subsequent structural-mechanical work in the ASPIRE work package 1.1 was therefore to provide the necessary realism in structural behavior for the important aerodynamic design of the fan stage so that this design would be permissible as a basis for further use in this and other Clean Sky 2 projects. A fan design, however sophisticated aerodynamically, would compromise all further results achieved with it, if it would have been based on unrealistic structural-mechanical assumptions or could not support the operating loads.

4.1 Assumptions and Boundary Conditions

The task at DLR BT therefore began with the design of a disk geometry for this aero design. The tight schedule restrictions mentioned above led to the adaption of a "single shot" approach. The design tools used at the Institute consist of both commercial and self-developed tools and software solutions. Hyperworks environment—Hypermesh and Hyperview from Altair—are used here as standard tools for pre-/post-processing [3]. The standard structural solver is PERMAS [4] from INTES GmbH in Stuttgart. Self-developed tools and scripts are used for the predesign of the disk, the meshing of the structures, the interpolation of the aerodynamic pressure loads on the FE mesh, for the hot-to-cold computation and the A $\times$ f evaluation. A number of simplifications as well as (geometric) boundary conditions and assumptions were defined for the design phase of the disk for the given blade geometry (see later also Fig. 6):

- blisk-design for reason of simplicity in modeling and simulation:
  - initial disk design with a 3-web disk, taken from fan disks in current Rolls-Royce engines;
  - approximate position of the shaft connection, also taken from current Rolls-Royce fan disks;
  - inner disk diameter limited and hub contour of the given blade design fixed.
- simulation of a segment only, consisting of blade and associated disc piece:
  - blade and disc finite element mesh tied with Multi-Point-Constraint (MPC, see [5], page 58) in PERMAS;
  - cyclic-symmetrical boundary conditions or fixed degrees of freedom on the segment flanks of the disc (depending on the simulation);
  - axial and circumferential movement suppressed at shaft connection.
- radial expansion of disk limited at the front and rear edges.
- strength criteria defined for the disc itself (disc burst at overspeed) and for the blade under consideration of the disc influence.
- operational loads: aerodynamic pressure loads applied onto blade suction and pressure side and rotational loads for the blisk, defined for 8 operating points (OP) in the Top-Level requirements: Red Line (RL), Top of Climb (ToC), Maximum Continuous Thrust (MCT), Cruise/Aero Design Point (ADP), Sideline (SDL), Maximum Take-Off (MTO), Cut-back (CB), Approach (APR) [21].

A conventional Ti6Al4V alloy was used as material. The material data and strength allowables were taken from the “Werkstoffleistungsblatt 3.7164”. No hollow blades, no modern Ti- or Al-alloys and no fiber-reinforced plastics were considered; this was in order to keep the effort low, to use existing data and tools which were capable of designing conventional structures during the optimization, and to stay within the scope of current experience during the analyses. This restriction had already been made before the start of the initial aerodynamic design. A further simplification was to omit a bird strike or foreign object damage (FOD) analysis of the design and no investigations regarding costs and weight were carried out. As the disk area below the flow channel was not relevant for the analyses and investigations planned for the further course of this and other CS2 projects, the scope for a disk design was only slightly restricted here. From the initial 3-web design, a 1.5-web disc was ultimately developed with the help of topology optimization and subsequent shape optimization, considering the above-mentioned strength and deformation criteria for disk and blades. Figure 5 shows the result with a contour plot of the radial displacements for the ToC operating point. This disk design would be considered as a preliminary

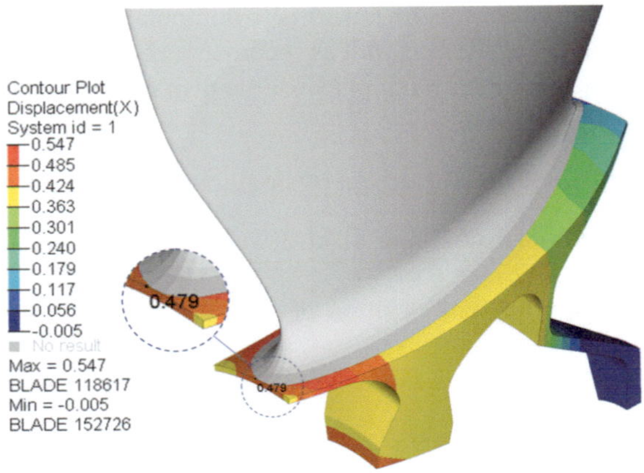

Fig. 5 ASPIRE fan rotor disk with "1.5-web" design and contour plot for radial displacements (in mm)

result only with need for further improvement but it had to be used here as no further iterations could take place due to time constraints.

4.2 Performed Analyses and Results

The first calculation step within this next part of the structural design process, just as during optimization, is a so-called "hot-to-cold" transformation of the geometry. As the initial geometry is a blade in operation, which represents the condition at a defined design point (here ToC), it differs from the "cold" production geometry. The latter represents the unloaded state of the component and is determined using an iterative process [12, 13]. The "hot" geometry is subjected to the operating load and the resulting deformation is then subtracted from the "hot" geometry. This result represents the first "cold" state. This geometry is again subjected to the operating load, leading to a geometry which is normally still significantly different from the desired "hot" state. So the deviation is subtracted again and so on until a "cold" geometry is reached which, subjected to the operating load, matches the "hot" geometry sufficiently well. This "cold" geometry is the starting point for all further structural simulations and can thus also be used for gaining shapes for aerodynamic "off-design" calculations. The loads and boundary conditions for the structural simulations are applied to the structure as shown in Fig. 6. The results of the structural simulations were then determined for each operating point. All the results presented below represent the status of the ASPIRE design review carried out in October 2016. Adjustments and modifications made later are not shown here. These results are:

- deformations (see Fig. 7)
- stresses or strains (here: stresses, see Fig. 8)
- tip gap behavior across the speed range (see Fig. 9)

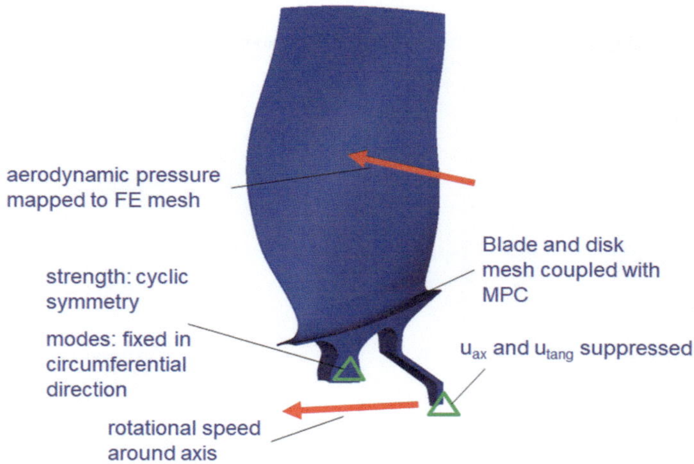

Fig. 6 Application of loads and boundary conditions for CSM calculations

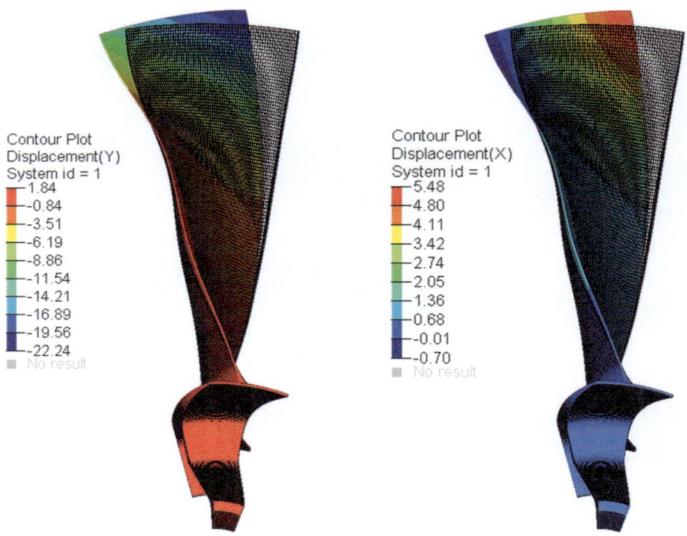

Fig. 7 Deformations (in mm) at the ToC operating point. Tangential displacement (left figure) and radial displacement (right figure). A kink in the blade due to bending stresses is spotted at approximately 70% of blade height, leading to a large variation of tip gap (see also Fig. 9). Displacements shown are exaggerated

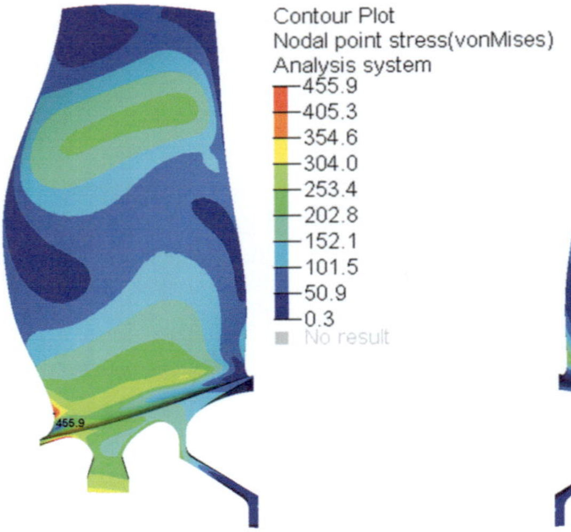

Fig. 8 Static stresses (v. Mises stresses in MPa) for Red Line (110%) conditions. The blade root area was not investigated during the initial optimization due to the application of boundary constraints here, so this area faces stress values above material limits. Blade balancing needs also to be improved

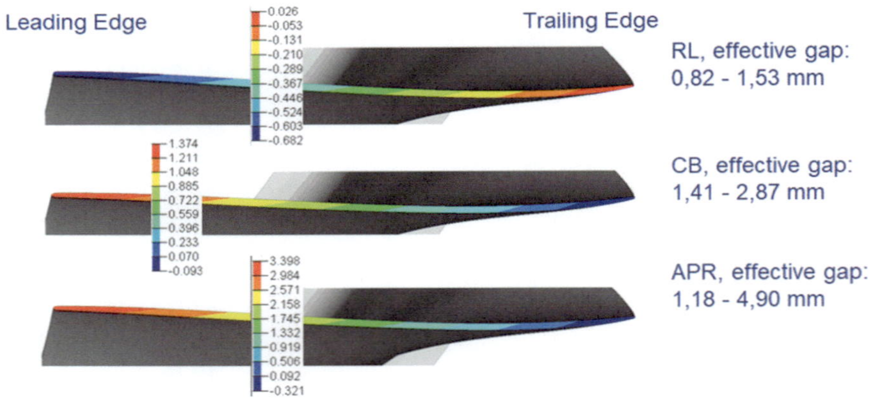

Fig. 9 Tip gap (in mm) at different operating points. Plots show deviation from nominal gap at ToC (1.5 mm). The tip gap minimum is located at leading edge for small gap widths (at RL), which is not the desired behavior as this involves the risk of the front edge digging into the housing in the event of contact

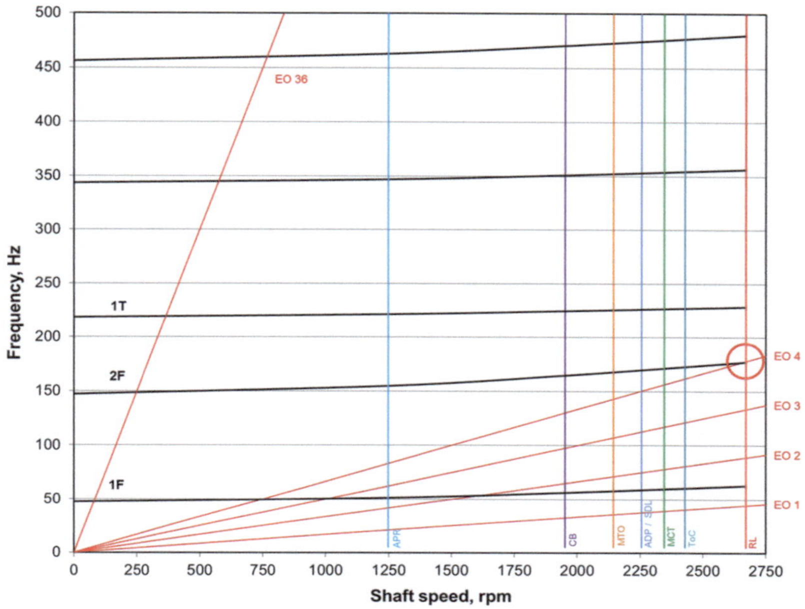

Fig. 10 Campbell diagram of the ASPIRE fan rotor. A crossing of Engine Order 4 (EO 4) with mode 2F is visible at Red Line conditions, which is an inadmissible property since excitation of this eigen mode is to be expected in stationary operation at Red Line with risk of structural overload

- eigenfrequencies over the speed range, including mode shapes up to engine order 36 due to OGV blade count, the evaluation is mainly carried out by use of a Campbell diagram (see Fig. 10)
- A × f analysis (Amplitude × frequency)—the determination of the dynamic residual load-bearing capacity for a given static load on the basis of a Goodman diagram (see Fig. 11)
- allowable tip amplitude for each mode
- risk of fan flutter within the operating range of the compressor (see Figs. 12 and 13)

In addition, extensive and time-consuming aeroelastic simulations have been carried out on the basis of these structural results at DLR AE Göttingen [23]. Typical results here are assemination of risk for fan flutter within the operating range of the compressor (see Figs. 12 and 13), or determination of structural response under forced excitation. Depending on these outcomes, new design iterations would follow in order to minimize or eliminate the recognized weaknesses. Due to the very early stage of the ASPIRE design, the specific results and values will not be presented here in detail. The aim is rather to give an overview of the evaluation possibilities of the process and to give an impression how the time restriction influenced the finding of a satisfactory result.

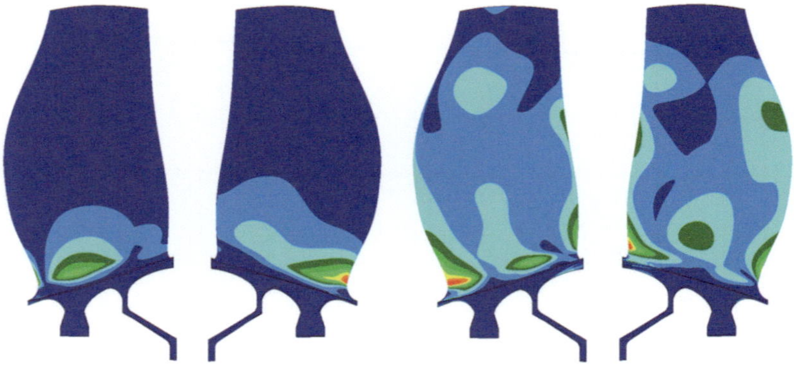

Fig. 11 A × f analysis: relative dynamic utilization for Red Line (110%) conditions for mode 1, first bending (left, pressure and suction sides) and mode 3, first torsion (right, pressure and suction sides). Red areas exceed material limits. Calculation of allowable tip amplitudes is based on the A × f values and has then to be compared against forced response values to be provided by DLR AE, Göttingen

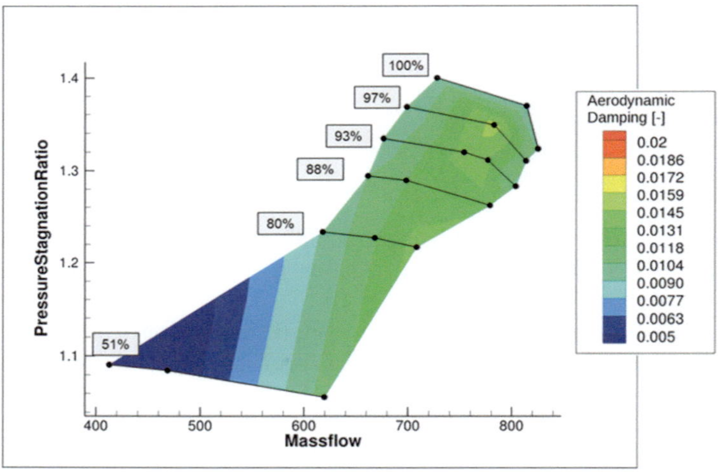

Fig. 12 Compressor map with contour plot for aerodynamic damping. Positive values show flutter free operation is possible over the entire operating range of the fan. Damping values close to 0 at part speed and near the stability limit of the fan were observed

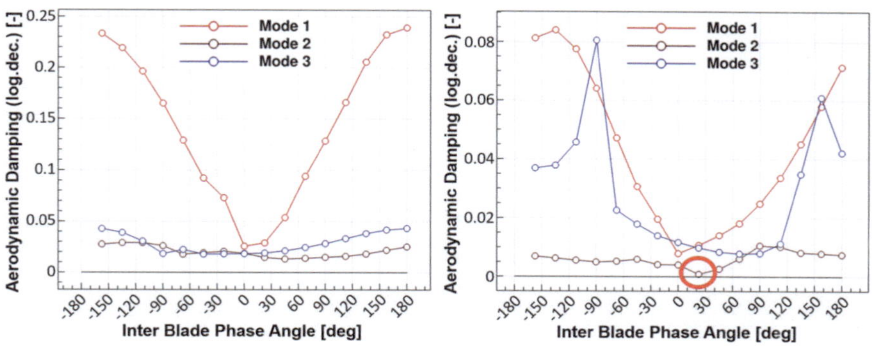

Fig. 13 Damping diagram for the ASPIRE Fan for ADP (left) and APR (right) operating point, showing very low values for aerodynamic damping at part speed conditions

In addition, analyses were carried out by DLR during the project to determine the extent to which an ASPIRE fan blade made from carbon fiber-reinforced plastic (CFRP) would be possible and what changes would result from such a design. These preliminary studies by DLR-BT were continued later in the ADEC project and extended by ONERA and NLR on the basis of the structural models and data to produce a feasible CFRP design [22]. The FE model of the ASPIRE fan and the associated data were transferred to ONERA and NLR for this purpose [1].

5 Summary and Outlook

In the course of the ASPIRE design review at the end of the fan/OGV design phase, it was noted that "all aero requirements were met with high levels of fan efficiency at ADP, (...) improvements were highlighted" [7]. The feasible aero design was the main achievement and of high importance for the project, so this can be seen as a success of DLR's activities. For the structural analysis and the work carried out here, the result is less satisfactory. The disk design and thus the blisk design of the entire fan can only be considered as preliminary. When looking at the results of the in-depth structural-mechanical analysis shown in Sect. 4.2, it can be seen in almost all areas that targets were not achieved or limits exceeded. These include the violation of the static and dynamic stress limits, the inadequate balancing of the blade, the necessary displacement of natural frequencies and the unsatisfactory splitting behavior.

What is noticeable on the process side is that due to the limited time available design freedom was restricted and a number of analyses were not carried out at all. Also, no iteration of the design could be carried out despite the unsatisfactory results. However, upon closer examination of the processes itself, it is noticeable that the established multidisciplinary collaboration between the institutes still involves much manual intervention and many adjustments and individual sometimes recurring work

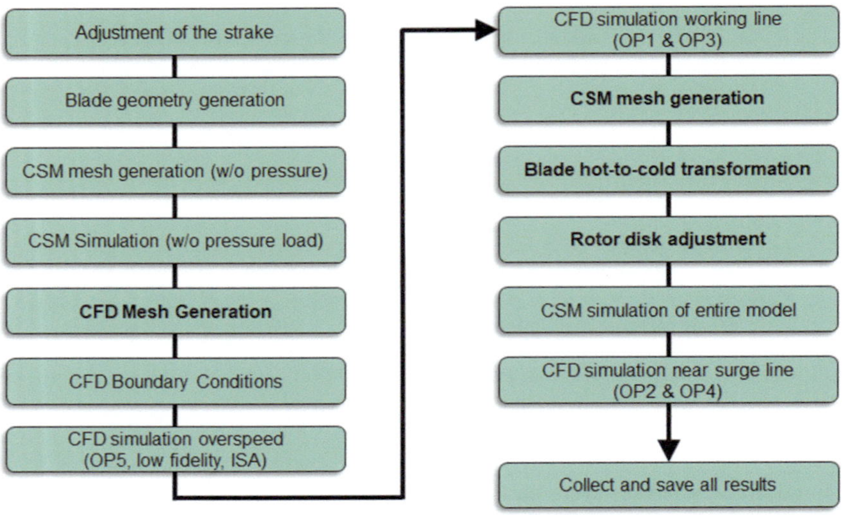

Fig. 14 State-of-the-Art MDO process chain [16]. Each member of the optimizations goes through the entire process chain. Sections marked in bold font are different for the high-fi and low-fi process chains

in many places, despite already existing interfaces and known and partly automated processes. As a result, the whole design, optimization and assessment process is still time-consuming and complex; important automated processes such as the provision of a suitable disk for blade geometry supplied by the aerodynamic specialist (with known data formats and boundary conditions) do not even exist and therefore require lots of manual work. In addition, these procedures were also prone to errors due to the low level of automation. An obvious development goal was therefore to ensure that these essential structural activities are still covered, but do not have to be carried out by "valuable" scientific staff. As these structural assessments are generally recurring and therefore basically automatable processes, routines can be developed that give scientists the freedom to turn their attention to new topics and not to work through recurring pure engineering tasks.

The methods presented briefly in Sect. 3 show the status from around 2016, at the time of ASPIRE. These methods have been further refined and expanded in subsequent years. A good example is the multidisciplinary optimization of a two-and-a-half-stage low-pressure compressor [14, 16], which has taken the MDO process to the extreme in terms of complexity, at least for the present, (see Fig. 14), but also has long runtimes and, above all, requires an intensive preparation phase. This process already includes fully automated individual preparation and adjustment (e.g. automated disk design) and high-fidelity CSM and CFD analysis for different operating points for every geometry considered out of the optimization.

Another approach that has been pursued intensively at DLR for several years is the bringing together of all relevant DLR institutes in the "Virtual Engine". This

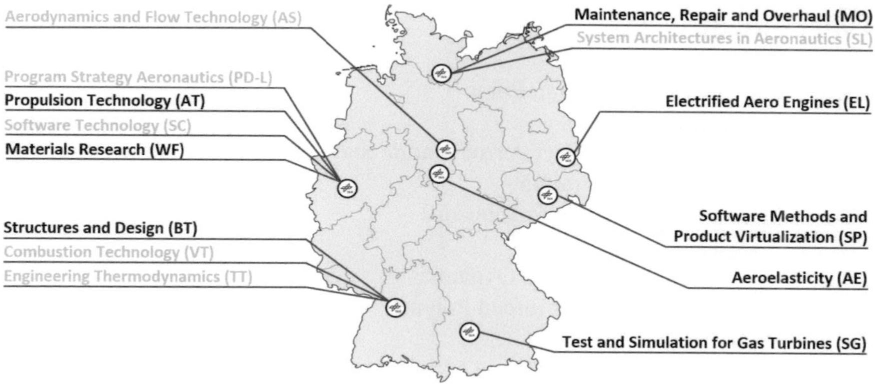

Fig. 15 Overview of DLR Institutes and units involved into the Virtual Engine Framework. Institutes marked in grey are not yet integrated but will be in the near future

is a collaborative, multidisciplinary and cross-institute framework for (pre-)design, simulation and optimization of engines and their essential components [10, 18]. Figure 2 suggests that the range of requirements and boundary conditions cannot be met by the simple approach outlined in Fig. 3 and by the three institutes involved to this date alone. DLR has grown steadily in recent years and new institutes have also emerged in the field of aeronautical propulsion and digitalization, which will gradually close these gaps in cooperation with the established propulsion research institutes. The aim is to jointly develop a cross-institute research strategy and to consistently strengthen cooperation between all participating institutes (see Fig. 15) by interlinking and increasingly automating standard processes. DLR's "Virtual Engine" serves to answer engine-specific questions and should ultimately be able to describe engines, test benches or parts of them based on simulation and data over the entire product life cycle and also in interaction with their respective environment. The most important components of the Virtual Engine Framework are the Gas Turbine Laboratory (GTlab) design and simulation framework [9] and its central data model for propulsion systems, the AutoOpti optimization package [26] as well as HyperGate and FlowSimulator [17] for Fluid-Structure coupling and for High Performance Computing (HPC) environments.

The contribution of the Institute of Structures and Design to this Virtual Engine Framework is the provision of standardized and automated or automatable structural mechanics processes for engines, gas turbines and propulsion structures which are summarized in the "Structural Mechanics GTlab Module" (SMM) [15, 19]. These tools and tool chains are available for all users of DLR's GTlab package in the Virtual Engine Framework and will provide fast and reliable process solution for multidisciplinary assessment with respect to structural mechanics, this being missing in ASPIRE. These methods are already in use in several DLR undertakings and projects for: optimization of radial compressors [20] or fast, automated bird strike investigations for engine (pre)design [19], to name just two examples.

Abbreviations

ADP	Cruise/Aero Design Point
APR	APRoach
ASPIRE	CleanSky 2 Project Aerodynamic and acouStic for high-byPass ratIo tuRbofan intEgration
CAA	Computational AeroAcoustics
CB	CutBack
CFD	Computational Fluid Dynamics
CFRP	Carbon Fiber Reinforced Polymer
CS2	CleanSky 2
CSM	Computational Structural Mechanics
DLR	Deutsches Zentrum für Luft- und Raumfahrt (German Aerospace Center)
DLR AE	DLR Institute of Aeroelasticity
DLR AT	DLR Institute of Propulsion Technology
DLR BT	DLR Institute of Structures and Design
EO	Engine Order
EU	European Union
FE	Finite Element
FEA	Finite Element Analysis
FOD	Foreign Object Damage
FPR	Fan Pressure Ratio
GTF	Geared TurboFan
JU	Joint Undertaking
MCT	Maximum Continuous Thrust
MDO	Multidisciplinary Design and Optimization
MPC	Multi-Point-Constraint
MRO	Maintenance, Repair and Overhaul
MTO	Maximum Take-Off
NLR	Nationaal Lucht- en Ruimtevaartlaboratorium (Aerospace Center of the Netherlands)
OGV	Outer Guide Vane
ONERA	Office National d'Etudes et de Recherches Aérospatiales (French Aerospace Center)
OP	Operating Point
RANS	Reynolds Averaged Navier Stokes equations
RL	Red Line
SDL	SiDeLine
SMM	Structural Mechanics Module
ToC	Top of Climb
TsaGI	Tsentral'nyy Aerogidrodinamicheskiy Institut (Russian National Aerodynamic Center)
u_{ax}, u_{tang}	Clamping condition u (axial and tangential)

UHBR Ultra High Bypass Ratio engine
VHBR Very-High Bypass Ratio engine

Acknowledgements The results presented in this paper were partly carried out with funding from the Clean Sky 2 Joint Undertaking (JU) under grant agreement No. 681856 (ASPIRE). The JU receives support from the European Union's Horizon 2020 research and innovation program, and also from Clean Sky 2 JU members who are not EU members.

Furthermore, support from colleagues at Airbus, NLR and ONERA and several other DLR Institutes are gratefully acknowledged for providing feedback and the always collegial and open cooperation. Colleagues from the DLR Institute of Propulsion Technology supported the activities in ASPIRE and are mainly responsible for driving the activities around the "Virtual Engine".

References

1. ASPIRE Engineering Coordination Memo (2016) ECM ASPIRE 0008 DLR FE Model and Eigen Vectors
2. Bigger engines, bigger challenges. (clean-aviation.eu). https://www.clean-aviation.eu/bigger-engines-bigger-challenges-clean-skys-aspire-resolves-them. Accessed 14 Sep 2023
3. Driving More Design with Simulation | Altair HyperWorks. https://altair.com/hyperworks. Accessed 12 Oct 2023
4. Intes Stuttgart. https://www.intes.de/k_permas/k_service/more_information/download. Accessed 12 Oct 2023
5. Permas product description v19 (intes.de). https://www.intes.de/k_service/more_information/download/product_description/infoE_v19.pdf. Accessed 08 Nov 2023
6. SA^{2}FIR Project Fair Stand. https://cleansky.virtualfair.be/hangar. Accessed 14 Sep 2023
7. WP 1 Fan/OGV Design Review/WP 2-3 Status Update and Planning. DLR Cologne, 06.10.2016
8. Aulich M, Siller U (2011) High-dimensional constrained multiobjective optimization of a fan stage. In: Proceedings of the ASME 2011 turbo expo: turbine technical conference and exposition. Volume 7: Turbomachinery, Parts A, B, and C. Vancouver, British Columbia, Canada, pp 1185–1196. https://doi.org/10.1115/GT2011-45618
9. Becker RG, Reitenbach S, Klein C, Otten T, Nauroz M, Siggel M (2015) An integrated method for propulsion system conceptual design. In: Proceedings of the ASME turbo expo 2015: turbine technical conference and exposition. Volume 1: aircraft engine; fans and blowers; Marine. Montreal, Quebec, Canada. https://doi.org/10.1115/GT2015-43251
10. Birschmann N, Heil J (2022) DLRmagazine 170 - Compute before flight. https://www.dlr.de/en/media/publications/magazines/2022_dlrmagazine-170-compute-before-flight. Accessed 02 Nov 2023
11. Boden F, Barth HP, Hein S, Heitmann D, Kirmse T, Klein C, Krüger W, Lemarechal J, Norambuena M, Schäfer D, Sinha K, Theiß A, Weiss A (2023) Recent developments of common numerical methods and common experimental means within the framework of the large passenger aircraft program. CEAS Aeronaut J 14:281–293. https://doi.org/10.1007/s13272-022-00635-w
12. Forsthofer N, Reiber C (2016) Structural mechanic and aeroelastic approach for design and simulation of CFRP fan blades. In: Deutscher Luft- und Raumfahrtkongress 2016. Braunschweig, Germany. documentID: 420132

13. Görke D, Denmat AL, Schmidt T, Kocian F, Nicke E (2012) Aerodynamic and mechanical optimization of CF/PEEK blades of a counter rotating fan. In: Proceedings of the ASME turbo expo 2012: turbine technical conference and exposition. Volume 7: structures and dynamics, parts A and B. Copenhagen, Denmark, pp 21–33. https://doi.org/10.1115/GT2012-68797
14. Hemmert-Pottmann S, Forsthofer N, Voß C, Schneider T (2019) High-dimensional multi-fidelity optimisation of a 2.5 stage low pressure compressor. In: International society for air breathing engines—ISABE conference 2019. Canberra, Australia
15. Kunc O, Broecker M (2024) Integration of large data based on HDF5 in a collaborative and multidisciplinary design environment, Part B: application to structural mechanics of gas turbine engines. In: 2024 AIAA science and technology forum and exposition. Orlando, Florida, USA. https://doi.org/10.2514/6.2024-0383
16. Pottmann S (2022) Multidisziplinäre Auslegung eines zweieinhalbstufigen Niederdruck-verdichters an seiner aerodynamischen Belastungsgrenze. Ph.D. thesis, Ruhr-Universität Bochum. https://doi.org/10.13154/294-8720
17. Reimer L (2015) The FlowSimulator—a software framework for CFD-related multidisciplinary simulations. In: European NAFEMS conference computational fluid dynamics (CFD)—beyond the solve. Munich, Germany
18. Reitenbach S (2023) Compute before flight—Erst rechnen, dann fliegen. In: 2. Virtual engine Kolloquium. Augsburg, Germany
19. Ritt SA, Kunc O, Forsthofer N, Schwämmle M, Reitenbach S (2023) Multi-disciplinary fan assessment with respect to bird strike. In: Deutscher Luft- und Raumfahrtkongress 2023. Stuttgart, Germany. https://doi.org/10.25967/610231
20. Schaffrath R, Nicke E, Forsthofer N, Kunc O, Voß C (2023) Gradient-free aerodynamic optimization with structural constraints and surge line control for radial compressor stage. In: Turbo expo 2023 turbomachinery technical conference & exposition. Boston, USA. https://doi.org/10.1115/GT2023-101593
21. Schnell R, Goldhahn E, Julian M (2019) Design and performance of a low fan-pressure-ratio propulsion system. In: International society for air breathing engine—ISABE conference 2019. Canberra, Australia
22. Schnell R and Trost, M (2026) Design and Integration of UHBR Propulsion Systems An Overview of DLR Activities within the Clean Sky 2 ASPIRE and SAÂ²FIR Projects. In: Boden, F et.al, Numerical Methods and Experimental Means for the Integration of Future Propulsion Systems, Springer Nature Switzerland, Cham, pp 71–91. https://doi.org/10.1007/978-3-031-97725-1
23. Schuff M, Reisberg J (2018) Flutter analysis of a flexible UHBR fan at different flight conditions. In: Proceedings of the ASME turbo expo 2018: turbomachinery technical conference and exposition. Volume 7C: structures and dynamics. https://doi.org/10.1115/GT2018-76930
24. Siller U, Aulich M (2010) Multidisciplinary 3D-optimization of a fan stage performance map with consideration of the static and dynamic rotor mechanics. In: Proceedings of the ASME turbo expo 2010: power for land, sea, and air. Volume 7: turbomachinery, parts A, B, and C. Glasgow, UK, pp 317–328. https://doi.org/10.1115/GT2010-22792
25. Siller U, Voß C, Nicke E (2009) Automated multidisciplinary optimization of a transonic axial compressor. Orlando, Florida, USA. https://doi.org/10.2514/6.2009-863
26. Voß C, Aulich M, Kaplan B, Nicke E (2006) Automated multiobjective optimisation in axial compressor blade design. In: Proceedings of the ASME turbo expo 2006: power for land, sea, and air. Volume 6: turbomachinery, parts A and B. Barcelona, Spain, pp 1289–1297. https://doi.org/10.1115/GT2006-90420, https://api.semanticscholar.org/CorpusID:109499590

Parametric Structural Modelling for Aeroelastic and Aeroacoustic Analyses of an Aircraft Fuselage

Wolf R. Krüger and Thomas Klimmek

Abstract This chapter describes results of work in the Cross-Demonstrator-Capabilities (XDC) activities in the project Large Passenger Aircraft on modelling and load analysis. As a result of this work, a simulation model of a 150-seat aircraft, used as a reference configuration in several investigations in the work package, has become available. The data includes a structural (finite-element) model and an aerodynamic (VLM/DLM) model. The model has been distributed to project partners for studies of aeroelastic stability and vibroacoustics. In this chapter, the modelling process is described and the simulation model is presented along with selected results.

Keywords Structural modelling · Aeroelasticity · Vibroacoustic analysis · Parametric modelling

1 General Objectives

In the project Large Passenger Aircraft (LPA), Platform 1, technologies for future transport aircraft are being studied. Specifically, the integration of novel engine concepts into future airframes is under investigation. Specific questions being addressed include performance, implications of novel engine concepts for the airframe structure, the design of innovative tail sections, and problems of airframe noise resulting from the engine integration [1]. All these studies require reliable airframe models with a suitable degree of detail. This is made more complicated by the fact that in many cases no converged airframe design is available for a certain investigation, or that a design is proprietary to a manufacturer and can thus not be freely distributed in the project. Sometimes, basic models are available, but lack the necessary level of detail for a specific investigation; this is often true for vibroacoustic investigation.

The main objective of the work described here, carried out between 2017 and 2019 in the so-called Cross-Demonstrator (XDC-) activities, was thus the generation of representative aircraft models for aeroelastic and vibroacoustic investigations, to

W. R. Krüger (✉) · T. Klimmek
German Aerospace Center (DLR), Institute of Aeroelasticity, Göttingen, Germany
e-mail: Wolf.Krueger@dlr.de

F. Boden (ed.), *Numerical Methods and Experimental Means for the Integration of Future Propulsion Systems*, Springer Aerospace Technology,
https://doi.org/10.1007/978-3-031-97725-1_6

be freely exchangeable among the project partners for the tasks to be performed in the XDC and the support of the LPA demonstrators.

2 Background and Considerations Dynamic Reference Aircraft Model and Rear End Load Analysis

For the development of numerical methods for aeroelastic and vibration analyses of future aircraft configurations, as is the aim of this project, representative reference models for all disciplines (here mainly aerodynamics and structures) are necessary. Only very few models with a sufficient level of detail are publicly available. One well-known example is the Common Research Model (CRM) of NASA, for which an outer geometry is available from the Drag Prediction Workshop. However, focus of this model has been on the aerodynamics of the wing, and no structural design of wing, fuselage or empennage has been available. Recently, structural (i.e. finite-element) models for aeroelastic analysis or for multi-disciplinary optimization (MDO) have been provided [7], but none of them had been intended to be used for vibroacoustic analysis.

Thus, a reference model has been developed and studied in the following analysis tasks in the framework of LPA:

- Load analysis, with a focus on rear end loads,
- Vibroacoustic analysis,
- Aeroelastic stability analysis of the tail, and
- Aeroelastic investigations of the influence of under-wing mounted very large UHBR engines, featuring an AFC system, on the flutter speed [13].

For a representative aircraft model, a proper modelling of global mass and stiffness characteristics is needed. The modelling process used by DLR in LPA is MONA [8], consisting of the in-house development ModGen and the commercial finite element software MSC NASTRAN. Focus of that process is the parametric modelling of wing structures, allowing fast variation of design parameters, thus providing a basis for optimization. MONA has been successfully used in a number of European, German (Lufo-) and DLR internal projects.

The exact but fast determination of loads acting on the aircraft is driving the aircraft development, especially due to the sizing of the airframe structure. DLR has developed an integrated, comprehensive loads process in the DLR-internal project iLOADS [9], and also in subsequent projects. The overall loads process at DLR has been the subject of verification as well as validation on different aircraft configurations, including comparisons to flight test results, see also [9]. This process is the basis for the loads-related activities in LPA.

The selected reference configuration was the result of extensive discussions in the project. When setting up the original LPA work program, a transport aircraft with rear-mounted engines had been envisaged for the investigations, either of CROR- or

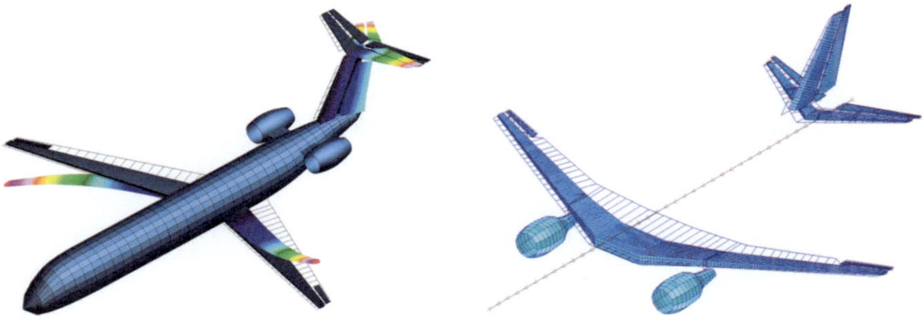

Fig. 1 Aeroelastic simulation models from MONA process available at start of the XDC activities: Left: ALLEGRA-S/Right: DLR-D150

of UHBR-type. This configuration would most likely be equipped with a T-tail. The aeroelasticity of such an aircraft has thus been investigated in LPA, see [14]. DLR-internally, the MONA process had been applied previously to a related configuration, the so-called ALLEGRA-S aircraft, based on the LamAir configuration, see [11, 16], and Fig. 1, left.

However, during the technical discussions in the work package, it became evident that for most investigations in the project the relevant configuration would be a mid-range aircraft with classical under-wing mounted engines. For this purpose, it was decided to base the modelling activities on the so-called DLR-D150 model, an A320-size configuration for which basic data and a model focusing on the investigations of wing loads was available at DLR [5, 9], and which could be extended to include structural details relevant for the current investigations. The extensions are mainly driven by the needs of local load analysis, aeroelastic analysis and of requirements for vibroacoustic investigations. The activities thus include an improved modelling of the engine attachments, of the fuselage and the tail section, and of the joining between fuselage, wings and tail.

2.1 Model Generation Process

The DLR modelling process is MONA, combining the in-house development ModGen [7] and the commercial finite element software MSC NASTRAN. The MONA process has been developed to perform an aeroelastic structural design for different aircraft configurations. The main application is the generation of models representing the global aeroelastic properties of an aircraft, the structural model can therefore be considered to be a global finite element (GFEM) model.

Three main steps are conducted: the parametric model set-up, followed by an extensive load analysis campaign, and finally a structural optimization of the wing structures (this includes the main wings and the tailplane). Aeroelastic requirements such as sufficient control surface efficiency are taken into account. MONA can be

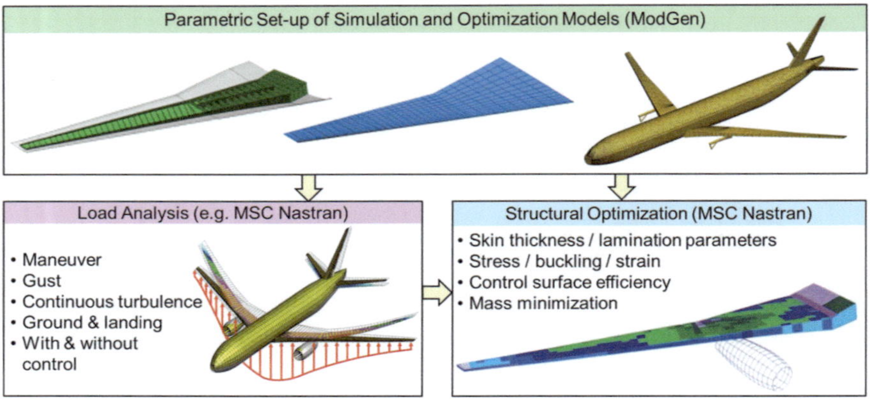

Fig. 2 Overview of the MONA process—ModGen and NASTRAN

used as stand-alone tool (as for the work in LPA) or integrated into high-fidelity MDO chains [8].

The name MONA indicates the two main computer programs involved in that process chain. First to mention is ModGen: ModGen is an inhouse computer program, developed at the DLR Institute of Aeroelasticity to parametrically build up the aeroelastic model. The model set-up within ModGen is based on a parametrically defined geometry of the wing-like components and the fuselage. Figure 2 gives an overview of the various components of that model. This comprises the outer geometry generated by the defined airfoils and their position on the wing, and all significant elements of the inner load carrying structure like the spars, ribs and stringers. In addition, ModGen formulates the optimization problem used in the structural sizing step.

The models generated by ModGen are used and processed within NASTRAN, the second computer program involved in the MONA process. MSC NASTRAN is a finite element software well established in aviation. It is applied here for the comprehensive load analysis and the subsequent structural sizing of the airframe (by means of structural optimization).

2.2 *Simulation Model*

In the following section, the structural model of the reference configuration will be presented. The original model of the D150 available at the project start, see Fig. 1, right, has been extended in several steps. First, a parametric fuselage model was developed, shown in Fig. 3. The topology of the fuselage is similar to that of an A320. The dynamics of the aircraft are very much dependent on the interconnection between fuselage and wings. For the detailed analysis of the rear end, a good representation

Fig. 3 Fuselage model of reference configuration

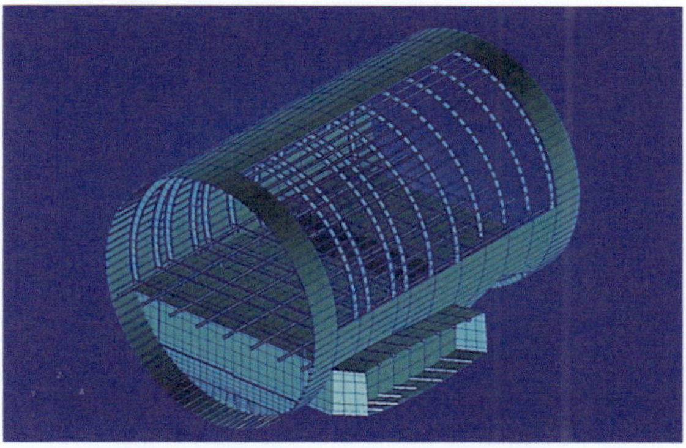

Fig. 4 Modelling elements: Center section

of the rear fuselage and the vertical and horizontal tail, as well as of the connecting element, is mandatory.

The next step was thus the modelling of the so-called center section, i.e. the joint between fuselage and the main wings, see Fig. 4. A similarly detailed topology is needed for the joint between fuselage and empennage, including the pressure bulkhead, see Fig. 5. In addition, an improved pylon model was developed, see Fig. 6. All of these elements are essential for analysis of loads, vibroacoustic and aeroelastic stabilities.

Finally, the transfer of these models for vibroacoustic analysis has been performed in cooperation with the respective project partners to ensure the validity of the model for the investigations performed in their work package. It quickly became evident that the vibroacoustic models need to be of a significantly finer discretization than the GFEM models usually generated by the MONA process. Thus, the model has been extended to include more details in those components essential for vibroacoustic analysis. A sample finite element model set-up for vibroacoustics has been obtained from project partners active in the instrumentation for vibration measurements and

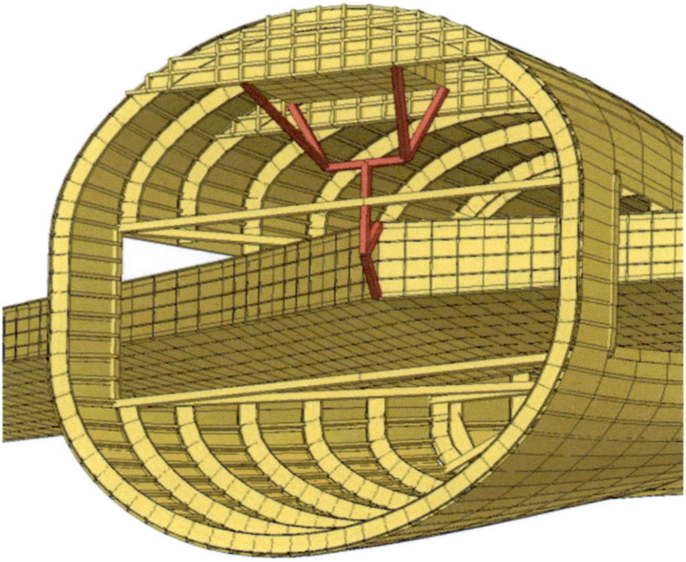

Fig. 5 Modelling elements: Empennage attachment

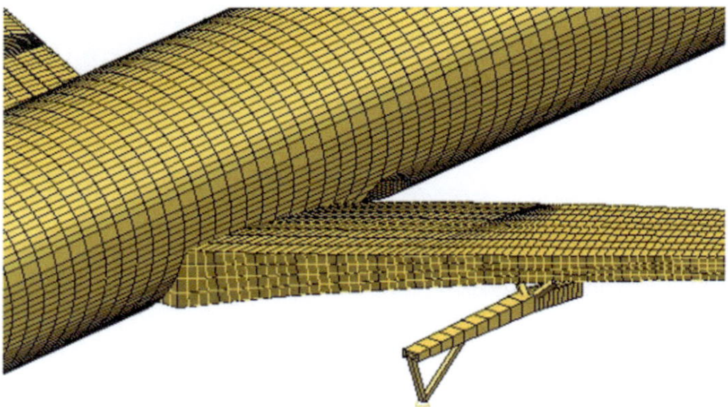

Fig. 6 Modelling elements: Pylon

vibration studies, to ensure the compatibility of the structural models between the activities. That fuselage reference is the so-called FlightLab model, a finite element model of an Airbus A320 fuselage which represents the target level of detail for the vibroacoustic analyses, see Fig. 7, right [3]. The model shown at the left of Fig. 7 is the corresponding NASTRAN model output from the MONA process.

After the validation of the MONA process with the FlightLab reference, a full rear section of the aircraft, following the requirements of the vibroacoustic rear fuselage analysis, has been set up and delivered to the project partners, see Fig. 8.

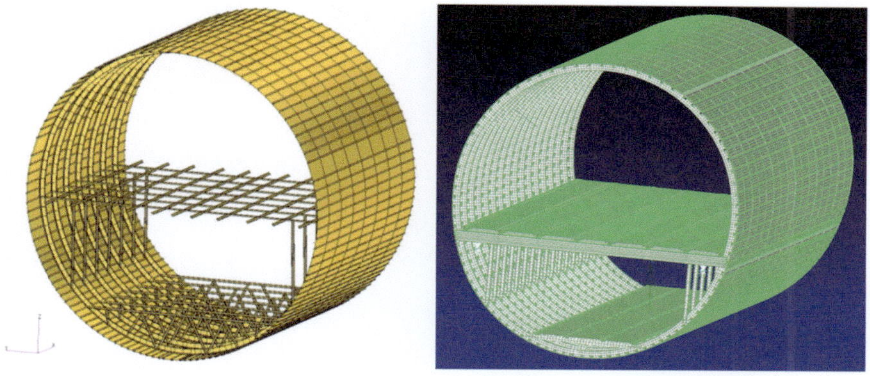

Fig. 7 Vibroacoustic model of center section from MONA process (left) and vibroacoustic reference (right)

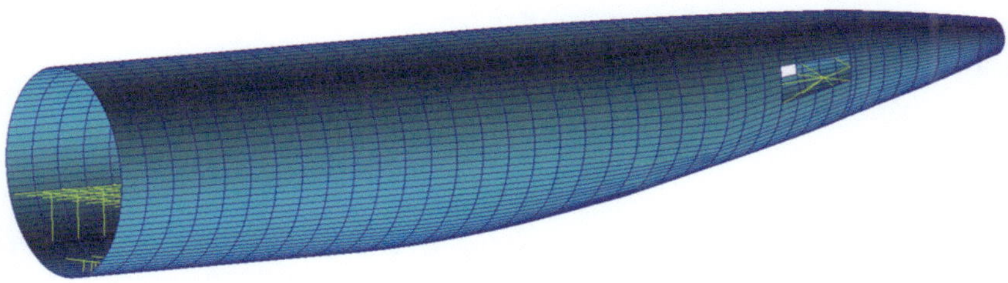

Fig. 8 Fuselage model from DLR MONA process for vibroacoustic analysis

Finally, a full aircraft model is assembled, see Fig. 9. For the representation of the aerodynamics, not required for vibroacoustic analysis but, of course, needed for aeroelastic investigations, the full aircraft model comes with a grid used for potential flow, i.e. Vortex-Lattice- and Doublet-Lattice- (VLM/DLM) aerodynamics. A definition of the coupling between structural model and aerodynamic model is also included.

2.3 Load Analysis

To demonstrate the suitability of the model for relevant investigations, load analyses on the parameterized aircraft model have been performed. Simulations include maneuver loads, gust loads, and ground loads. Some selected results are presented in the following section.

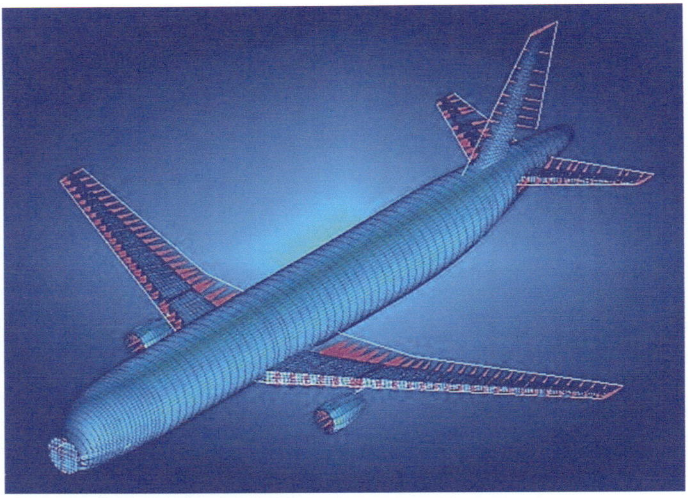

Fig. 9 Structural model of the reference aircraft developed in XDC

2.3.1 Gust Loads

Figure 10 shows an investigation of gust loads determined by the so-called Pratt gust and a dynamic excitation, the so-called 1-cos-gusts. The Pratt equation [12] is a quasi-static approximation of a gust excitation, independent of a specific gust length. In comparison, a 1-cos-gust is a dynamic excitation defined according to CS-25 certification standards (see CS 25.341 [4]). These gusts are defined using a gust amplitude and length, which are chosen according to the flight point.

In Fig. 10 results for the equivalent load factor Δn_z, monitored at the center of gravity (CG) of the aircraft, are compared. It can be seen that the Pratt assumption (dashed black line), despite being a quasi-static approximation, still yields loads equivalent to the peak loads for different moderate gust lengths of the 1-cos-gust definition (magenta, blue and orange lines, T representing the gust length in meters or chords). The comparison supports the classical assumption, found in relevant textbooks on load analysis like [6], that a 1-cos-gust length of 25 wing chords (blue line, c signifies chord in Fig. 10) gives the equivalent load factor Δn_z of the Pratt gust for the same flight point.

2.3.2 Tail Loads

In the work package, loads on the tail are of high interest. In Fig. 11, the design load envelope at root of the horizontal tail plane (HTP) is represented which results from dynamic 1-cos gust and quasi-steady maneuver simulations. It is apparent that the largest bending moment (M_x) and torsion moment (M_y) are reached during maneu-

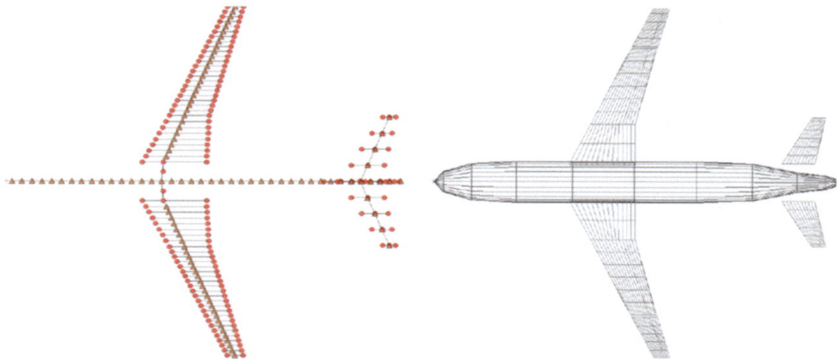

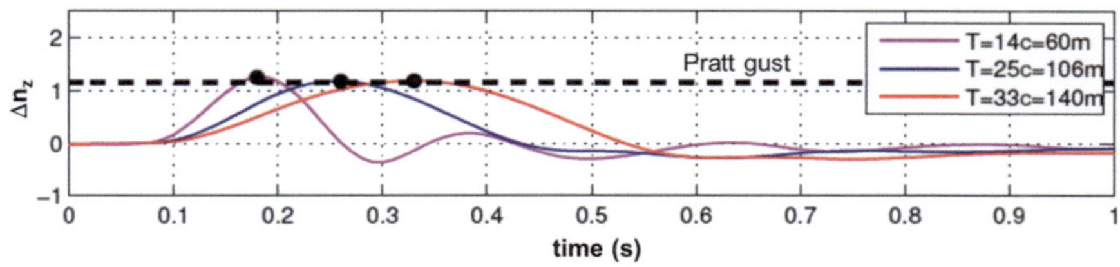

Fig. 10 Comparison of gust loads (load factor Δn_z around steady flight) according to Pratt equation (dashed line) and dynamic simulations for three gust lengths (colored lines)

Fig. 11 Design load envelope at the HTP root

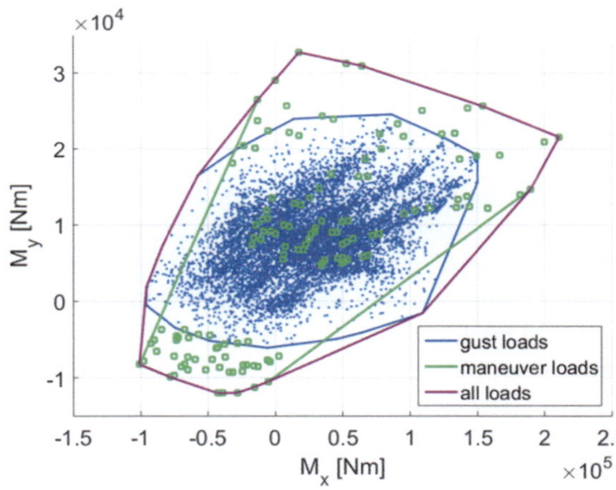

vers; however, there are also points on the overall load envelope which result from gust encounters.

For structural sizing purposes, mainly the maximum amplitude of loads at the structural components are relevant. Next to the maximum loads, however, the frequency of occurrence of all load levels is of interest, especially for fatigue analysis of the structure. For this purpose, investigations based on turbulence spectra or load

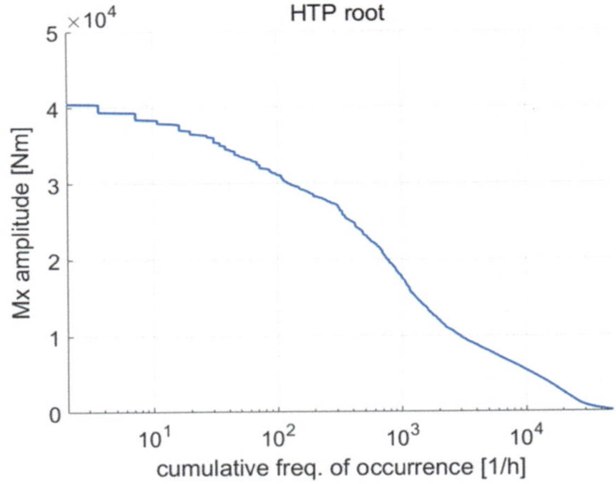

Fig. 12 Bending moment collective at the HTP root in turbulence

spectra are performed. Figure 12 shows an example of a bending moment collective at the HTP root in continuous turbulence during the climb phase. As common in standard applications, the number of exceedances of load increments increases for smaller loads. In this case, the von-Kármán spectrum with a turbulence scale of 2,500 ft (762 m), as prescribed in CS-25, is considered. The root mean square value of the vertical wind speed is selected such that the probability of occurrence is 10^{-3} which is classified as moderate turbulence according to MIL-STD-1797A [15].

2.3.3 Ground Loads

As a final demonstration of the model capabilities, a bi-directional interface of the finite element model to a multibody simulation (MBS) tool was established to simulate loads on the rear fuselage from a typical landing impact [10]. First, a multibody simulation applying the commercial MBS tool SIMULIA/SIMPACK was performed, using a model including a comprehensive, fully non-linear landing gear model and a linear (modal) flexible aircraft structure. Attachment load histories for all gears are calculated and saved. The data is used as force input for a transient solution inside MSC NASTRAN where loads on all components represented in the full finite element model can be generated. Figure 13 shows a comparison of the acceleration at CG of the MBS analysis (blue) and the subsequent finite element analysis (red). Acceleration levels are comparable, but a phase shift between both plots can be seen which can be attributed to the differences in the modelling of the load path between the modal model of the structure used in the MBS and the full finite element model used in the NASTRAN transient analysis.

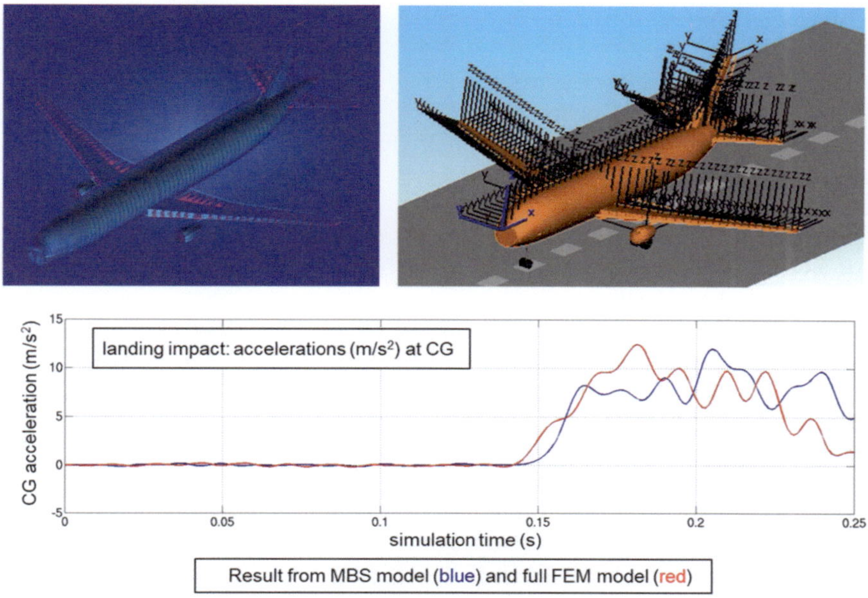

Fig. 13 Coupled multibody and finite-element simulation for analysis of landing loads

3 Conclusion

In the project Large Passenger Aircraft, LPA, technologies for future transport aircraft have been studied. For the investigation of novel engine concepts into future airframes, with a focus on the rear fuselage, a reference model has been set up and made available for the project partners. The model is representative of a short- and medium range aircraft of A320-size, but can be freely distributed. In the project, focus was on the further development of modelling capabilities to closer represent realistic load paths for flight and ground load analysis, as well as flutter analysis, and to extend the level of detail of the model for vibroacoustic analysis.

The vibroacoustic model is in use in the Clean Sky 2 framework for the quantification of power flow input into the fuselage in dependence of the engine installation point on the wing [17]. For this purpose, numerical investigations are carried out on FEM models of an Airbus A320 wing generated with the parametric model generator described in this chapter.

The capabilities developed in the XDC to generate representative parametric models for loads and aeroelastic analysis are widely applied in follow-on projects, an example is the so-called DLR-F25 configuration investigated in the Clean Aviation project UP Wing [2].

Abbreviations

AFC	Active Flow Control
CG	Center of Gravity
CS-25	Certification Specification 25 (for Large Aeroplanes)
c	Chord
CS2	CleanSky 2
CRM	Common Research Model
CROR	Counter Rotating Open Rotor
Δn_z	Load Factor
DLR	Deutsches Zentrum für Luft- und Raumfahrt e.V. (German Aerospace Center)
DLM	Doublet-Lattice-Method
FEM	Finite Element Method
GFEM	Global Finite Element Model
HTP	Horizontal Tail Plane
LPA	Large Passenger Aircraft (one of the three Innovative Aircraft Demonstration Platforms (IADP) of CS2)
MBS	Multi-Body Simulation
MDO	Multi-Disciplinary Optimization
MIL-STD	United States Military Standard
MONA	DLR Modelling Process based on ModGen and NASTRAN
M_x	Bending Moment
M_y	Torsion Moment
NASA	National Aeronautics and Space Administration of the USA
T	Gust Length
UHBR	Ultra High Bypass-Ratio Engine
VLM	Vortex-Lattice-Method

Acknowledgements This project has received funding from the Clean Sky 2 Joint Undertaking under the European Union's Horizon 2020 research and innovation program under grant agreement N° CS2-LPA-GAM-2020-2023-01.

References

1. Boden F, Barth HP, Hein S, Heitmann D, Kirmse T, Klein C, Krüger W, Lemarechal J, Norambuena M, Schäfer D, Sinha K, Theiß A, Weiss A (2023) Recent developments of common numerical methods and common experimental means within the framework of the large passenger aircraft program. CEAS Aeronaut J 14:281–293. https://doi.org/10.1007/s13272-022-00635-w , Springer

2. Cumnuantip S, Schulze M, Krüger WR (2024) Assessment of the aeroelastic stability of high aspect ratio wing aircraft during the preliminary design. In: International forum on aeroelasticity and structural dynamics, IFASD 2024. Den Haag, The Netherlands
3. Dewald RD, Klimmek T, Algermissen S, Hesse C, Winter R (2023) Angepasste Modellierungsvorschriften für vibroakustische Untersuchungen von Flugzeugrümpfen. In: 49. Jahrestagung fur Akustik (DAGA). https://elib.dlr.de/194901/
4. European Aviation Safety Agency (2019) CS25—Certification specifications and acceptable means of compliance for large aeroplanes—Amendment 13
5. Handojo V, Lancelot P, De Breuker R (2018) Implementation of active and passive load alleviation methods on a generic mid-range aircraft configuration. In: AIAA aviation forum. https://doi.org/10.2514/6.2018-3573
6. Hoblit F (1988) Gust loads on aircraft: concepts and applications. American Institute of Aeronautics & Astronautics, AIAA Education Series
7. Klimmek T (2014) Parametric set-up of a structural model for Fermat configuration for aeroelastic and loads analysis. J Aeroelast Struct Dyn 3(2):31–49. https://doi.org/10.3293/asdj.2014.27
8. Klimmek T, Schulze M, Abu-Zurayk M, Ilic C, Merle A (2019) cpacs-MONA—an independent and in high fidelity based MDO tasks integrated process for the structural and aeroelastic design for aircraft configurations. In: International forum on aeroelasticity and structural dynamics 2019, IFASD 2019. Savannah, GA, USA
9. Krüger W, Ciampa PD, Geier M, Kier T, Klimmek T, Kohlgrüber D, Ohme P, Risse K, Schwinn J (2018) A comprehensive loads process at the DLR—definition, analysis, experimental evaluation. Aerospace Lab 14. https://aerospacelab.onera.fr/al14/comprehensive-load-process-at-the-dlr-definition-analysis-and-experimental-evaluation, ONERA Scientific Information Department
10. Krüger W, Cumnuantip S (2019) A hybrid approach for the analysis of aircraft ground loads. In: International forum on aeroelasticity and structural dynamics, IFASD 2019. Savannah, GA, USA
11. Krüger W, Klimmek T, Liepelt R, Schmidt H, Waitz S, Cumnuantip S (2014) Design and aeroelastic assessment of a forward-swept wing aircraft. CEAS Aeronaut J 5:419–433. https://doi.org/10.1007/s13272-014-0117-0. Springer
12. Pratt KG, Walker WG (1954) A revised gust-load formula and a re-evaluation of v-g data taken on civil transport airplanes from 1933 to 1950. Tech. Rep. NACA-TN-2964, National Advisory Committee for Aeronautics (NACA), Washington D.C., USA
13. Schulze M, Neumann J, Klimmek T (2021) Parametric modeling of a long-range aircraft under consideration of engine-wing integration. Aerospace 8(1):1–20. https://doi.org/10.3390/aerospace8010002, https://www.mdpi.com/2226-4310/8/1/2
14. Schäfer D (2021) T-tail flutter simulations with regard to quadratic mode shape components. CEAS Aeronaut J 12:621–632. https://doi.org/10.1007/s13272-021-00524-8, Springer
15. US Department of Defense (2004) MIL-STD-1797A—flying qualities of piloted aircraft notice 3
16. Waitz S (2017) The aeroelastic behaviour of a forward-swept wing configuration with focus on engine gyroscopics and T-tail flutter. In: International forum on aeroelasticity and structural dynamics, IFASD 2017. Como, Italy
17. Zettel S, Winter R, Norambuena M, Klimmek T, Böswald M (2022) Numerical investigation of power flow input into a fuselage due to jet engine induced wing vibrations. In: Internoise 2022—51st international congress and exposition on noise control engineering. INTER.NOISE 2022. Glasgow, Scotland, pp 22–248. https://elib.dlr.de/187993/1/zettel_2022.pdf

Flutter Assessment of a Passenger Aircraft with Standard and UHBR Engines

Jürgen Arnold and Jens Neumann

Abstract Focus is on the stability assessment of an aircraft equipped with the currently introduced high-efficiency engines with ultra-high bypass ratios (UHBR) and to compare the critical flutter onset to the same aircraft using standard engines with a smaller fan diameter. The D150 configuration modelled in-house by DLR was chosen, since it is similar to the Airbus A320 in terms of size and dynamic behavior. The setup of the aeroelastic model is shown and results from the flutter assessment clarify differences between the investigated standard and UHBR engine types with respect to flutter onset. Here, the sensitivity study includes varying locations of the UHBR engine in front of the D150 wing and is based on unsteady aerodynamic forces using the doublet lattice method (DLM). In addition, a methodology for the use of CFD based aerodynamic forces is developed and applied to the AMP wing model which allows the comparison to measured flutter data. The aeroelastic models are analyzed with the commercial ZAERO aeroelastic software package.

Keywords D150 configuration · UHBR engine integration · Flutter analysis

1 Introduction and General Remarks

This contribution summarizes the aeroelastic modelling and flutter assessment of a UHBR-powered aircraft featuring an active flow control (AFC) system that requires the development of a methodology for the utilization of unsteady CFD-based generalized aerodynamic forces in the flutter analysis. The objective was to give a stability assessment of an aircraft equipped with the currently introduced high-efficiency engines with ultra-high bypass ratio (UHBR) and to compare the critical flutter onset with the same aircraft using standard engines with a smaller fan diameter. Here, the D150 configuration [1] modelled in-house by DLR was chosen, since it is similar to the Airbus A320 in terms of size and dynamic behavior.

J. Arnold (✉) · J. Neumann
German Aerospace Center (DLR), Institute of Aeroelasticity, Göttingen, Germany
e-mail: Juergen.Arnold@dlr.de

F. Boden (ed.), *Numerical Methods and Experimental Means for the Integration of Future Propulsion Systems*, Springer Aerospace Technology,
https://doi.org/10.1007/978-3-031-97725-1_7

When it became clear during the investigation that active flow control is switched off for those airspeeds and Mach numbers of D150 where flutter occurs, the AFC system was classified to be no longer relevant for the flutter assessment of this aircraft. As a consequence, the decision was taken to study the difference between the standard and UHBR engine types with unsteady aerodynamic forces using the doublet lattice method (DLM). This would speed up the computation and hence, allow a more detailed sensitivity study with varying locations of the engine in front of the D150 wing. The parameter variations for the flutter sensitivity study with the standard CFM56 engine include location and pylon wall thickness, whilst the variation of UHBR parameters is rather limited due to integration aspects related to size and required ground clearance of this enlarged engine type. The utilized aeroelastic model has the following features:

- NASTRAN dynamic finite element model, providing different payload-fuel cases as well as pylon and engine models according to the studied engine type,
- ZAERO unsteady aerodynamic forces from DLM for lifting surfaces, as well as body definitions for fuselage and engine nacelles.

In parallel, the required methodology for the extraction of CFD-based generalized aerodynamic forces (GAF), as originally foreseen for the AFC modelling, has been developed, introduced into flutter analyses and finally applied to the AMP wing model [2, 3] for which measured flutter data are available for comparison. The available GAF-matrices of the aerodynamic approaches comprising DLM for the D150 aircraft and CFD for the AMP wing were included as part of the aeroelastic models and analyzed with the commercial ZAERO aeroelastic software package [4] to predict the related flutter onsets.

2 Influence of UHBR Engines on Flutter Behavior

2.1 D150 Flutter Assessment with Standard CFM56 Engine

The technology projection, in terms of an aeroelastic assessment, of the A320 equipped with large UHBR engines and related active flow control (AFC) requires both the integration of rotating effects in the dynamic aircraft model and the setup of a flutter model to account for large UHBR engine integration and AFC system effects. All investigations are based on the generic D150 configuration which is available from the former DLR project VAMP and similar to the A320 in planform as well as typical dynamic behavior.

The model input was a NASTRAN finite element model of the generic D150 configuration, where its condensed model includes four mass cases comprising different distributions of payload (0, 50%) and fuel (0, 50, 100%). The initial model contained rigid substructure definitions for pylon, engine and nacelle considering mass but neglecting related stiffness. Also, the panel definitions for pylon and nacelle

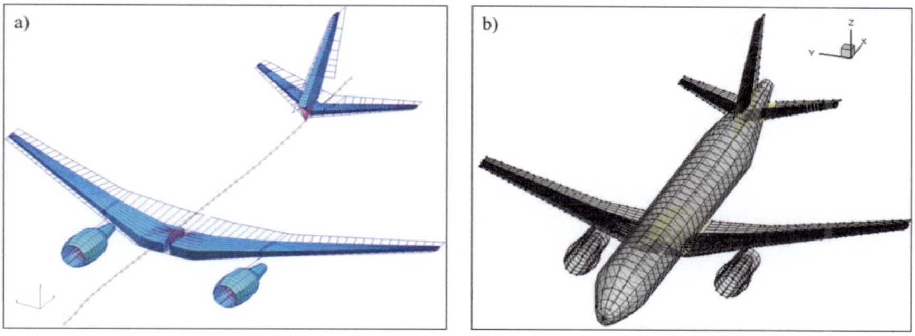

Fig. 1 D150 dynamic (**a**) and aerodynamic model (**b**) with CFM56 engine

were not available. Hence, the technical work focused on the setup of an aeroelastic model that could form the reference for the flutter analyses with standard and larger UHBR engine types. This required the implementation of flexible structures for pylon, engine and nacelle into the dynamic model, as well as the aerodynamic model setup for the new substructures. Here, two alternative modelling techniques in ZAERO – comprising the BODY7 (displacing body) and CAERO7 (lifting body) elements – were investigated with pylon and nacelle with respect to differences in the flutter results. Figure 1 shows the D150 dynamic NASTRAN model with leading and trailing edge nodes (left figure), as well as the aerodynamic ZAERO model using CAERO7 (lifting body) elements for wing, empennage, pylon, engine and BODY7 elements (displacing body) for the fuselage (right figure). The ZAERO model was used to perform an initial flutter analysis for the D150 with the standard CFM56 engine type including the flexible engine attachment to the wing and the aerodynamic influence of pylon and nacelle. The following flutter results are derived for the empty aircraft without payload and fuel representing a total mass of approximately 44,500 kg using a non-matched flutter analysis. Figure 2 depicts the eigenfrequencies and damping versus true airspeed using BODY7 elements (displacing body) to model pylon and nacelle, whilst Fig. 3 gives the results for CAERO7 elements (lifting body). Aircraft flutter is indicated by positive damping values. The critical true airspeeds (TAS) are at 324.7 m/s and 330.4 m/s for the two modelling approaches, respectively. The corresponding flutter modes show large nacelle contributions, which can be seen for flutter mode 23 in Fig. 4.

In a next step, a methodology has been setup and implemented to consider rotating effects for under-wing engines and engine thrust in the ZAERO flutter computation. This consideration introduces unsymmetrical modes due to gyroscopic coupling of mode shapes and so-called follower forces related to the thrust vector. In the following, flutter analyses for the D150 reference aircraft with standard engine were performed to assess the influence of the rotating effects on flutter onset. Again, pylon and nacelle were modelled with BODY7 and CAERO7 elements, respectively. All depicted damping results are derived from non-matched flutter analyses for the empty aircraft without payload and fuel, representing a total mass of around 44,500 kg. Here,

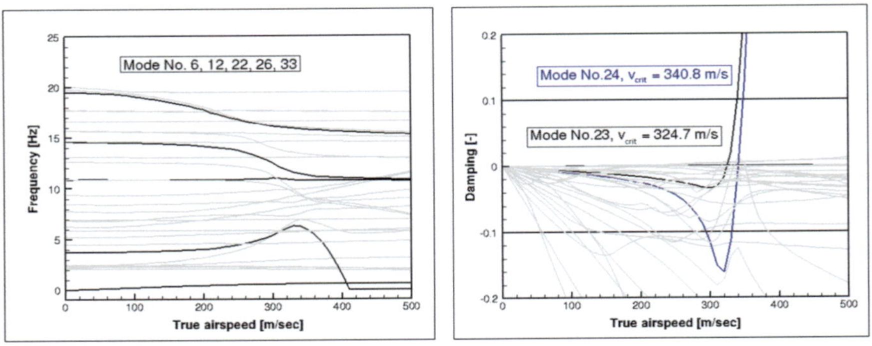

Fig. 2 Flutter diagrams for D150 aeroelastic model (BODY7 elements)

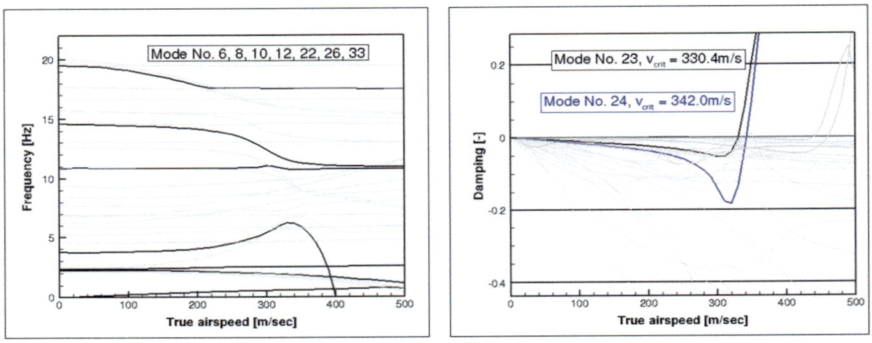

Fig. 3 Flutter diagrams for D150 aeroelastic model (CAERO7 elements)

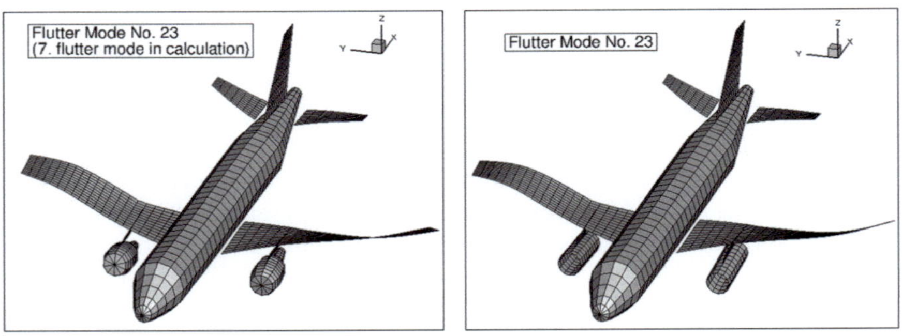

Fig. 4 Critical flutter mode No. 23 with BODY7 (left) and CAERO7 elements (right)

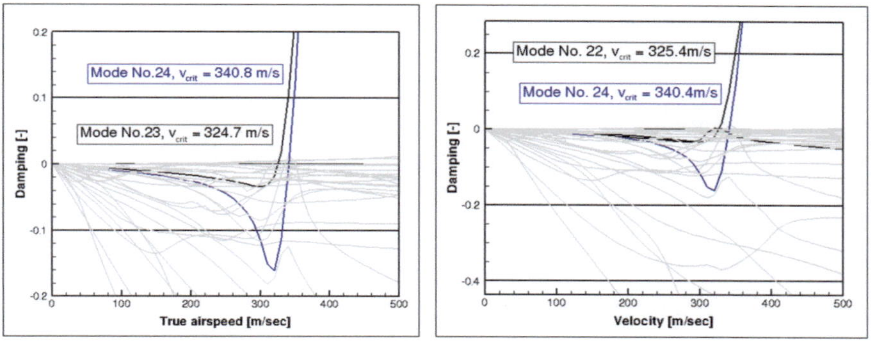

Fig. 5 D150 with standard engines using BODY7 – Damping for engines switched off (left) and on (right)

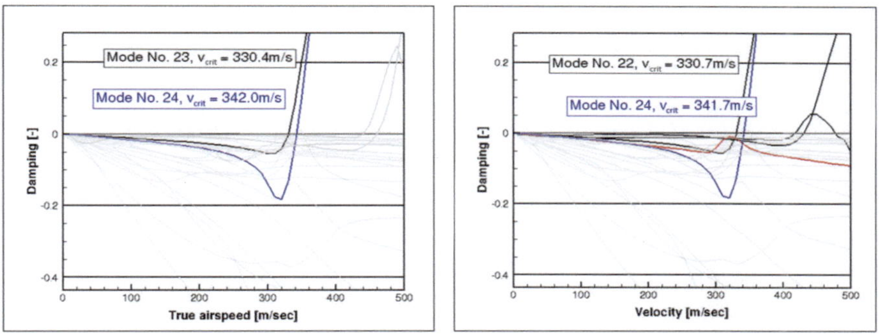

Fig. 6 D150 with standard engines using CAERO7 – Damping for engines switched off (left) and on (right)

Fig. 5 depicts the resulting damping values versus true airspeed for engines being switched off and on, showing the influence of gyroscopic coupling and thrust on the respective flutter onset using BODY7 elements, whilst Fig. 6 gives the results for CAERO7 elements. For the D150 with standard engine, the influence of the considered engine effects is found to be rather small. An increase of the critical airspeed for flutter onset of 0.7 m/s (+0.2%) and 0.3 m/s (+0.1%) is obtained for the two modelling approaches, respectively. A summary of the presented flutter results with due account given to the aerodynamic nacelle model and gyroscopic engine effects is available in Table 1.

Finally, the sensitivity of the engine position on flutter onset was studied. The analyses were performed for the D150 reference aircraft with standard engines under consideration of basic constraints which are applicable to the integration of an UHBR-engine. They comprise the same spanwise position for both engine configurations as well as a forward and/or upward shift of the geometrically larger UHBR-engine. The ZAERO model was used to perform non-matched flutter analyses for the D150

Table 1 Summary of flutter results considering pylon and engine aerodynamic models (BODY7, CAERO7) and engine effects (gyroscopic coupling, thrust)

ID	D150 VAMP-configuration	Flutter mode	v_{crit} [m/s]	Flutter f [Hz]	Payload (%)	Fuel (%)
c01	Engine-BODY7	23	324.7	9.82	0	0
c01	Engine-BODY7 with Gyroscopy, Thrust	22	325.4	9.80	0	0
c01	Engine-CAERO7	23	330.4	9.62	0	0
c01	Engine-CAERO7 with Gyroscopy, Thrust	22	330.7	9.60	0	0

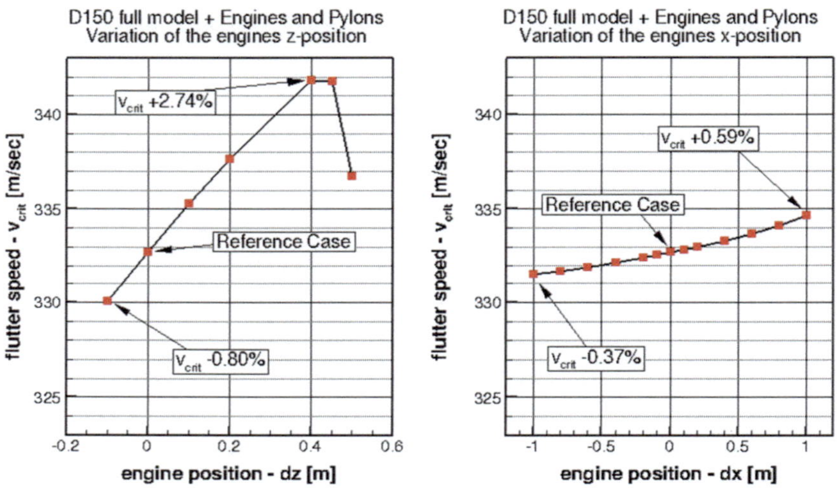

Fig. 7 D150 with standard CFM56 engines – Influence of engine vertical (left) and longitudinal installation position (right) on flutter speed

with standard engine type CFM56 to gain reference results for comparison with the UHBR-configuration. Figure 7 depicts the influence of the engine installation position on the critical speed for flutter onset, whilst Fig. 8 shows the influence of the pylon wall thickness. First, it is found that the influence of changing the vertical installation position of the engine from its reference position by dz on critical speed (−0.80 to +2.74%) is much larger than for the same longitudinal variation dx (−0.37 to +0.59%). Second, the influence of pylon wall thickness on critical speed is of a similar magnitude (−2.16 to +2.55%) as found for changes in the vertical engine installation position. Here, the increase of wall thickness above the reference value of 15 mm introduces a larger pylon stiffness and thus a stiffer connection of the engine to the wing structure.

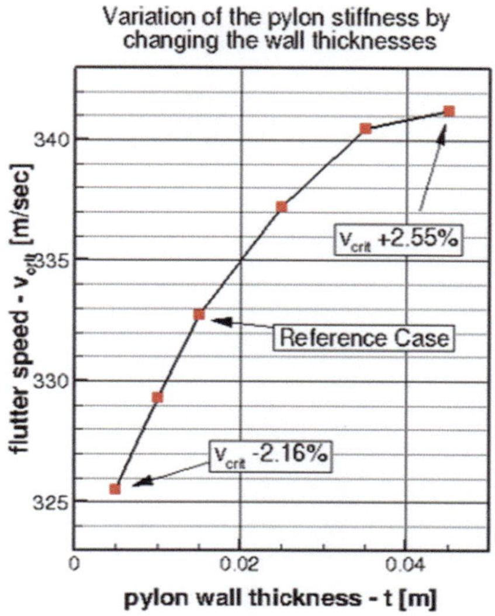

Fig. 8 D150 with standard CFM56 engines – Influence of pylon wall thickness on flutter speed

2.2 D150 Flutter Assessment with UHBR Engine

The change from the standard to UHBR engine type required modifications to the dynamic and aerodynamic model for pylon (geometry, stiffness, mass), engine (mass) and nacelle (geometry, mass). Here, the UHBR mass was assumed to increase to 120% of the standard engine for which CFM56 data is used for engine and nacelle. The assumed UHBR mass could be verified against DLR internal data which consider the CFM56 standard engine and a geared turbofan representing a modern NEO type of engine. The provided data gives factors of approximately 1.20 and 1.28 for the increase in mass and diameter, respectively. These factors are the same as used in the flutter model and confirm the implemented engine masses and nacelle geometry for the standard and UHBR engines in the D150 model. The new reference aircraft considers basic constraints applicable for the integration of an UHBR engine which comprise the same spanwise position for both engine configurations as well as a backward and upward shift of the larger UHBR engine according to the engine position used within the project. This shift also moves the center of gravity (c.g.) of engine and nacelle in longitudinal and vertical direction ($\Delta x = 0.44m/\Delta z = 0.25m$). Related geometries of the engine configurations and spatial coordinates are given in Fig. 9, whilst Fig. 10 shows the D150 aeroelastic models with structural parts (blue) and aerodynamic surfaces (grey) using CAERO7 (lifting body) elements for wing, empennage, pylon, engine and BODY7 elements (displacing body) for the fuselage.

The model setup was followed by non-matched flutter analyses with the aeroelastic software ZAERO for the two engine configurations at a flight altitude of

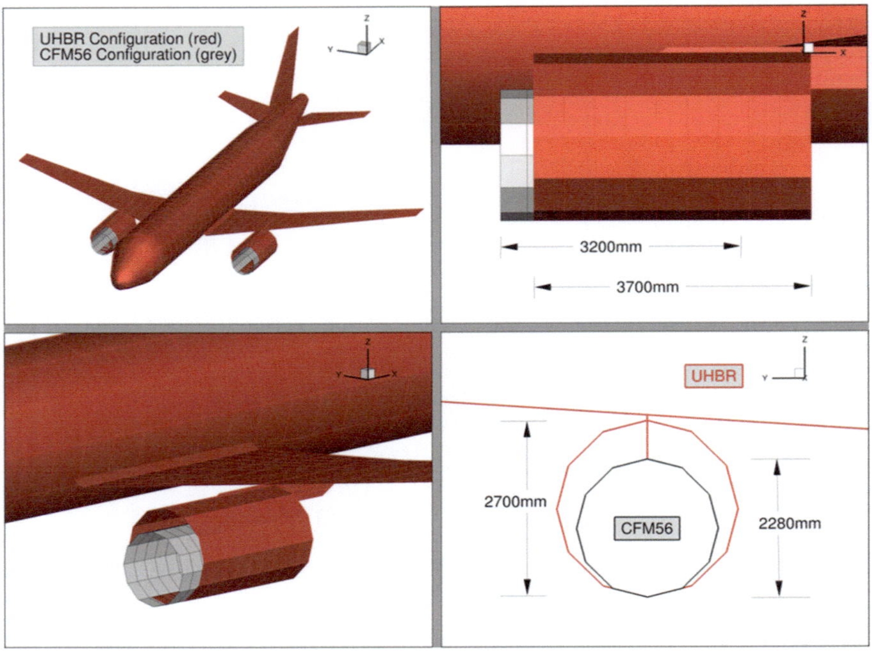

Fig. 9 Comparison of final D150 nacelle geometries for UHBR (red) and standard engine (grey) configuration

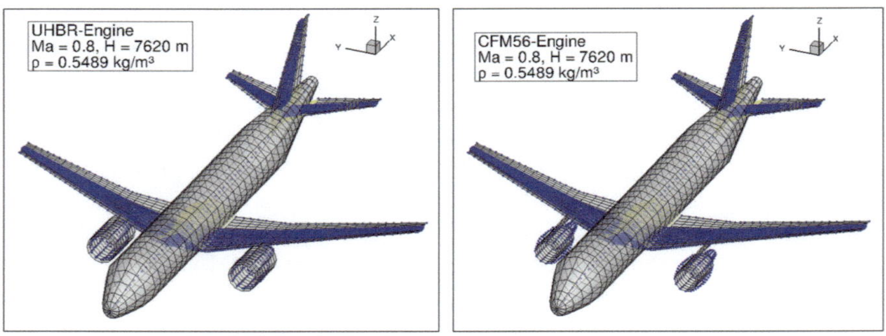

Fig. 10 D150 aeroelastic model with UHBR (left) and standard engines (right)

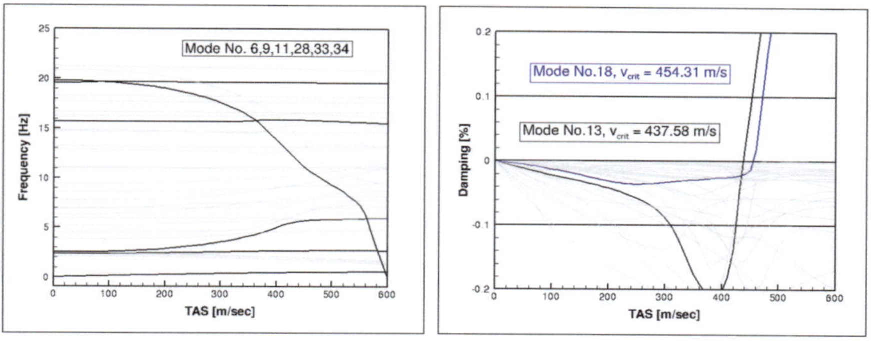

Fig. 11 D150 with CFM56 engines – Frequency and damping curves (FL250)

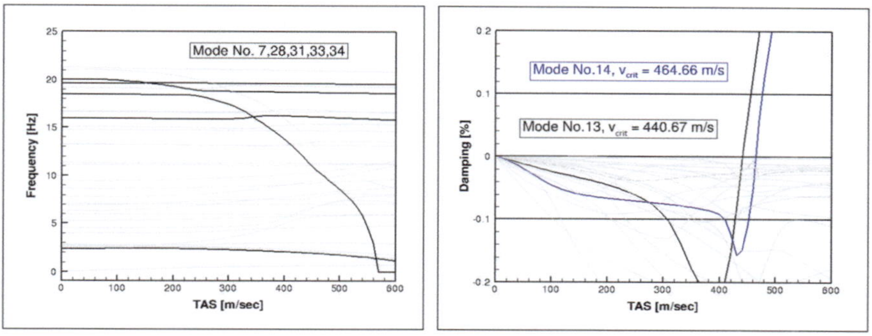

Fig. 12 D150 with UHBR engines – Frequency and damping curves (FL250)

7,620 m, which represents a corner point (FL250, $M_{MO} = 0.82$) in terms of flight level FL and maximum operating Mach number M_{MO} in the A320 flight envelope. Figure 11 depicts the flutter curves (frequency, damping) for D150 with standard CFM56 engines, whilst Fig. 12 gives the same curves for the aircraft in UHBR configuration. It is found that for the chosen installation position of the UHBR engine and mass model assumption, the critical speed for flutter onset is more or less unchanged in comparison to the aircraft with standard engines.

Finally, the extension of the D150 flutter assessment to the complete aeroelastic envelope required the setup of the according V_D/M_D and $1.15 \cdot V_D/M_D$ envelopes according to CS 25.629 [5] and AC 25.629 [6]. A large number of non-matched flutter analyses were performed with the aeroelastic software ZAERO for the reference configuration with standard CFM56 and UHBR engines to investigate both the influence of Mach number and flight altitude on flutter stability with respect to the derived envelope. Figure 13 depicts the relation of flutter speed (defined in knots, Equivalent Air Speed EAS) as a function of Mach number for the two engine configurations, whilst Fig. 14 shows the change of flutter speed with flight altitude (defined in feet). It can be summarized that for the chosen installation position of the UHBR engine

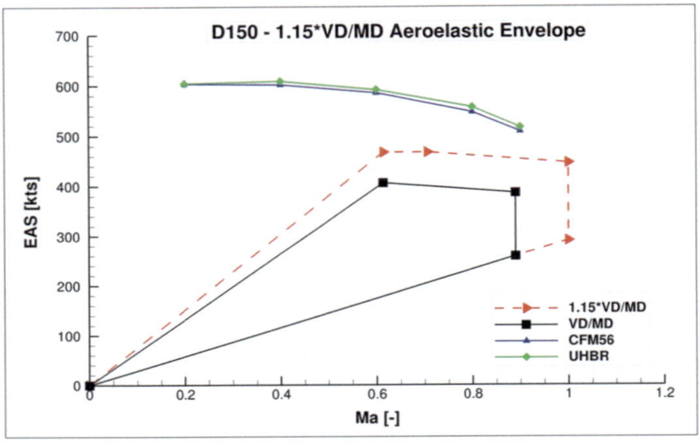

Fig. 13 D150 aeroelastic envelope (influence of Mach number on flutter speed)

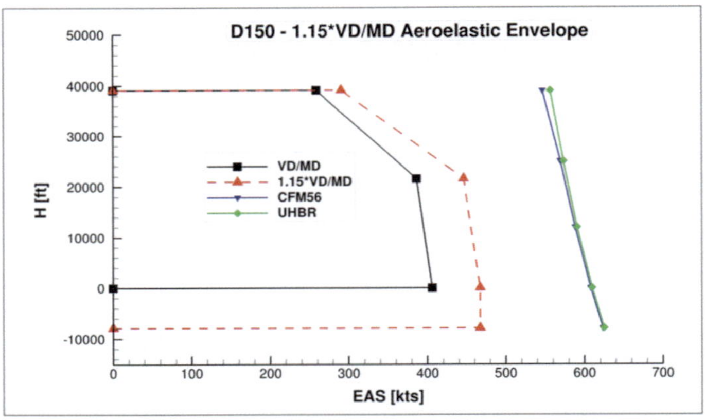

Fig. 14 D150 aeroelastic envelope (influence of flight altitude on flutter speed)

and mass model assumption, the replacement of the standard CFM56 with the UHBR engine does not deleteriously affect the flutter stability. All computed flutter points are found well outside the $1.15 \cdot V_D/M_D$ aeroelastic envelope.

3 Influence of CFD Aerodynamics on Flutter Behavior

3.1 AMP Wing Flutter Assessment with Standard DLM Aerodynamics

So far, the aeroelastic envelope has been assessed by using the standard flutter methodology based on DLM unsteady aerodynamics. The aeroelastic assessment of the A320 aircraft equipped with large UHBR engines mandates the integration of active flow control (AFC) system effects and hence, the consideration of unsteady CFD forces in the flutter analysis. The assessment of flutter stability for aircraft in the transonic flight regime with strong system changes requires the consideration of unsteady CFD forces in the flutter analysis. In order to reduce the complexity of the whole aircraft system and to investigate new CFD methods like the linearized frequency domain TAU solver [7], all investigations have been performed on the clean wing configuration of the well-known AMP wing [2, 3].

The basic setup of the aeroelastic model considering the structural dynamic NASTRAN finite element model and the ZAERO DLM aerodynamic model is depicted in Fig. 15. The ZAERO model of the AMP half wing was used for non-matched flutter simulations based on unsteady DLM aerodynamics to study the influence of the wind tunnel walls on flutter onset. Figure 16 depicts the related flutter curves comprising the development of frequencies and damping versus true airspeed. Here, flutter is found for positive damping values.

3.2 AMP Wing Flutter Assessment with High-Fidelity CFD Aerodynamics

Complementary to the basic ZAERO model, a high-fidelity fluid structure coupled CFD-CSM simulation using PyCSM [8] was set up. Figure 17 shows the TAU mesh

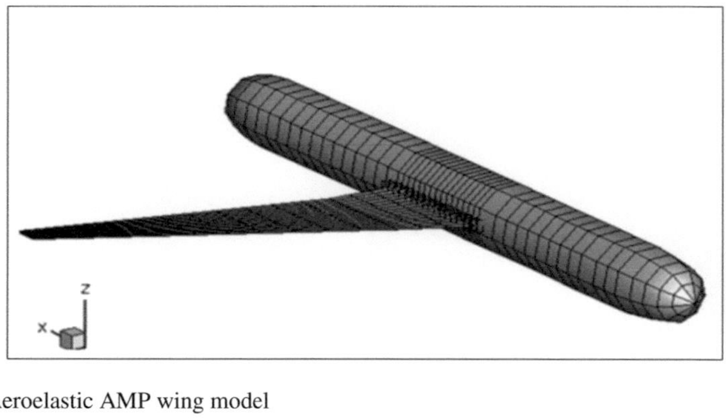

Fig. 15 Aeroelastic AMP wing model

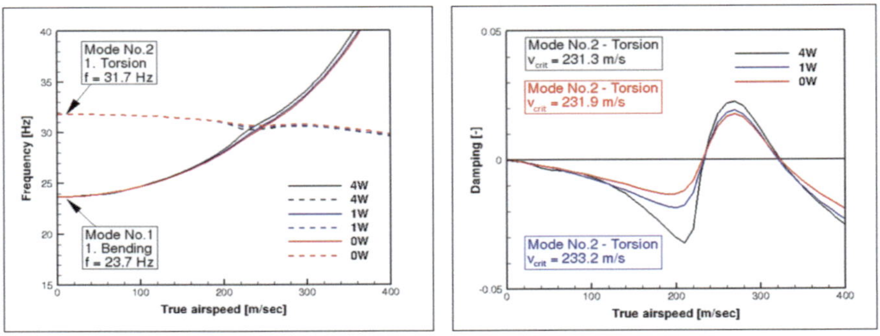

Fig. 16 AMP flutter sensitivity with respect to wind tunnel wall model at Mach 0.7 (no wall: 0 W; wall in symmetry plane: 1 W; four walls: 4 W)

and NASTRAN finite element model utilized for both the static and dynamic simulations. Since the flutter results from Fig. 16 show only a low sensitivity with respect to the wind tunnel wall modelling, the walls have been neglected in the further studies.

In order to compare the results of classical flutter analysis using DLM with those calculated with fully non-linear CFD-CSD coupled simulations in the time domain, simulations for the AMP clean wing configuration were performed in a Mach number range between 0.4 and 0.9 using PyCSM in RANS mode. The flutter results obtained with the classical linear ZAERO/ZONA6 method and the non-linear CFD-CSD coupled method are depicted in Fig. 18. It can be seen that the non-linear PyCSM result is less conservative for Mach numbers beyond 0.75 and furthermore, that the fluid-structure interaction (FSI) clearly shows the expected transonic dip phenomenon.

Additionally, in order to compare the results of the flutter analysis using ZAERO/ZONA6 with flutter results based on generalized aerodynamic forces (GAF) from the linearized frequency domain TAU-LFD solver, simulations for the AMP

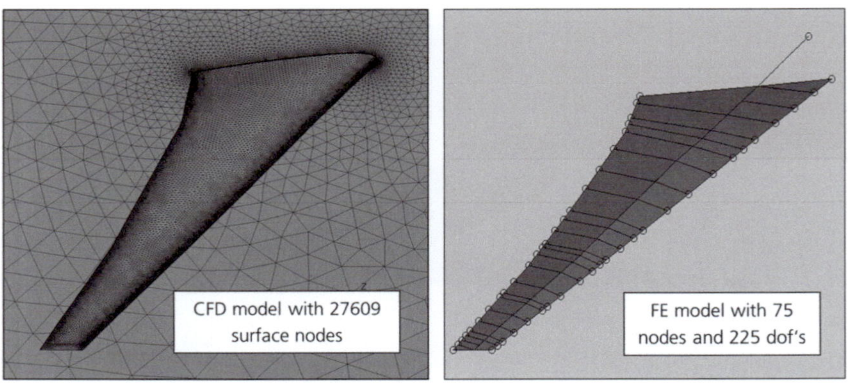

Fig. 17 CFD mesh and FE model of the AMP clean wing configuration

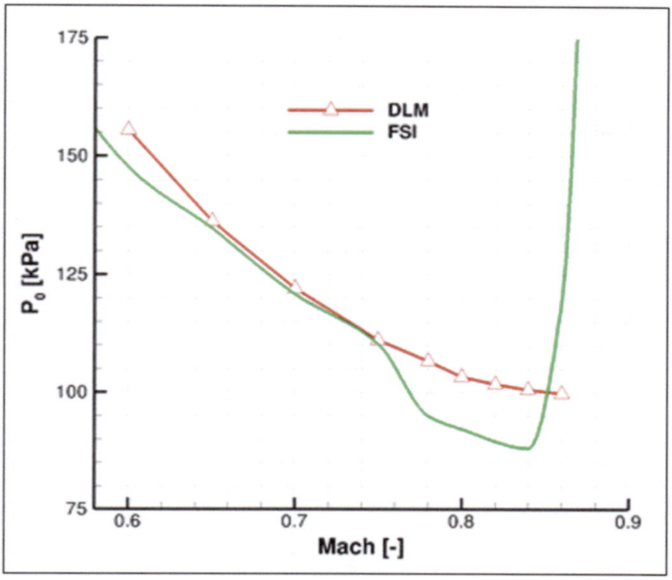

Fig. 18 Comparison of AMP flutter boundaries for linear DLM and non-linear CFD-CSD coupled simulations

clean wing configuration were performed in the Mach number range between 0.60 and 0.87. Here, the performed work addressed the derivation of generalized aerodynamic forces from TAU-LFD in Euler and URANS mode, using for both the FSForcedMotion module which is implemented in the TAU FlowSimulator simulation environment. The generalized aerodynamic forces were calculated in terms of magnitude and phase in the frequency domain for each combination of Mach number, reduced frequency, structural mode shape and dynamic pressure. Due to the large number of required calculations, an automated process chain was built up and used. In a final step, the improved GAF results were included in the ZAERO aeroelastic model to replace the generalized DLM aerodynamic forces as calculated by the ZAERO/ZONA6 method in the flutter analysis of the AMP clean wing configuration.

Finally, the obtained results for the AMP wing permit a comparison of the predicted flutter onset and resolution of the transonic dip phenomenon with unsteady aerodynamic force input based on linear potential flow theory (ZONA6) or linearized CFD (TAU-LFD) used with ZAERO in the frequency domain. These flutter results are complemented by fully non-linear CFD-CSD coupled simulations with CFD (TAU) using the PyCSM environment in a time domain simulation. Corresponding results are depicted in Fig. 19 together with experimental results from the ONERA S2MA wind tunnel.

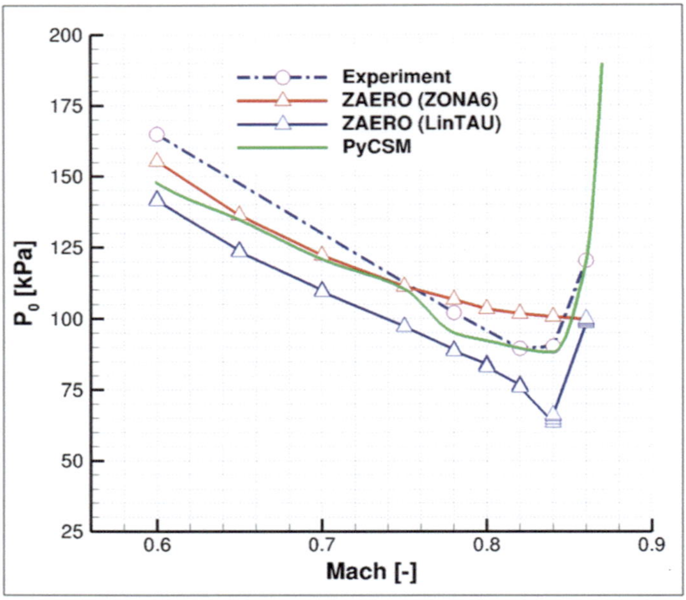

Fig. 19 Comparison of flutter boundaries from ZAERO (ZONA6, LinTAU) and PyCSM (TAU) with experimental data for the AMP clean wing configuration

4 Summary and Conclusions

A stability assessment of an aircraft equipped with the currently introduced high-efficiency engines with ultra-high bypass ratio (UHBR) was performed and compared to the critical flutter onset of the same aircraft using standard engines with a smaller fan diameter. The D150 configuration was chosen, since it is similar to the Airbus A320 in terms of size and dynamic behavior. The setup of the aeroelastic model was shown and results of the flutter assessment clarified differences between the investigated standard and UHBR engine types with respect to the flutter onset. Here, the sensitivity study included varying locations of the UHBR engine in front of the D150 wing and was based on unsteady aerodynamic forces from the doublet lattice method (DLM). In addition, a methodology for the use of CFD based aerodynamic forces was developed and applied to the AMP wing model, thereby providing a comparison with measured flutter data. The aeroelastic models were analyzed with the commercial ZAERO aeroelastic software package.

Abbreviations

AFC	Active Flow Control
AMP	Aeroelastic Model Programme
BODY7	ZAERO Body Macroelements
CAERO7	ZAERO Wing Macroelements
CFD	Computational Fluid Dynamics
CS-25	Certification Specification 25 (for Large Aeroplanes)
CSD	Computational Structural Dynamics
CSM	Computational Structural Mechanics
DLM	Doublet-Lattice-Method
DLR	Deutsches Zentrum für Luft- und Raumfahrt e.V. (German Aerospace Center)
dx, Δx	Longitudinal Variation of Engine Position
dz, Δz	Vertical Variation of Engine Position
FEM	Finite Element Method
FL	Flight Level
FSI	Fluid-Structure-Interaction
GAF	Generalized Aerodynamic Forces
H	Flight Altitude
Ma, M	Mach Number
MD, M_D	Design Dive Mach Number
M_{MO}	Maximum Operating Mach Number
NASTRAN	NAsa STRuctural ANalysis System; Commercial Software Tool for Multidisciplinary Structural Analysis
NEO	New Engine Option
ONERA	Office National d'Etudes et de Recherches Aérospatiales (French Aerospace Center)
PyCSM	DLR in-house Fluid-Structure-Interaction Software
ρ	Density of Air
TAS	True Airspeed
TAU	DLR CFD software package
TAU-LFD; LinTAU	TAU Linearized Frequency Domain Solver
UHBR	Ultra High Bypass-Ratio Engine
URANS	Unsteady Reynolds-Averaged Navier-Stokes
v_{crit}	Critical True Airspeed Flutter Speed
VAMP	DLR project "Virtual Aircraft Multi-Disciplinary Analysis and Design Process"
VD, V_D	Design Dive Speed
ZAERO	Commercial Software Suite for Aeroelastic Design and Analysis
ZONA6	ZAERO Engineering Module for Subsonic Unsteady Aerodynamics

Acknowledgements This project has received funding from the Clean Sky 2 Joint Undertaking under the European Union's Horizon 2020 research and innovation program under grant agreement N° CS2-LPA-GAM-2020-2023-01.

References

1. Neumann J (2020) Flutter assessment of DLR-D150-Vamp configuration w.r.t. UHBR engine integration. Tech. Rep. DLR-IB-AE-GO-2020-106, DLR Internal report
2. Arnold J, Einarsson G, Krüger W (2009) Multibody simulation of an oscillating aeroelastic wing model. NAFEMS Int J CFD Case Stud 8:5–18. https://doi.org/10.59972/ghx4dqen
3. Neumann J, Arnold J, Krüger W (2006) Numerische Simulation komplexer aeroelastischer Phänomene und Vergleich mit Windkanalexperimenten. In: Deutscher Luft- und Raumfahrtkongress 2006. Braunschweig, Germany
4. ZAERO Theoretical Manual (2016), Version 9.2, Zona 02-12.4
5. CS-25 amdt. 28, Appendix K. https://www.easa.europa.eu/en/document-library/certification-specifications/cs-25-amendment-28
6. FAA Advisory Circular AC 25.629-1B. https://www.faa.gov/documentLibrary/media/Advisory_Circular/AC_25_629-1B.pdf
7. Thorman R (2018) Accurate and efficient, time-linearized flutter analysis of transport aircraft. Phd thesis, TU Braunschweig
8. Neumann J, Friedewald D, Hennings H (2018) The influence of fixed transition modeling on aeroelastic simulations in comparison to wind tunnel experiments. CEAS Aeronaut J 9:491–503.https://doi.org/10.1007/s13272-018-0299-y

Experimental Means

Temperature-Sensitive Paint Method for the Investigation of Stationary Crossflow Instability Induced Boundary Layer Transition

Jonathan Lemarechal, Benjamin Daniel Dimond, Christian Klein, and Hans Peter Barth

Abstract Laminar-turbulent boundary-layer transition in the Swept flat PlatE Crossflow TRAnsition (SPECTRA) configuration is investigated by means of the Temperature-Sensitive Paint method and Hot-Wire Anemometry. The results of both measurement techniques are compared in detail far upstream of the transition and also in the transition region in order to investigate the capabilities of Temperature-Sensitive Paint to obtain the wavelength of stationary crossflow (CF) instabilities in the laminar flow, to ascertain the detectability of the transition onset and to determine the transition location. The Temperature-Sensitive Paint method captures the footprints of stationary crossflow vortices and resolves the dominant crossflow wavelength already at low amplitudes of the crossflow disturbance. The chordwise transition location, as detected by the Temperature-Sensitive Paint method, agrees with the maximum in the chordwise evolution of the root-mean square of the high-pass filtered velocity fluctuations. Furthermore, the detected chordwise location of the onset of the secondary instabilities from Hot-Wire Anemometry measurements coincides with the local surface temperature gradient observed by the Temperature-Sensitive Paint method. Finally, the Temperature-Sensitive Paint method was used to investigate the influence of a backward-facing step of 0.45 mm height on the transition location.

Keywords Laminar-turbulent transition · Crossflow instability · SPECTRA · Backward-facing step · Temperature-sensitive paint (TSP) · Hot-wire anemometry

1 Introduction

Friction drag is the dominating source of drag for aircraft [51]. Thus, efforts are underway to develop aircraft with an increased portion of laminar flow on wings, fuselage, and horizontal and vertical tail planes [45]. Increasing the portion of lam-

J. Lemarechal (✉) · B. D. Dimond · C. Klein · H. P. Barth
Institute of Aerodynamics and Flow Technology, German Aerospace Center (DLR), Göttingen, Germany
e-mail: Jonathan.Lemarechal@dlr.de

F. Boden (ed.), *Numerical Methods and Experimental Means for the Integration of Future Propulsion Systems*, Springer Aerospace Technology,
https://doi.org/10.1007/978-3-031-97725-1_8

inar flow directly leads to a reduction of the friction drag, which in turn results in energy savings and reduced fuel consumption [1]. The three-dimensional laminar boundary layer on swept wings of transport aircraft is prone to different boundary layer instabilities: crossflow instability, Tollmien-Schlichting instability, centrifugal instability, and attachment line instability [48]. For sweep angles larger than $\varphi \geq 15°$ and large Reynolds numbers, the crossflow instability dominates and limits the extent of laminar flow [50]. The extent of laminar flow can be severely reduced by the presence of two-dimensional imperfections, which are, for example, steps caused by panel interfaces and high-lift devices on aircraft wings [18, 25]. Early experimental investigations on the influence of a backward-facing step (BFS), led to the development of empirical criteria which could be utilized to describe their influence on the transition location. A critical Reynolds number based on the step height and the boundary layer edge velocity was used to describe the beginning of an upstream movement of the transition location, which occurred when the critical Reynolds number value is exceeded [19, 40, 43]. Recent experiments showed that for sub-critical conditions the unsteady disturbances initially grow downstream of the BFS but they decay before reaching critical levels [20]. For critical step heights the amplitude of steady and unsteady disturbances grow stronger compared to the case without a BFS. The experiments observed that unsteady disturbances downstream of the BFS led to the transition but these were not observed in the case without the BFS [20]. However, latest research observed an increased portion of laminar flow caused by the presence of a forward-facing step of suitable height and position [46]. Hence, knowledge about the critical step height is crucial for the design of future wings and horizontal and vertical tail planes to increase the proportion of laminar flow; this has been addressed in the Clean Sky 2 initiative.

The so-called Cross-Capability Demonstrators (XDC) of the Clean Sky 2 initiative are included in the Large Passenger Aircraft Platform 1. The XDC are also to be used to mature and demonstrate experimental methods for aerodynamic testing relevant for the development of future aircraft [9]. One of these XDC—the SPECTRA (Swept flat PlatE Crossflow TRAnsition [6, 7]) configuration—is utilized as a testbed for studying crossflow-dominated transition and the influence of steps on the crossflow dominated transition. These investigations require a fast, i.e., with respect to acquisition and evaluation time, non-intrusive measurement technique to measure the transition location. The Temperature-Sensitive Paint (TSP) method, a thermographic measurement technique utilized for flow measurements [21, 36], fulfills these requirements. Futhermore, it has already been applied in flows where the transition mechanism is dominated by crossflow instability [22, 32].

The boundary layer of a swept wing with a pressure gradient produces a three-dimensional velocity profile with a crossflow component, i.e. a flow component perpendicular to the inviscid streamline [8]. The crossflow is unstable for stationary or travelling oblique waves [8, 49, 53]. Freestream turbulence interacting with the surface roughness close to the attachment line determines whether travelling or stationary crossflow instabilities dominate the transition process [16, 23]. At aircraft cruise flight conditions, stationary crossflow instabilities are typically dominant [8, 49]. The stationary crossflow instability is associated with co-rotating vortices inside the laminar boundary layer. These vortices are at a fixed spanwise position with respect to the model [39]. Initially, the growth of the disturbance amplitude of these vortices is exponential, but the amplitude of the vortices already saturates far upstream of the location of the final breakdown to turbulence. Subsequently, the breakdown of the laminar flow is driven by secondary instability modes [38], of which different types have been identified at three distinct locations in the stationary crossflow vortex [53]: type-I [30, 52], type-II [37, 52], and type-III [52]. The type-I secondary instability mode is observed in many wind tunnel experiments [4, 30, 52, 53]. This secondary instability is characterized by high-frequency velocity fluctuations (as compared to the frequency of the primary traveling instabilities) in the updraught part of the stationary crossflow vortices. This very localized development of the secondary instability with large growth rates leads to a local wedge-shaped onset of turbulence and the characteristic saw-tooth pattern is observed.

In this chapter, the measurement capabilities of the TSP method for the investigation of transition dominated by crossflow instability are evaluated in detail. This type of evaluation had previously been done for TSP and transition dominated by Tollmien-Schlichting instability: This included detailed TSP and velocity measurements to determine the structures visible in TSP [33], comparison of the determined transition location [13, 21, 33], and an assessment of the influence of a TSP measurement on the flow [12, 24]. The conditions of the SPECTRA configuration provide the opportunity to directly compare the TSP results with velocity data from Hot-Wire Anemometry (HWA) measurements, due to the controlled conditions of the flow of the SPECTRA configuration. These data are used to further investigate the structures appearing in the TSP measurements and to compare the transition location detected by TSP with the development of the secondary instabilities, which are obtained from the HWA measurements; this leads to an improvement in the interpretation of TSP flow measurements where boundary layer transition is dominated by crossflow instability. This comparison also takes account of the non-adiabatic conditions which prevail during TSP data acquisition. Finally, the capabilities of TSP are demonstrated in an investigation of the change in transition location in the presence of a BFS.

2 Experimental Methods

2.1 1-Meter Wind Tunnel Göttingen

The experiments were carried out in the one-meter wind tunnel Göttingen (1MG) of the German Aerospace Center (DLR) Göttingen, Institute of Aerodynamics and Flow Technology. The 1MG is a low-speed wind tunnel of the so-called Göttingen-type. The SPECTRA configuration is installed in the long open test section of 2.4 m length. The cross section at the nozzle exit is 1 m wide and 0.74 m high. For this test section a turbulence level Tu $\approx 0.13\%$ at 25.8 m/s in the frequency range of 2 Hz to 2 kHz has been measured upstream of the SPECTRA configuration [5]. The flow velocity U_{1MG} is derived by measuring the pressure drop between the settling chamber and the nozzle of the 1MG.

2.2 SPECTRA Configuration

The SPECTRA configuration [6, 7] is a modification of the original swept flat plate arrangement utilized by Bippes et al. [8, 41] for an isolated investigation of crossflow instability. Compared to the original swept flat plate arrangement, this SPECTRA configuration has an increased chord length of $c = 600$ mm and a modular setup. The main components of the SPECTRA configuration are a flat plate and a displacement body, see Fig. 1. The modular design of the SPECTRA configuration enables the installation of different displacement bodies or flat plates in order to investigate different pressure distributions and to integrate different measurement techniques. Figure 2 shows the model chord c and the three adopted coordinate systems: The model fix coordinate system (x_c, y_c), the vortex coordinate system (x_v, y_v) and a spanwise coordinate with shifted origin (x_c, y_c') are used for data evaluation and presentation. The y_c' coordinate represents the spanwise distance to the line connecting all pressure taps, i.e. $y_c'(x_c) = y_c - y_{c,p} + x_c \tan(\varphi)$ with $y_{c,p}$ denoting the spanwise coordinate at which the line connecting all pressure taps intersects the leading edge.

The model is installed at a sweep angle of $\varphi = -45°$ to the incoming flow. The blockage effect of the model in the 1MG test section is non-negligible, so that a nozzle adapter and contoured side walls are required to approximate the desired infinite-swept wing flow condition [5, 39]. In order to avoid flow separation on the displacement body, a slat has been installed, see Fig. 1.

The displacement body imprints a pressure distribution onto the upper side of the flat plate, which stabilizes the boundary layer with respect to Tollmien-Schlichting

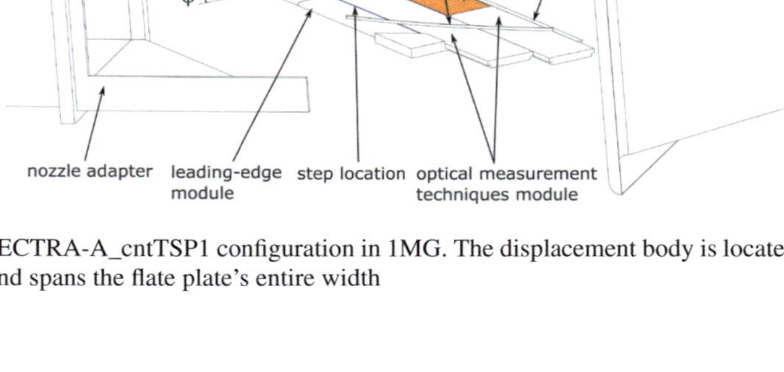

Fig. 1 SPECTRA-A_cntTSP1 configuration in 1MG. The displacement body is located above the flat plate and spans the flate plate's entire width

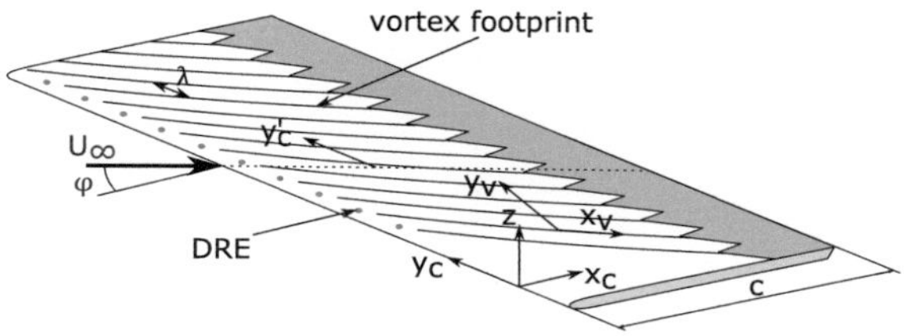

Fig. 2 Definitions of the used coordinate systems, sweep angle φ, location of the DRE (discrete roughness element), wavelength of the stationary crossflow vortices λ, free stream velocity U_∞, and chord length c

instablities. In addition, other design choices of the SPECTRA configuration [7] stabilize the boundary layer with respect to all typical primary instabilities of a swept-wing configuration, with the exception of crossflow instabilities. Two different displacement bodies were utilized in the present investigation: The SPECTRA-A configuration is similar to the original pressure distribution designed by Bippes et al. [41], whereas the SPECTRA-B [6, 7] configuration provides characteristics suitable for the investigation of the transition control technique called "upstream flow deformation" [6, 44, 47, 52].

The flat plate is made of aluminium and consists of either three or four segments: The leading-edge segment ($x_c/c \leq 0.25$), one or two segments for optical measurement techniques ($0.25 \leq x_c/c \leq 0.96$), and the trailing-edge segment ($0.96 \leq x_c/c \leq 1$), see Fig. 1. The flat plate features 40 pressure taps arranged parallel to the main axis of the free-stream velocity. Two segments have been specifically designed for the TSP method: The first for measurements from $0.61 \leq x_c/c \leq 0.96$ and the second from $0.25 \leq x_c/c \leq 0.96$.

The modular design of the flat plate provides the possibility to install forward- or backward-facing steps at $x_c/c = 0.25$, as indicated in Fig. 1. Adjustment of the vertical (z-direction) position of the leading-edge segment relative to the downstream segments, leads to the formation of a step. In the presented work, the flat plate is set up with $z = 0$ mm (no step, reference condition) or a BFS of 0.45 mm.

The turbulence level of the 1MG leads to similar amplitudes of traveling and stationary crossflow instabilities in the investigated boundary layer. In order to ensure that the laminar–turbulent transition is predominantly caused by stationary crossflow instabilities and to improve the reproducibility, stationary crossflow instabilities are artificially excited with a spanwise-equidistant arrangement of discrete roughness elements (DREs) [31, 44]. The DREs are installed by applying rub-on dots with diameter $d_{DRE} = 2.5$ mm and height $h_{DRE} \approx 6\,\mu$m or $h_{DRE} \approx 10\,\mu$m. The chordwise position of the DRE $x_{c,DRE}/c$ and the spanwise spacing λ_{DRE}, as listed in Table 1, were chosen based on an accompanying linear stability analysis to ensure that the excited crossflow instability mode is amplified over the entire range of planned experimental conditions. The transition on the lower side is fixed close to the leading edge.

Table 1 DRE geometry parameters

Configuration	λ_{DRE} [mm]	d_{DRE} [mm]	h_{DRE} [μm]	$x_{c,DRE}/c$
SPECTRA-A_cntTSP1	10	2.5	30	0.068
SPECTRA-A_cntTSP2	9.5	2.5	6	0.066
SPECTRA-B_cntTSP3	9	2.5	6	0.067

2.3 Temperature-Sensitive Paint Method

The TSP method is an optical, non-intrusive, reversible temperature measurement technique relying on the temperature-dependent emission of light from luminophores embedded in a binder. These luminophores are first excited by light of a suitable wavelength, whereupon the excited luminophores can return to their ground state mainly by one of two processes: a radiationless process (called thermal quenching) or by the emission of light at a larger wavelength. Since the process of thermal quenching becomes more likely with increasing temperature, the light emission is temperature dependent. Further details on the photophysics of TSP is given by Liu et al. [36]. In order to acquire the temperature distribution on a body, a surface coating of a transparent binder (a clear coat) with embedded luminophores is required in the area of interest. For data acquisition an excitation light source and a photosensitive device are needed [36]. The TSP method has been used, among other applications, for transition measurements and surface flow measurements [21, 36] in wind and water tunnels.

In order to measure surface temperature changes due to different boundary-layer states by thermographic imaging techniques like TSP, an artificial heat flux between model surface and fluid is required at low flow speeds [21]. Several methods have been developed to generate this artificial heat flux: The temperature-step method [2]; electrical current carrying layers of carbon nanotubes (CNT) [27] or carbon fibers [33]; and infrared laser heating [28]. Applying an artificial heat flux results in a temperature difference between the surface temperature T_S and fluid temperature T_∞ of $\Delta T = T_S - T_\infty$. The temperature ratio $T_S/T_\infty = (T_\infty + \Delta T)/T_\infty$ of the model surface temperature to the fluid temperature needs to be balance between sufficient contrast of the TSP measurement and not influencing the boundary layer and transition location to ensure that the TSP results are obtained at the required adiabatic conditions [13, 32].

Heating and TSP integration

The flat-plate segments prepared for TSP measurements utilize CNT to generate the heat flux between the model surface and the flow, see [27] for details; from here on the combination of CNT heating and TSP will be called cntTSP. A pocket is milled into the segment to incorporate the cntTSP flush with the remaining surface. Before applying the CNT, a polyurethane coating is applied in order to electrically insulate

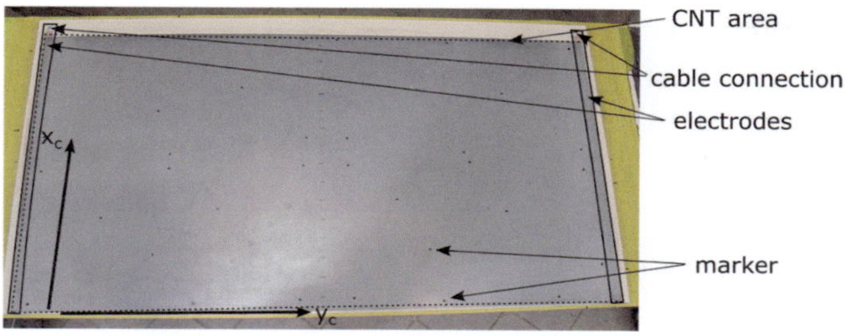

Fig. 3 cntTSP coated pocket of the cntTSP2 optical measurement technique segment with labeling of the cntTSP (gray colored region), the electrodes and electrical connections, and the marker pattern

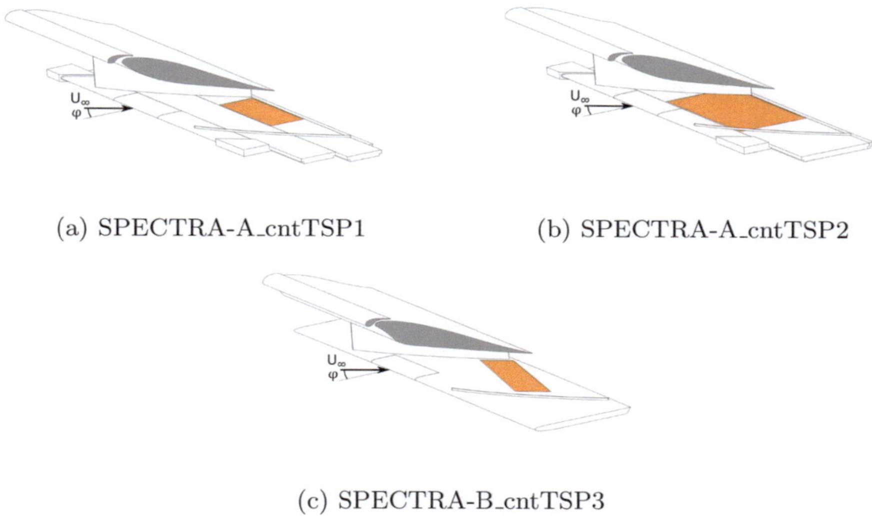

Fig. 4 Location, orientation and the SPECTRA configuration of each manufactured cntTSP configuration

the CNT from the underlying aluminum model. Subsequently, the copper electrodes for connecting the CNT to cables are installed and the CNT layer is applied by spray gun. Afterwards, the white screen layer is coated onto the CNT and a dot pattern (markers) is applied for image registration during the data evaluation, see Fig. 3. Finally, the TSP layer is applied by spray gun. After curing, the junction between TSP surface and the uncoated aluminum model surface is polished to achieve the flush integration of the cntTSP. For the entire project a TSP based on an Europium-complex with high temperature sensitivity ($-3.6\%/\mathrm{K}$ at 30 °C) [42] is utilized. This TSP is excited in the near-UV range ($\lambda_{ex} \approx 400$ nm) and the emitted luminescence is red light ($\lambda_{em} = 615$ nm).

Throughout the project three optical measurement technique segments of the SPECTRA configuration were coated with cntTSP, see Fig. 4. The dimensions of the

Table 2 cntTSP parameter

Configuration	cntTSP area [mm^2]	TSP LE x_c/c
SPECTRA-A_cntTSP1	200 × 580	0.61
SPECTRA-A_cntTSP2	400 × 800	0.26
SPECTRA-B_cntTSP3	200 × 250	0.56

rectangular shaped cntTSP areas and the location of the leading edge (LE) are listed for each cntTSP-configuration in Table 2.

TSP Setups

In the framework of the project two significantly different setups for TSP data acquisition were implemented and tested to investigate the capabilities of TSP regarding the investigation of CF instabilities and the influence of BFS on the transition location. The first setup is called the external setup, where a scientific-grade camera and LEDs are placed outside of the open test section. For the second setup, which is called the integrated setup (described in more detail in [34]), miniaturized cameras and LEDs are installed in the displacement body of the SPECTRA-A configuration.

External TSP setup

The camera and LEDs of the external setup are placed outside of the open test section at its downstream end, see Fig. 5a. The resulting distance between the camera and the TSP surface on the SPECTRA model in horizontal direction is approximately 2.0 m, as indicated in Fig. 5b. The cameras and LEDs and the position of the equipment is the same for both SPECTRA configurations. The location for the camera and LEDs is chosen to optimize the field of view and also enable the install of the equipment outside of the zone influenced by the flow and deflected flow outside of the wind tunnel. The monochromatic 14-bit CCD camera PCO.4000 with 10.8 MPx resolution in combination with a 180 mm lens is utilized. The achieved spatial resolution and the set exposure times t_{exp} are given in Table 3. The camera is equipped with an optical band-pass filter (630 nm central wavelength and 75 nm band width). Because of the small distance of 93 mm between the flat plate and the displacement body trailing edge, this setup has an acute viewing angle between the optical axis and the TSP surface: An indication of the view onto the TSP surface as seen by the camera is provided in Fig. 5c. The illumination of the TSP surface is also restricted, since the excitation is less efficient due to the large distance and the acute angle [34]. It is therefore necessary to employ, four high-power LEDs (IL-105/6X Illuminator, HARDsoft), see Fig. 5a. Each LED is equipped with a low-pass filter (cut-off wavelength 550 nm) and a lens to focus the light onto the TSP surface.

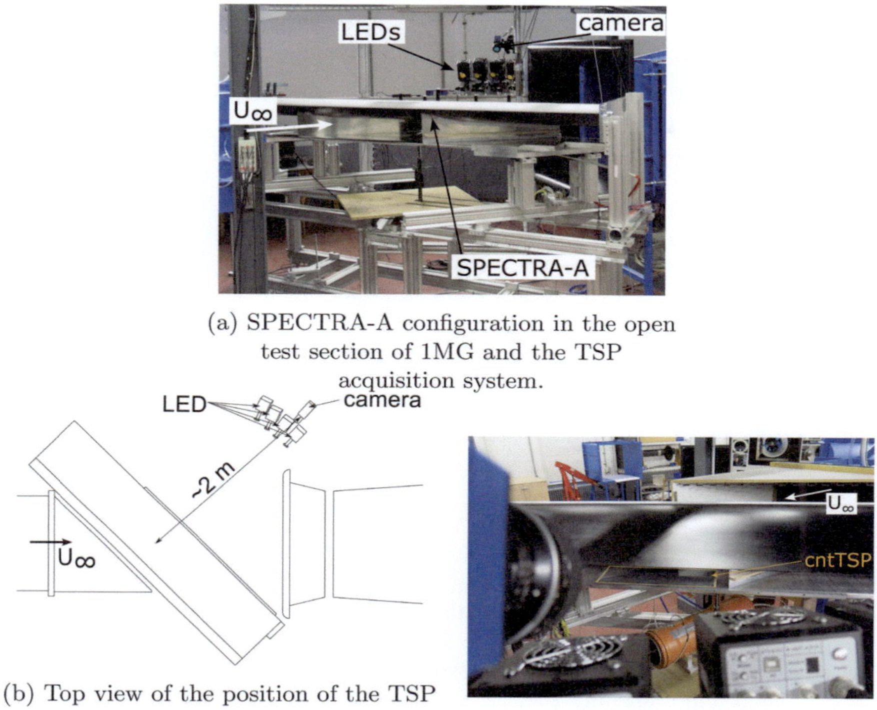

(a) SPECTRA-A configuration in the open test section of 1MG and the TSP acquisition system.

(b) Top view of the position of the TSP acquisition system.

(c) Camera perspective.

Fig. 5 External TSP setup shown for the SPECTRA-A_cntTSP1 configuration

Table 3 TSP acquisition parameters

Configuration	TSP setup	avg. res. [px/mm]	t_{exp} [ms]	$\dot{q}_{el}$ [W/m^2]	t_h [s]	T_S/T_∞
SPECTRA-A_cntTSP1	External	0.3–0.7	105	4800	180	1.01
SPECTRA-A_cntTSP2	Integrated	0.2–0.4	130	600	300	1.01
SPECTRA-B_cntTSP3	External	0.2	180	4000	73	1.01

Integrated TSP setup

In this setup miniaturized cameras and LEDs are integrated into the displacement body of the SPECTRA-A configuration, see Fig. 6. The machine vision camera VC MIPI IMX296 is a monochromatic CMOS camera with 1.6 MPx, 10-bit resolution, a global shutter and requires only $22 \times 23.5 \times 10\,\text{mm}^3$ of space (without lens). The suitability of this camera for TSP flow measurements and transition measurements

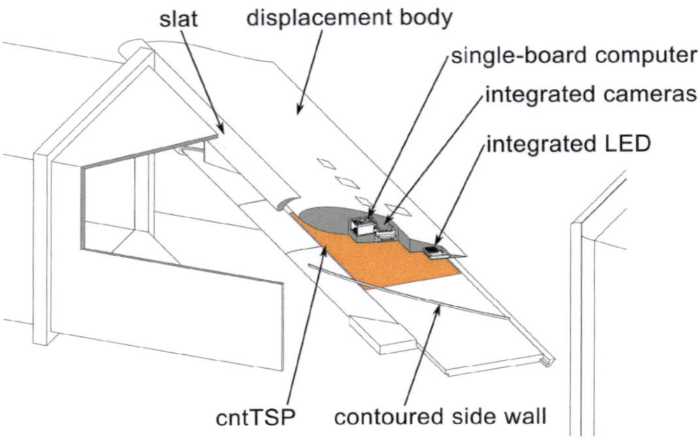

Fig. 6 Isometric view of the TSP setup integrated into the SPECTRA-A displacement body with the cntTSP2 measurement segment

had been previously demonstrated—see [34]. Four cameras inside of two cavities of the SPECTRA-A displacement body with lenses of 3 mm focal length are necessary to capture the entire cntTSP area of the SPECTRA-A_cntTSP3 configuration. The cavity is closed flush with the contour of the displacement body using a N-BK7 glass window with anti-reflection coating (wavelength range of 400 to 700 nm). Additionally, the cameras are equipped with a colored glass high-pass filter (cut-off wavelength 550 nm). The TSP is excited by five Luminus UV CBT-120-UV LEDs (emission peak of 385 nm), which are installed with a low-pass filter (cut-off wavelength of 550 nm) in individual cavities, see Fig. 6. Values for the average camera resolution and exposure time t_{exp} are given in Table 3.

Data Acquisition and Evaluation

For both TSP setups, the TSP data acquisition requires recording images in three different phases: dark, reference and run phases. These phases and the acquisition procedure in general are described in the following. After preparing and cleaning the installed model, the wind tunnel is adjusted to the first velocity to be tested. At first, a number of so-called dark images are acquired to measure the background light. Subsequently, reference images are acquired with the LEDs switched on but with the artificial heat flux switched off. Afterwards, the model is heated up with the CNT heating until an over temperature of $\Delta T \approx 2$ to 4 K is reached, which corresponds to a temperature ratio of surface to flow of $T_S/T_\infty \approx 1.01$. For the different measurements the adopted electrical heat flux $\dot{q}_{el}$ was varied between 600 and 4800 $\mathrm{W/m^2}$, see Table 3, where the heat flux is presented as the set electrical power per CNT area, so that losses due to heat conduction into the flat plate are included. Finally, run images are acquired, after which the next wind tunnel velocity is set. Up to five velocities are acquired consecutively, then the acquisition procedure is restarted after a cool down phase to re-establishing thermal equilibrium between model and flow. Values of the acquisition parameters are listed in Table 3, and further details of TSP data acquisition with machine vision cameras is provided in [34].

The TSP data evaluation (camera image processing) is performed with the DLR in-house software nToPas [26]. The first step is to average all acquired images of each phase in the image space. If a misalignment of the model markers is observed, which occurred due to deformation of the model or displacement body, an image registration between reference and run images is performed. Subsequently, the reference images are divided by the run images (I_{ref}/I) and projected onto a structured grid of the flat plate with physical coordinates in the plate-oriented coordinate system x_c, y_c. Finally, the spatial intensity ratio distribution on this grid is converted to surface temperate maps T_S by means of a previous calibration, which is acquired with a coating sample. More details on the data evaluation are provided in [11] and on the integrated setup in [34].

2.4 Hotwire Anemometry

HWA is used to investigate the flow field and the CF instabilities in the boundary-layer above the model surface. A V-type boundary-layer probe, manufactured by DANTEC Dynamics, is operated with a DISA 55M constant-temperature anemometer. The anemometer output voltage is sampled and recorded using a National Instruments "NI-USB-6363" analog-to-digital converter at a sampling rate of $2^{15} = 32768$ Hz. From the HWA measurement the velocity component in the direction of the boundary-layer edge streamline u_s is derived.

The hot-wire probe is mounted onto an extension, which is moved by a system of stepper motors mounted above the wind tunnel test section—see Fig. 7. In this way it can be accurately positioned in the boundary layer on the upper side of the flat plate of SPECTRA over a wide range of chordwise-, spanwise and wall-normal locations. HWA measurements were performed in different planes: First, spanwise scans at a constant wall-normal distance are performed in the range of $x_c/c \geq 0.6$ and $x_c/c \leq 0.9$, in which the breakdown to turbulence is observed. Secondly, planar wall-normal and spanwise scans at constant x_c/c are measured to acquire spatially resolved data of the crossflow vortices and the secondary instabilities. All presented HWA measurements were performed above the cntTSP coating, see Fig. 7, and the CNT heating was switched off.

3 Results and Discussion

In this section, the TSP visualization and measurement capabilities are discussed and compared with the HWA measurements. Throughout this paper, the TSP data of the utilized heating configuration is processed in such a way that lighter areas (light

Fig. 7 Joint demonstration of both measurement techniques: Hot-wire probe and probe mounting above excited TSP surface showing the visible part of the utilized excitation light (blue) and the emitted light (red). Flow from the left

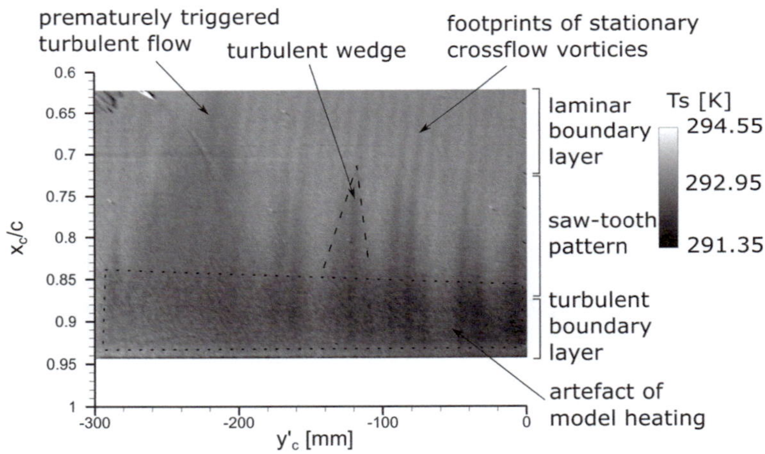

Fig. 8 TSP surface temperature T_S for the SPECTRA-A_cntTSP1 configuration with $U_{IMG} = 23.5\,\mathrm{m/s}$; characteristic phenomena are labelled. The flow direction is from top to bottom

grey) correspond to areas with a locally higher temperature, which can correspond to the lower skin friction of the laminar boundary layer. Conversely, darker areas represent lower temperatures, which can be caused by the larger skin friction of a turbulent boundary layer.

A typical TSP surface temperature map of the laminar-turbulent boundary-layer transition dominated by stationary crossflow instability is shown in Fig. 8 for $U_{IMG} = 23.5\,\mathrm{m/s}$. The result was acquired with the external TSP acquisition setup (see Fig. 5) and the SPECTRA-A_cntTSP1 configuration (see Fig. 4a). The result is presented in the x_c/c, y'_c coordinate system with the flow direction being from top to bottom.

In Fig. 8 the characteristic regions of the crossflow-dominated transition process are visible in the region $-200\,\text{mm} \leq y_c' \leq 0\,\text{mm}$, where the visible flow structures are consistent in spanwise direction, thus lending support to the applicability of the infinite-swept wing flow condition in this region. The following characteristics can be observed consecutively in flow direction: laminar flow with footprints of stationary crossflow vortices; beginning of the turbulent flow at $x_c/c \approx 0.72$ in the form of wedge-shaped regions (saw-tooth pattern) of turbulent flow; and finally turbulent flow downstream of the conversion of the turbulent wedges at $x_c/c \approx 0.88$. The footprints of the stationary crossflow vortices are visible as parallel medium-gray lines in the laminar region approximately aligned with the direction of the inflow. The breakdown to turbulence, as observed by TSP, occurs over an extended region, due to its onset being in the form of individual turbulent wedges, which form the typical saw-tooth pattern [49]. The upstream tip of turbulent wedges is located at particularly pronounced crossflow vortex footprints.

Besides the visualization of the crossflow vortex footprints and laminar and turbulent boundary layer, there are two further observations which stand out. First, an artefact of reduced surface temperature is visible between $0.85 \leq x_c/c \leq 0.95$ in Fig. 8. It is assumed that this is caused by a variation in the heat flux of the model heating or the heat conduction properties inside of the model below the heating layer. Second, a patch of prematurely triggered turbulent flow is visible between $-300\,\text{mm} \leq y_c' \leq -200\,\text{mm}$. Inside of this patch, structures comparable to those of the turbulent wedges of the saw-tooth pattern are still visible. This leads to the assumption that this patch of turbulent flow is triggered by a local surface imperfection upstream, which leads to an increase of the initial amplitudes of the crossflow modes locally.

In the following, a detailed comparison between TSP and HWA results is given. First, the footprint of the stationary crossflow vortices in the TSP results and the derivable characteristics are discussed and compared to HWA results in Sect. 3.1. Secondly, a detailed comparison of the observable development in the vicinity of the transition location is conducted in Sect. 3.2. Thirdly, this setup has also been used to investigate the flow with and without a BFS of 0.45 mm using the TSP method, to be discussed in Sect. 3.3. Finally, an uncertainty analysis of the TSP application for the investigation of crossflow-dominated transition is provided in Sect. 3.4.

3.1 TSP Observations of Stationary Crossflow Vortices

The stationary crossflow vortices in the laminar flow region result in a spanwise pattern of parallel lines aligned with the flow direction, which is well captured by TSP, as shown in Fig. 8. The relationship between the footprints of this pattern and the flow field structures of stationary crossflow instability is analyzed at three locations with respect to the saw-tooth pattern visualized by TSP.

The first location for the comparison is far upstream of the location of breakdown to turbulence, which has been achieved at a flow speed of $U_{IMG} = 18.0\,\text{m/s}$ for the

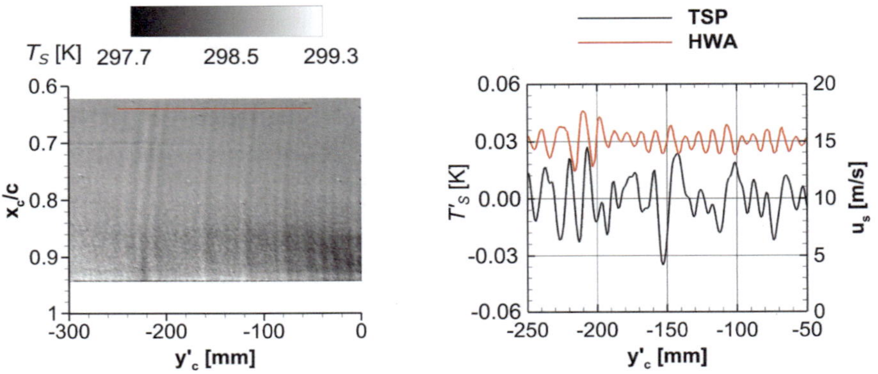

(a) Surface temperature map T_S and HWA measurement location (–).

(b) u_s and T'_S at $x_c/c = 0.64$.

Fig. 9 TSP and HWA results for the SPECTRA-A_cntTSP1 configuration at $U_{IMG} = 18.0\,\mathrm{m/s}$

SPECTRA-A_cntTSP1 configuration. This inflow velocity results in a completely laminar boundary layer over the entire flat plate, see Fig. 9. A Gaussian filter is applied to the TSP results in order to remove noise and enhance the visibility of the footprints of these stationary crossflow vortex footprints. The second and third locations for analyzing the relationship between the stationary crossflow pattern in TSP and the flow field measured by HWA are located just upstream of the transition, see Fig. 10. These were acquired with the SPECTRA-B_cntTSP3 configuration and the external TSP acquisition setup at $U_{IMG} = 27.0\,\mathrm{m/s}$. The TSP results in Fig. 10a, b do not require additional filtering because the signal is more pronounced close to the transition. The spanwise cuts of the velocity u_s and deviation from mean surface temperature T'_S at the location indicated by a red line in Figs. 9a and 10a, b are shown in Figs. 9b and 10c, d, respectively. The presented velocity data from the HWA measurement are located at $z = 0.6\,\mathrm{mm}$ above the model surface. The TSP data shown in Fig. 10c required a spanwise shift of $y'_c - 4\,\mathrm{mm}$ to align with the HWA data. The shift was determined based on transition onset as discussed in Sect. 3.2 and is shown in Fig. 13a.

The stationary crossflow vortices transport low-momentum fluid from the vicinity of the wall upwards and high-momentum fluid from larger wall-distances towards the wall which causes the spanwise modulation of the time-averaged flow field [8, 49] clearly visible in the velocity measurements as shown in Figs. 9b and 10c, d. At the measurement location presented in Fig. 9b, where the crossflow amplitude is small, the velocity modulation has already a regular spanwise wavelength with some variation in the amplitude. The spanwise velocity modulation correlates to a spanwise modulation of the wall-shear stress. Adding the homogeneous heat flux for the TSP measurements results in a modulation of the convective cooling, which in turn results in a modulation of the surface temperature. At the most forward location (Fig. 9b), the velocity modulation of approximately $\pm 1\,\mathrm{m/s}$ causes a deviation from

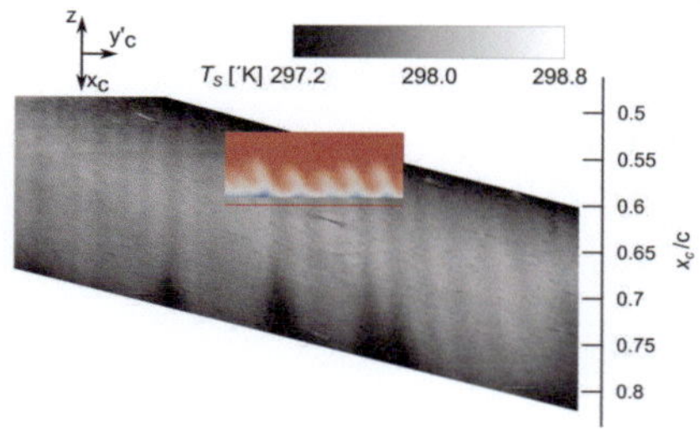

(a) Surface temperature T_S, HWA measurement location at $x_c/c = 0.60$ (–), and velocity component u_s from HWA.

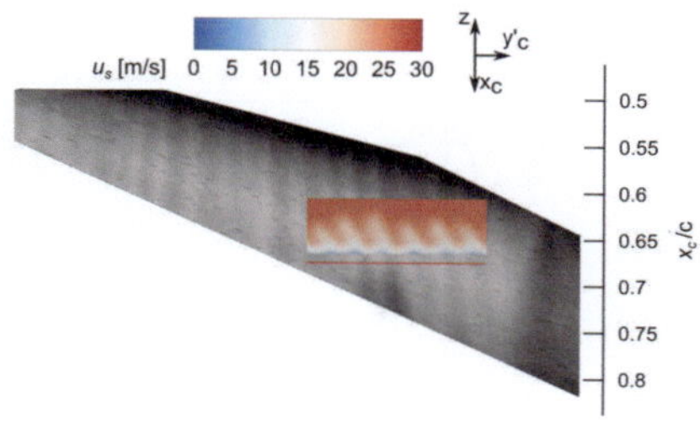

(b) Surface temperature T_S, HWA measurement location at $x_c/c = 0.67$ (–), and velocity component u_s from HWA.

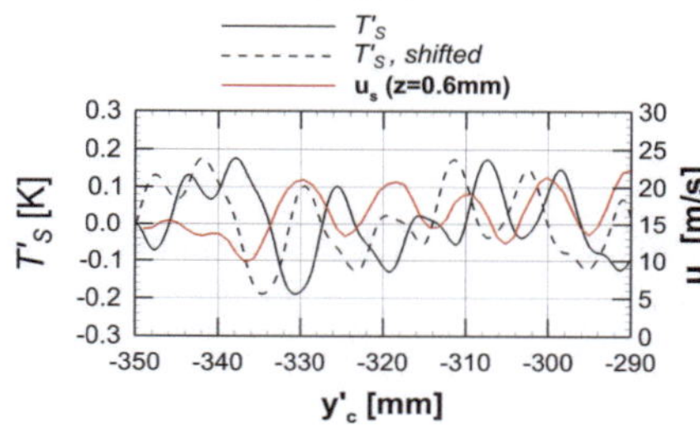

(c) u_s and T'_S at $x_c/c = 0.60$.

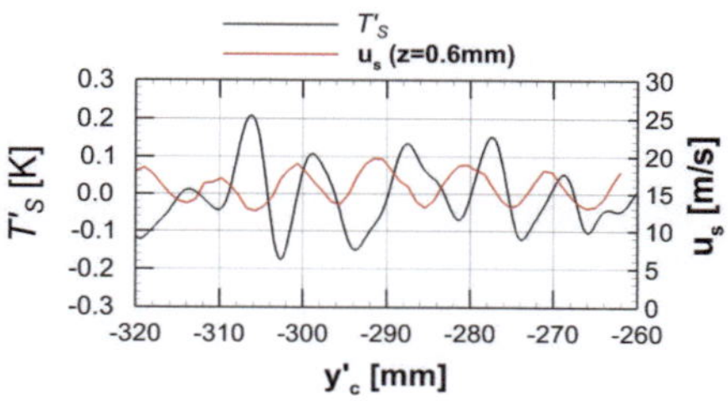

(d) u_s and T'_S at $x_c/c = 0.67$.

Fig. 10 TSP and HWA results for the SPECTRA-B_cntTSP3 configuration acquired with the external TSP acquisition system at $U_{IMG} = 27\,\text{m/s}$

the mean surface temperature of approximately $\pm 0.02\,\text{K}$, which demonstrates the outstanding sensitivity of the TSP method.

The TSP method provides a measure of the variation of the wall-shear stress by measuring the variation of heat transfer and its ensuing surface temperature change. Thus, the medium-gray lines of the stationary crossflow vortices in Figs. 9a and 10a, b are expected to be located below the largest wall-shear stress. In order to further elucidate the relationship between the TSP measurement and the velocity field, the velocity modulation of u_s at $z = 0.6\,\text{mm}$ and the spanwise surface temperature deviation from the mean value T'_S are compared in Fig. 9b. Here the local maxima of the deviation from mean surface temperature T'_S are observed to coincide approximately with the maxima of the velocity u_s. This coincidence is also seen in Fig. 10c, d. However, due to the inclination in spanwise direction of the crossflow flow structure and the remaining distance between the hot-wire measurement location closest to the wall and the surface, the exact location of the the crossflow vortex structure above the TSP surface temperature map cannot be concluded from the current measurements. From published velocity fields it is to be expected that the core of the crossflow vortex is approximately located at every second changeover from dark to bright lines in the TSP temperature map. The spanwise shift between the near-

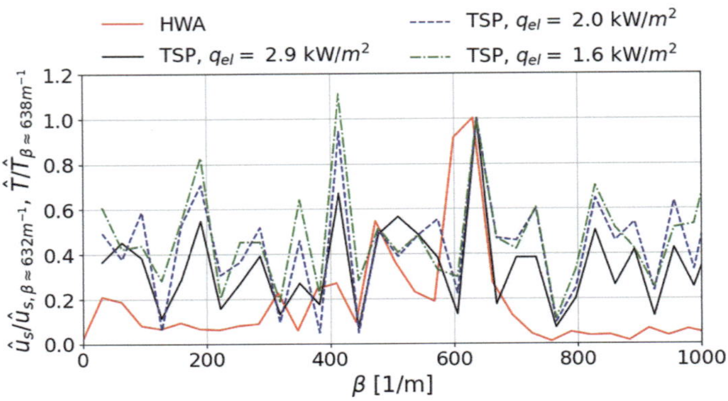

Fig. 11 Wavenumber spectrum derived from TSP measurements compared to HWA results for a variation of the electrical heating power density $\dot{q}_{el}$

wall high speed fluid and the lowest surface temperature, both being associated with the largest wall-shear stress, is attributed to conducting the HWA measurement at a wall distance of $z = 0.6$ mm and the flow structures being inclined in the boundary layer. Also the differences in regions $y_c' \approx -175$ mm and $y_c' \approx -100$ mm in Fig. 9b or $-350\,\text{mm} \leq y_c' \leq -335$ mm in Fig. 10a, is attributed to the HWA measurement being conducted at a certain distance from the wall.

The observed amplitudes of the spanwise velocity variation and the deviation from mean surface temperature increase from the measurement location far upstream in Fig. 9b relative to the locations closer to transition, as shown in Fig. 10c, d. This increase is particularly pronounced for the TSP result, where the increase is by more than a factor of ten, in spite of the lower heat flux, see Table 3. These observations lead to the conclusion that the TSP method is able to qualitatively measure the variation of the crossflow vortex strength. However, TSP setups are typically optimized for measuring the transition location in terms of measurement region, heat flux, and number of image acquisitions, which limits the acquisition of the small influence on the surface temperature of low amplitude crossflow vortices. Another observation supporting the conclusion that the vortex strength is captured qualitatively, is that the vortices in which the transition initiates are wider and more pronounced stripes than neighboring vortices in the TSP results. This confirms also the observation in previous TSP measurements of crossflow instability dominated transition that the vortices where transition starts have a darker footprint [32]. However, the footprints of the neighboring vortices vanish in the region of the saw-tooth pattern ($0.72 \leq x_c/c \leq 0.88$ in Fig. 8).

The spanwise HWA and TSP data of the location presented in Fig. 9a are used to derive power density spectra. These are shown in Fig. 11 as the amplitude of the spanwise velocity modulation $\hat{u}_s$ normalized by the amplitude at $\beta = 632\,\text{m}^{-1}$ and the amplitude of the surface temperature modulation normalized by the temperature amplitude at $\beta = 638\,\text{m}^{-1}$ as a function of the wavenumber $\beta = 2\pi/\lambda$, respectively.

The power density spectrum of the TSP measurement was derived from the unfiltered surface temperature distribution. Apart from the electrical heating power density of $\dot{q}_{el} = 2.9\,\mathrm{kW/m^2}$ utilized for the data shown in Fig. 9a, lower heating power densities settings of 2.0 and $1.6\,\mathrm{kW/m^2}$ were also used during data acquisition. The amplitude peak of the HWA result is $\beta \approx 632\,\mathrm{m^{-1}}$, which corresponds to a wavelength of $\lambda \approx 9.9\,\mathrm{mm}$. Thus, the result is in excellent agreement with the wavelength of the DRE ($\beta = 628\,\mathrm{m^{-1}}$), as listed in Table 1 for the SPECTRA-A_cntTSP1 configuration. Also from the surface temperature map of the TSP measurements this stationary crossflow wavelength of $\beta_{TSP} \approx 638\,\mathrm{m^{-1}}$ ($\lambda \approx 9.8\,\mathrm{mm}$) is captured with all electrical heat flux settings.

The HWA result of the power density spectrum shows only one additional distinct peak, apart from the aforementioned main peak, with a larger wavelength of $\lambda \approx 13\,\mathrm{mm}$ ($\beta \approx 475\,\mathrm{m^{-1}}$). In comparison, the power density spectrum of the surface temperature data shows several additional peaks besides the one triggered with DRE and the signal-to-noise ratio is smaller than that of the HWA results. These secondary peaks, which were obtained at all heat flux settings, correspond to larger wavelengths of $\lambda \approx 15\,\mathrm{mm}$ ($\beta \approx 414\,\mathrm{m^{-1}}$) and $\lambda \approx 33\,\mathrm{mm}$ ($\beta \approx 192\,\mathrm{m^{-1}}$). Since, these peaks do not coincide with HWA results, it is expected, that these spanwise structures in the surface temperature signal are the result of minor large scale temperature variation artefacts caused by the model heating or a variation of heat conduction inside of the model. The amplitude of these variations decreases with increasing heat flux, this influence being most pronounced for the peak at $\beta \approx 414\,\mathrm{m^{-1}}$ in Fig.11. The signal-to-noise ratio of TSP is in general lower than that from the HWA measurement; for example, the secondary peak identified by HWA is not distinguishable from the noise floor in TSP. However, the signal-to-noise ratio improves when the heat flux is increased. In conclusion, TSP is suitable to measure the dominant crossflow wavelength, given that the heat flux is sufficiently high.

3.2 *TSP Observations of Crossflow-Dominated Transition*

The final breakdown to turbulence in crossflow-dominated transition is driven by secondary instability modes, which are observable by HWA measurements inside of the crossflow vortices [29]. The onset of the secondary instability (type-I) does not develop at the wall, but rather in the updraught of the crossflow vortex in the form of high-frequency velocity fluctuations [53]. This behaviour had also been observed in previous investigations with the SPECTRA-B configuration [6]. For the SPECTRA configuration the secondary instabilities are observable in the form of velocity fluctuations at frequencies larger than 1000 Hz at the free-stream conditions used in this investigation. Thus, it is of interest to compare the detected onset of the transition and the measured transition location at the wall by TSP with the velocity data from HWA measurements in the boundary layer. To this purpose, spanwise HWA measurements parallel to the flat-plate surface were conducted at a wall-normal

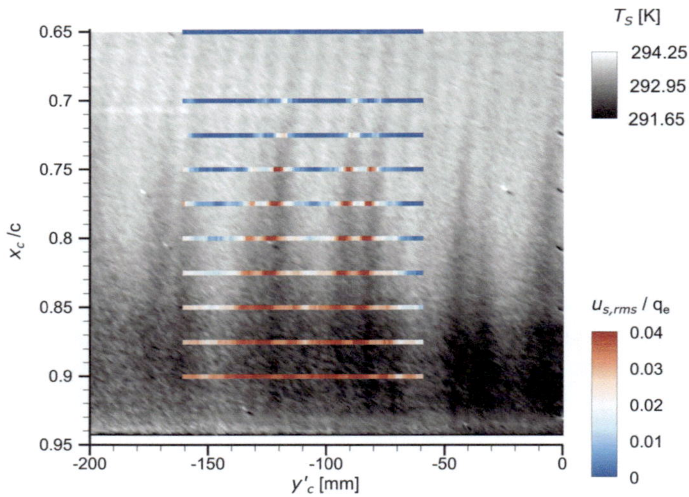

Fig. 12 TSP surface temperature T_S (extracted from Fig. 8) and disturbance amplitude $u_{s,rms}/q_e$ from the HWA measurement. SPECTRA-A_cntTSP1 configuration and external TSP acquisition setup at $U_{IMG} = 23.5\,\mathrm{m/s}$

distance of 0.7 mm in the region of the turbulent wedge visible in the TSP result, shown in Fig. 8.

Figure 12 shows the root-mean square (rms) of the high-pass filtered ($f >$ 1000 Hz) velocity component u_s normalized by the boundary-layer edge velocity q_e, namely $u_{s,rms}$, and the TSP surface temperature T_S. The agreement of the locations of increased $u_{s,rms}$ values with those of relatively low T_S values can be clearly seen. Thus, it can be concluded that the turbulent wedges in TSP are caused by the increased skin friction of the commencing turbulent flow, which is detected by the high-frequency velocity fluctuations of the secondary instability. The TSP measurement captures only the average of the skin friction field because the low acquisition frequency of the cameras and the frequency resolution of the utilized TSP [42] act effectively as low-pass filters, thus hindering a faithful measurement of these high-frequency velocity fluctuations (kilohertz regime) caused by the secondary instability. Hence, the TSP results are compared with the development of statistical velocity parameters like the mean velocity u_s or the rms of the high-pass filtered velocity fluctuations $u_{s,rms}$, as well as the stationary flow structures causing the footprints in the TSP results.

After the turbulent wedges have merged, the footprints of stationary crossflow vortices, from which the turbulent wedges arose, persist further into the turbulent boundary layer. This is also observed for neighboring vortices, where the transition occurs further downstream, for example at $y'_c \approx -35$ mm in Fig. 12. This observation leads to the conclusion that this crossflow vortex persists into the beginning of the turbulent boundary layer. The same observation is derived from an unrelated DNS study [52], where the primary crossflow vortex is only weakly modulated by the

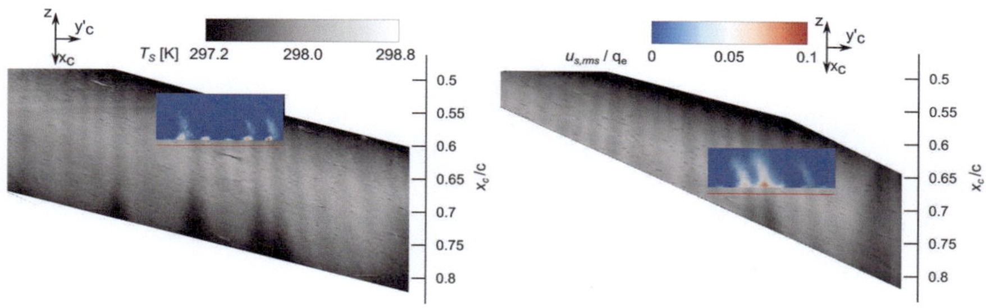

(a) $u_{s,rms}$ at $x_c/c = 0.60$. y_c' corrected by $\Delta y_c' = -4\,\mathrm{mm}$.

(b) $u_{s,rms}$ at $x_c/c = 0.67$.

Fig. 13 TSP and HWA results for SPECTRA-B showing the surface temperature T_S and the normalized disturbance amplitude $u_{s,rms}/q_e$ for $U_{IMG} = 27\,\mathrm{m/s}$ at two HWA measurement locations

secondary instability. Previous experimental investigations with surface based flow visualization techniques also showed this pattern (see TSP visualization [32], sublimating chemicals [15]). The visualization of the prematurely initiated transition in the region $-260\,\mathrm{mm} \leq y_c' \leq -200\,\mathrm{mm}$, see Fig. 8, shows that the crossflow vortex footprint persists for a significant distance, but then eventually vanishes. Thus, the turbulent mixing in the boundary layer disperses the typical stationary signs of the vortex.

A more detailed comparison of the flow structures in the region of turbulent wedges from HWA measurements and those from the TSP measurement is provided in Fig. 13. The velocity fluctuations in the updraught of the crossflow vortex, which can be identified in velocity distributions in Fig. 10a, b, reach the surface near the proximity of the near-wall tip of the secondary instability, which correspond to the low temperature in the surface temperature map in Fig. 13. A DNS study [52] analyzing the vortical structures relevant for the transition process showed that finger vortices originating at this location in the crossflow vortex were the first structures to reach the surface.

The comparison of the surface temperature with $u_{s,rms}$ measurements, shown in Fig. 12, shows already qualitatively that the TSP method is sensitive towards the flow structures related to crossflow dominated transition. The transition location is derived from TSP data by determining the largest temperature gradient [3, 13, 14, 21]. For this evaluation, the TSP data is transformed from model-fixed coordinates x_c, y_c into a coordinate system whereby surface temperature gradients from cutting crossflow vortices at an angle or shadow effects of the saw-tooth pattern can be avoided. These requirements are fulfilled by the model-fixed coordinate systems x_c, y_c' or the vortex coordinate system x_v, y_v, as shown in Fig. 2. Then cuts along the flow direction or the vortex axis are used to determine the largest gradient.

In Fig. 14, three examples of nearly streamwise surface temperature cuts at constant y_c' are shown. These have been extracted from the data presented in Fig. 12. First, in Fig. 14a, b, the constant y_c' cuts located in a spanwise region where the

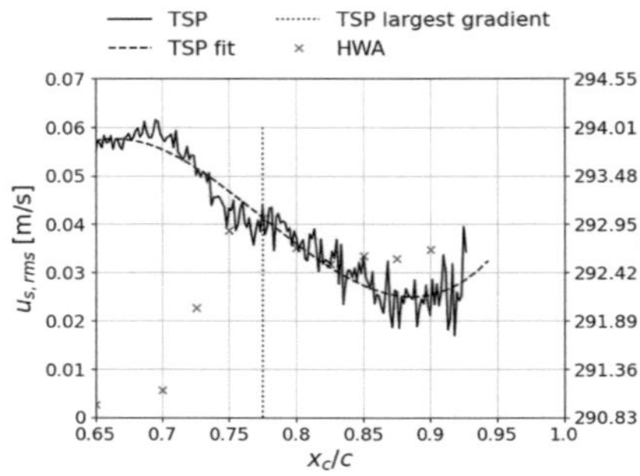

(a) $y_c' = -120\,\mathrm{mm}$.

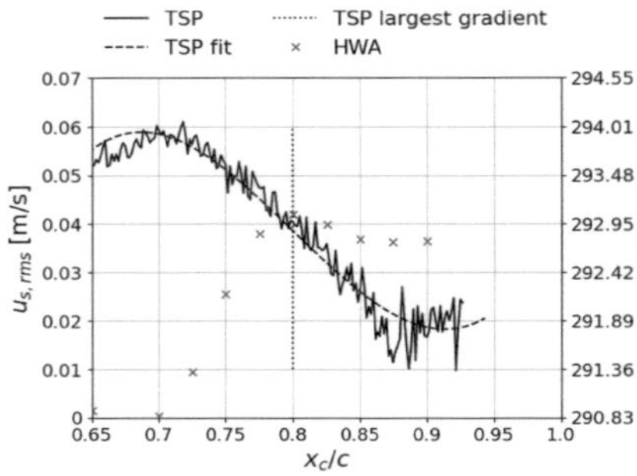

(b) $y_c' = -82\,\mathrm{mm}$.

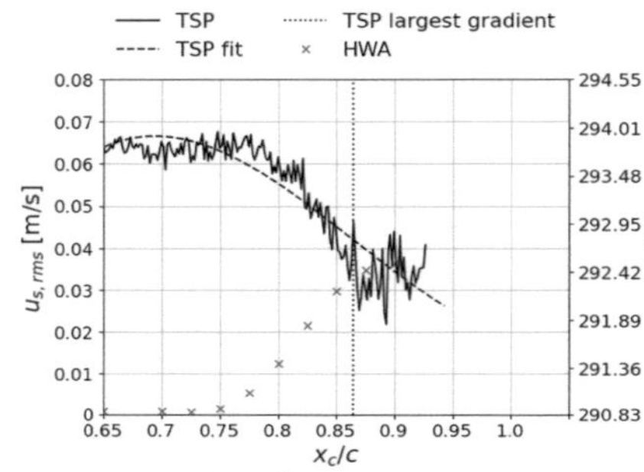

(c) $y_c' = -105\,\mathrm{mm}$.

Fig. 14 Surface temperature cuts T_S from TSP and the normalized disturbance amplitude $u_{s,rms}/q_e$ from HWA measurements for SPECTRA-A_cntTSP1 at three different y_c' locations with $U_{IMG} = 27\,\mathrm{m/s}$. See also Fig. 12

transition starts (namely below the updraught part of the crossflow vortex) are examined. The rms of the high-frequency velocity fluctuations $u_{s,rms}$ starts to increase at $x_c/c \approx 0.7$ in Fig. 14a. A sharp increase of the high-frequency velocity fluctuation $u_{s,rms}$ lies at $x_c/c \approx 0.74$. After reaching a maximum at $x_c/c \approx 0.775$, a plateau with lower values is reached at $x_c/c \approx 0.825$. The drop in surface temperature starts only a short distance later than the increase of $u_{s,rms}$ detected by HWA. The location of the largest gradient (dotted line in the figures) is in good agreement with the maximum of $u_{s,rms}$. Thus, the TSP method has detected transition at a location $\Delta x_c/c \approx 0.05$ to 0.10 further downstream than the sudden increase seen in $u_{s,rms}$.

In the region where HWA measures a plateau in $u_{s,rms}$, the development of the surface temperature is less clear due to effects from heating artefacts at the downstream end of the cntTSP area. The data presented in Fig. 14b lead to the same observations, only with transition occurring at a x_c/c value. Finally, the constant y_c' cut in a location between turbulent wedges is examined—see Fig. 14c. The rms of the high-frequency velocity fluctuation $u_{s,rms}$ starts increasing further downstream at $x_c/c \approx 0.75$ and the rms grows slower than in the updraught region of the crossflow vortex where transition started. Furthermore, the rms of the velocity fluctuations

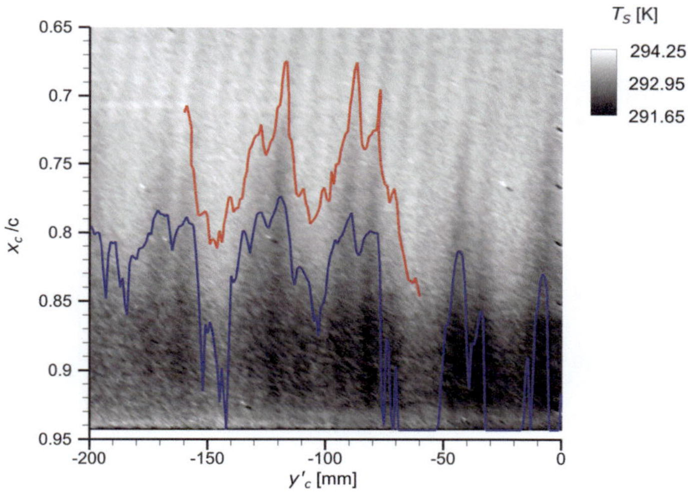

Fig. 15 TSP surface temperature map showing location of largest temperature gradient (–) and $u_{s,rms}/q_e = 0.01$ (–). SPECTRA-A_cntTSP1 configuration and external TSP acquisition setup at $U_{IMG} = 23.5\,\mathrm{m/s}$

do not reach a maximum. Moreover, the TSP method resolves the beginning of the surface temperature drop at a similar location where the HWA technique detects an increase of $u_{s,rms}$. The largest temperature gradient is located where the rms of the high-frequency velocity fluctuations have reached their largest value. Again, the location of the largest temperature gradient is located $\Delta x_c/c \approx 0.05$ further downstream than the sudden rise of the rms of the high-frequency velocity fluctuation. Both measurement techniques show that the transition is not completed before the downstream end of the measurement region is reached.

The comparison of the detected transition location to the locations of significant values of high-frequency velocity fluctuations over a larger spanwise extent is shown in Fig. 15, which is based on the data of Figs. 8, 12, and 14. The figure compares the location of the largest surface temperature gradient with $u_{s,rms}/q_e = 0.01$. The value of $u_{s,rms}/q_e = 0.01$ is chosen from Fig. 14 because this value represents the value where a large increase of the high-frequency velocity fluctuations has occured. From Fig. 15 the impression has been reinforced that the transition location detected by TSP is $\Delta x_c/c \approx 0.05$ to 0.10 downstream of the beginning of the first detectable increase of $u_{s,rms}$. Nevertheless, the onset of the increase of the high-frequency velocity fluctuations has been well resolved along x_c, see Fig. 14.

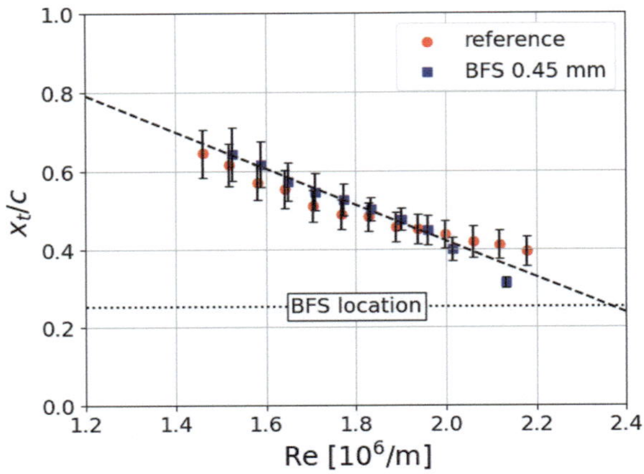

Fig. 16 Transition location with and without a BFS of 0.45 mm for SPECTRA-A_cntTSP2 with 3 DRE. The errorbars are the standard deviation of the detected transition location

3.3 Influence of BFS on the transition location

In the following, the TSP method is used to investigate the influence of a backward-facing step (BFS) of 0.45 mm height on the transition location. The tests were performed with the SPECTRA-A configuration and the 3 DRE configuration. The experiments were conducted with the cntTSP2 flat plate to provide sufficient space to track transition movement whilst the wind tunnel velocity was varied.

Figure 16 shows transition location data for the reference condition (no step) and with BFS. Additionally, temperature maps showing the transition location are presented in Fig. 17 in the vortex oriented coordinate system x_c/c, y_v. The spanwise average of the transition location x_t/c in the region $-100\,\text{mm} \leq y_v \leq 100\,\text{mm}$ is shown in Fig. 16 as a function of the unit Reynolds number $\text{Re} = U_{IMG}/\nu$ in the range of $1.4 \times 10^6\,\text{m}^{-1} \leq \text{Re} \leq 2.2 \times 10^6\,\text{m}^{-1}$. In addition, the standard deviation of the measured transition locations in the spanwise range is presented in the form of the error bars. In the reference case, the transition location gradually moves upstream until $\text{Re} \approx 1.8 \times 10^6\,\text{m}^{-1}$, whereafter, at larger Reynolds numbers, the rate of the transition movement is reduced. In contrast, for the case with the BFS, an upstream movement of the transition location is observed up to $\text{Re} \approx 2.0 \times 10^6\,\text{m}^{-1}$. In this case, the transition approaches the step location faster.

Interestingly, the average transition location with the BFS is observed approximately 5% downstream of the transition location of the reference condition for Re $\approx 1.8 \times 10^6\,\text{m}^{-1}$. Nevertheless, the standard deviation of the transition location, as well as the TSP results in Fig. 17 for $\text{Re} \approx 1.77 \times 10^6\,\text{m}^{-1}$, indicate that the portion of laminar flow shows little change due to the presence of the BFS.

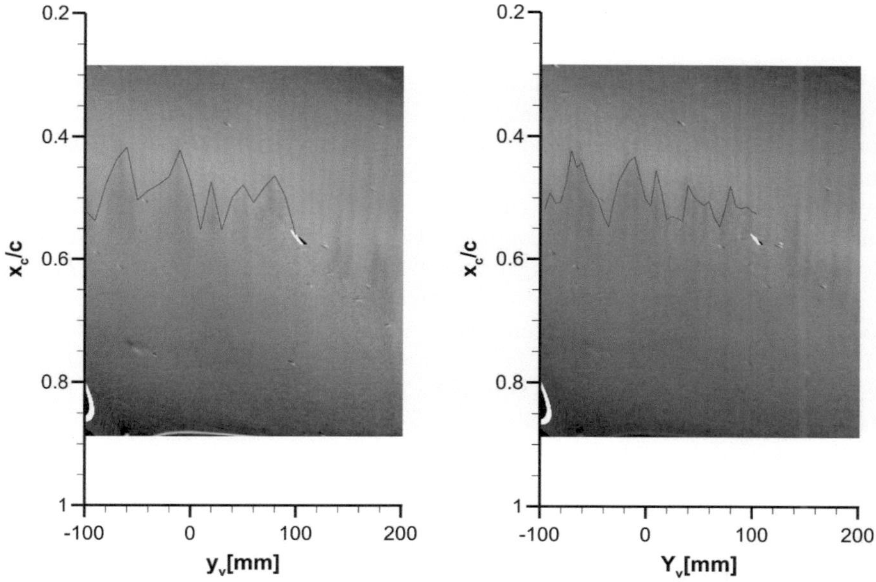

(a) Reference (no BFS). (b) BFS.

Fig. 17 TSP results for SPECTRA-A_cntTSP2 configuration at Re $\approx 1.77 \times 10^6\,\mathrm{m}^{-1}$ for the reference and BFS of 0.45 mm height. Location of the largest temperature gradient (–), namely the transition location, is indicated

3.4 Uncertainty Analysis

A detailed description of the uncertainty analysis for quantitative surface temperature measurements by means of TSP is provided by Liu et al [35] and Cattafesta et al [10]; this has been recently applied to measurements with the TSP utilized in this experiment, only with water rather than air as the test fluid [33]. The data evaluation in the current paper does not rely on the quantitative temperature measurement but an uncertainty of 10% for the temperature measurement is expected. Mapping of the camera images onto the structured surface grid is the main factor affecting the uncertainty in these measurements of transition location; both setups provide challenging conditions for the mapping. The external setup has an acute angle between the TSP surface and the optical axis of the camera, which results in deformed markers, which complicates the determination of their centers. The integrated setup utilizes lenses with a focal length of only 3 mm so that a nonlinear approach to model the lens parameters for the camera model is needed. Also, for both optical setups, the applied markers are larger than the optimal diameter of 10 px. Still, all of these factors lead to an uncertainty of the absolute position on the model surface that is better than

± 0.5 mm, which has been derived from mapped marker positions at the edge of the projection.

The influence of the model heating on the flow is more difficult to quantify. Apart from the influence of a non-adiabatic surface on the flow, an artefact of the model heating, has been visible between $0.85 \leq x_c/c \leq 0.95$ in Figs. 8 and 9a. This inhomogeneous temperature distribution is caused by a local variation of the CNT layer thickness, leading to an uneven model heating. The influence of a non-adiabatic surface on crossflow instability dominated transition has not yet been conclusively determined. When strongly heating the model surface at the leading edge to $T_S/T_\infty \approx 1.21$ in a wind tunnel ($Tu \approx 0.5\%$) experiment an upstream shift of the transition location was observed [17]. Recent experiments in a low turbulence wind tunnel ($Tu = 0.06\%$) by Barahona et al. [4] also showed an upstream shift of the transition location when heating the model surface to $T_S/T_\infty = 1.055$ and $T_S/T_\infty \approx 1.02$ in the middle region of the model and near the leading edge, respectively. Significantly stronger amplitudes of the type-I and type-III modes of the secondary instability were observed. However, a previous TSP investigation in a low turbulence wind tunnel (Tu = 0.02% at 30 m/s) with the model heating downstream of $x/c \approx 0.12$ and a temperature ratio of $T_S/T_\infty \leq 1.07$ in the middle region of the model [32] did not show an influence of the surface temperature on the transition location. The temperature ratio at the leading edge is expected to be significantly lower and a shift of the transition location had not been observed when switching off the model heating. For the current TSP investigation, the model heating starts downstream of $x_c/c = 0.26$ and the temperature ratio is minimized to limit the influence of the heated model surface on the crossflow instability. The agreement of the location of the transition location determined by TSP and the location of maximum $u_{s,rms}$, see Fig. 14, provides that the model heating does not influence the transition location, since the HWA data had been recorded without model heating.

4 Conclusion

The TSP method was successfully integrated into the SPECTRA configuration, one of the Cross Capability Demonstrators of the Clean Sky 2 initiative. This includes the design of two TSP acquisition setups, one external and one integrated into the SPECTRA-A displacement bodies, and the manufacturing of flat plate segments with integrated cntTSP. The comparison of TSP results with complementary HWA measurements provide a detailed assessment of the capabilities of TSP related to the investigation of crossflow dominated transition.

In the laminar region the TSP method resolves the footprint of the crossflow vortices with the most dominant wavelength. When the heat flux and number of acquired images are optimized, the wavelength can be determined also for low amplitudes of the crossflow instability. The crossflow vortices appear as alternating brighter and darker lines in the TSP visualization with the updraught of the crossflow vortex being located above the line of better cooling.

The onset of transition is initiated by secondary instabilities, which are recorded by HWA as high-frequency velocity fluctuations. The developing wedge-shaped turbulent flow increases the heat transfer, which leads to a variation of the average surface temperature, i.e., a cooler surface. The transition location as determined by the TSP method, which is identified based on the largest gradient of the intensity ratio in flow direction, is in good agreement with the maximum of the rms of the high frequency velocity fluctuations $u_{s,rms}$ derived from HWA. In these HWA measurements, the first occurrence of high-frequency velocity fluctuations reveals early signs of the turbulent breakdown upstream of the transition location as determined by the TSP method.

Finally, the TSP method was deployed to investigate the influence of a BFS on the transition location in the SPECTRA-A configuration. Development of the transition location with respect to wind tunnel velocity is well resolved over the entire parameter range. A slightly delayed transition is observed in the case of a BFS of height 0.45 mm for most investigated wind tunnel velocities.

The uncertainty analysis concludes that the mapping of the images onto the structured surface grid and the potential influence of the model heating on the crossflow instability are the most relevant sources of uncertainty. It was determined that the mapping error is small compared to the relevant flow structures and the dimensions of the model. An influence of model heating on the measured flow structures could not be ascertained.

In conclusion, the TSP method was able to deliver precise transition measurements and flow characterization of crossflow-dominated transition. The TSP flow measurements can be conducted for large areas with short acquisition and evaluation times. This provides an opportunity of using larger test matrices, with the concomitant identification of interesting locations on the model, which can then be investigated using more time-consuming measurement techniques such as HWA and Lagrangian particle tracking.

Abbreviations

1MG	One-meter wind tunnel Göttingen
BFS	Backward-facing step
c	Chord length
CCD	Charge-coupled device
CF	Crossflow
CNT	Carbon nanotubes
cntTSP	TSP coating combined with CNT heating
CMOS	Complementary metal-oxide-semiconductor
d_{DRE}	DRE diameter
DLR	German Aerospace Centre
DNS	Direct numerical simulation
DRE	Discrete roughness element

h_{DRE}	DRE height
HWA	Hot-wire anemometry
I	Measured TSP emission intensity during the run-phase
I_{ref}	Measured TSP emission intensity during the reference-phase
LE	Leading edge
LED	Light-emitting diode
$\dot{q}_{el}$	Electrical heating power density
q_e	Boundary-layer edge velocity
Re	Reynolds number
rms	Root-mean square
SPECTRA	Swept flat PlatE Crossflow TRAnsition
t_{exp}	Exposure time
ΔT	Over temperature
T_∞	Flow temperature
T_S	Surface temperature
T'_S	Surface temperature deviation from mean value
Tu	Turbulence level
TSP	Temperature-sensitive paint
U_{1MG}	Flow velocity of 1MG
u_s	Velocity component in streamline direction
$u_{s,rms}$	rms of high-pass filtered velocity fluctuation of u_s
x_c,y_c	Model fix coordinate system
$x_{c,DRE}$	Chordwise position of DRE
x_t	Transition location
x_v,y_v	Vortex coordinate system
y'_c	$y'_c = y_c - y_{c,p} + x_c \tan(\varphi)$
XDC	Cross-capability demonstrators
λ	Crossflow wavelength
ν	Kinematic viscosity
φ	Sweep angle

Acknowledgements This project has received funding from the Clean Sky 2 Joint Undertaking under the European Union's Horizon 2020 research and innovation program under grant agreement N° CS2-LPA-GAM-2020-2023-01.

The authors would like to thank U. Gerhard and S. Hein for the design of the displacement body; M. Weberschock and DLR-SHT for the manufacturing of the flat plate and displacement body; V. Ondrus for synthesizing the TSP; C. Fuchs, T. Kleindienst, K. Borchert for their support during the manufacturing and preparation of the model and setting up the experiment.

References

1. Arnal D, Archambaud J (2008) Laminar-turbulent transition control: NLF, LFC, HLFC. Tech. Rep. RTO-EN-AVT-15, NATO STO
2. Asai K, Kanda H, Kunimasu T, Liu T, Sullivan J (1997) Boundary-layer transition detection in a cryogenic wind tunnel using luminescent paint. J Aircr 34:34–42. https://doi.org/10.2514/2.2132
3. Ashill P, Betts C, Gaudet I (1996) A wind tunnel study of transition flows on a swept panel wing at high subsonic speeds. In: CEAS 2nd European forum on laminar flow technology, pp 10.1–10.17
4. Barahona Lopez M, van de Weijer A, Rius Vidales A, Kotsonis M (2024) Impact of a heated wall on the laminar-to-turbulent transition of crossflow vortices: an experimental study. In: Proceedings of the AIAA SCITECH 2024 forum. AIAA SciTech Forum and Exposition, 2024, American Institute of Aeronautics and Astronautics Inc. (AIAA), United States. https://doi.org/10.2514/6.2024-0695
5. Barth HP (2021) Beeinflussung des laminar-turbulenten Grenzschichtumschlags durch kontrollierte Anregung stationärer Querströmungsinstabilitäten. Ph.D. thesis, Georg-August-Universität Göttingen
6. Barth HP, Hein S (2021) Experimental investigation of spanwise-periodic surface heating for control of crossflow-dominated laminar-turbulent transition. In: IUTAM laminar-turbulent transition, 9th IUTAM symposium, London, pp 267–278. Springer International Publishing. https://doi.org/10.1007/978-3-030-67902-6_23
7. Barth HP, Hein S, Rosemann H (2017) Redesigned swept flat-plate experiment for crossflow-induced transition studies. In: Notes on numerical fluid mechanics and multidisciplinary design. Springer International Publishing, pp 155–165. https://doi.org/10.1007/978-3-319-64519-3_14
8. Bippes H (1999) Basic experiments on transition in three-dimensional boundary layers dominated by crossflow instability. Prog Aerosp Sci 35:363–412. https://doi.org/10.1016/S0031-8914(53)80099-6
9. Boden F, Barth HP, Hein S, Heitmann D, Kirmse T, Klein C, Krüger W, Lemarechal J, Norambuena M, Schäfer D, Schnell R, Sinha K, Theiß A, Weiss A (2023) Recent developments of common numerical methods and common experimental means within the framework of the large passenger aircraft program. CEAS Aeronaut J 14:281–293. https://doi.org/10.1007/s13272-022-00635-w
10. Cattafesta L, Liu T, Sullivan J (1998) Uncertainty estimates for temperature-sensitive paint measurements with charge-coupled device cameras. AIAA J 36(11):2102–2108
11. Costantini M (2016) Experimental analysis of geometric, pressure gradient and surface temperature effects on boundary-layer transition in compressible high reynolds number flows. PhD thesis, RWTH Aachen
12. Costantini M, Fey U, Henne U, Klein C (2012) Influence of non-adiabatic model surface on transition measurements using the temperature-sensitive paint technique in a cryogenic wind tunnel. In: 42nd AIAA fluid dynamics conference and exhibit. https://doi.org/10.2514/6.2012-2830
13. Costantini M, Henne U, Risius S, Klein C (2021) A robust method for reliable transition detection in temperature-sensitive paint data. Aerosp Sci Technol 113:106702. https://doi.org/10.1016/j.ast.2021.106702
14. Crawford B, Duncan G, West D, Saric W (2013) Laminar-turbulent boundary layer transition imaging using IR thermography. Opt Photon J 3:233–239. https://doi.org/10.4236/opj.2013.33038
15. Dagenhart J, Saric W, Mousseux M, Stack J (1989) Crossflow-vortex instability and transition on a 45 deg swept wing. In: 20th fluid dynamics, plasma dynamics and lasers conference, fluid dynamics and co-located conferences. AIAA-89-1892. https://doi.org/10.2514/6.1989-1892

16. Deyhle H, Bippes H (1996) Disturbance growth in an unstable three-dimensional boundary layer and its dependence on environmental conditions. J Fluid Mech 316:73–113. https://doi.org/10.1017/S0022112096000456
17. Dovgal A, Levchenko V, Timofeev V (1990) Laminar-turbulent transition. In: International union of theoretical and applied mechanics, chap. Boundary layer control by a local heating of the wall. Springer, Berlin, Heidelberg, pp 113–121. https://doi.org/10.1007/978-3-642-84103-3_9
18. Drake A, Westphal R, Zuniga F, Kennelly R, Koga D (1996) Wing leading edge joint laminar flow tests. Tech. Rep. NASA Technical Memorandum 4762, NASA
19. Duncan GT, Crawford B, Tufts MW, Saric WS, Reed HL (2013) Effects of step excrescences on swept-wing transition. In: 31st AIAA applied aerodynamics conference. https://doi.org/10.2514/6.2013-2412
20. Eppink JL, Wlezien RW, King RA, Choudhari M (2019) Influence of a backward-facing step on swept-wing boundary-layer transition. AIAA J 57(1):267–278. https://doi.org/10.2514/1.J057437
21. Fey U, Egami Y (2007) Springer handbook of experimental fluid mechanics. In: Transition-detection by temperature-sensitive paint. Springer, Berlin, Heidelberg, New York, pp 537–552
22. Fey U, Egami Y, Klein C (2007) Temperature-sensitive paint application in cryogenic wind tunnels: transition detection at high Reynolds numbers and influence of the technique on measured aerodynamic coefficients. In: 22nd International congress on instrumentation in aerospace simulation facilities, Pacific Grove, CA, 2007
23. Hunt L, Saric W (2011) Boundary-layer receptivity of three-dimensional roughness arrays on a swept-wing. In: 41st AIAA fluid dynamics conference and exhibit, 27–30 June 2011, Honolulu, Hawaii. AIAA 2011-3881. https://doi.org/10.2514/6.2011-3881
24. Jacobitz F, Sysyn I, Ryan J, Comfort J, Bonner P, Poole D, Lemarechal J, Costantini M (2025) Receptivity of a sept-wing boundary layer to micron-sized discrete roughness elements. Int J Heat Fluid Flows
25. Joslin RD (1998) Aircraft laminar flow control. Annu Rev Fluid Mech 30:1–29. https://doi.org/10.1146/annurev.fluid.30.1.1
26. Klein C, Engler R, Henne U, Sachs W (2005) Application of pressure-sensitive paint for determination of the pressure field and calculation of the forces and moments of models in a wind tunnel. Exp Fluids 39:475–483. https://doi.org/10.1007/s00348-005-1010-8
27. Klein C, Henne U, Sachs W, Beifuß U, Ondrus V, Bruse M, Lesjak R, Löhr M (2014) Application of carbon nanotubes (CNT) and temperature-sensitive paint (TSP) for the detection of boundary layer transition. In: 52nd Aerospace sciences meeting, 13–17 January 2014, National Harbor, Maryland. No. AIAA 2014-1482
28. Klein C, Yorita D, Henne U, Ondrus V, Hensch AK, Longo R, Gimbel A, von Deetzen S (2020) Application of temperature sensitive paint to investigate laminar-to-turbulent transition on nacelles. In: AIAA Scitech 2020 forum. https://doi.org/10.2514/6.2020-1608
29. Kohama Y, Onodera T, Egami Y (1996) Design and control of crossflow instability field. In: Duck P, Hall P (eds) IUTAM symposium on nonlinear instability and transition in three-dimensional boundary layers. Springer Netherlands, pp 147–156. https://doi.org/10.1007/978-94-009-1700-2_14
30. Kohama Y, Saric W, Hoos J (1991) A high-frequency, secondary instability of crossflow vortices that leads to transition. In: Proc. R. Aerosp. Soc. Conf. Bound.-Layer Transit. Control Conf. Peterhouse College, Cambridge, UK, pp 4.1–4.13
31. Kurz H, Kloker M (2014) Receptivity of a sept-wing boundary layer to micron-sized discrete roughness elements. J Fluid Mech 755:62–82. https://doi.org/10.1017/jfm.2014.425
32. Lemarechal J, Costantini M, Klein C, Kloker M, Wuerz W, Kurz H, Streit T, Schaber S (2019) Investigation of stationary-crossflow-instability induced transition with the temperature-sensitive paint method. Exp Thermal Fluid Sci 109:1–10. https://doi.org/10.1016/j.expthermflusci.2019.109848
33. Lemarechal J, Klein C, Henne U, Puckert D, Rist U (2019) Detection of lambda- and omega-vortices with the temperature-sensitive paint method in the late stage of controlled laminar-turbulent transition. Exp Fluids 60(91):1–14. https://doi.org/10.1007/s00348-019-2734-1

34. Lemarechal J, Dimond BD, Barth HP, Hilfer M, Klein C (2023) Miniaturization and model-integration of the optical measurement system for temperature-sensitive paint investigations. Sensors 23(16). https://doi.org/10.3390/s23167075
35. Liu T, Campbell B, Sullivan J (1995) Accuracy of temperature-sensitive fluorescent paint for heat transfer measurements. In: 30th AIAA thermophysics conference, 19–22 Juni 1995, SanDiego, CA, USA. AIAA 95-2042
36. Liu T, Sullivan J, Asai K, Klein C Egami Y (2021) Pressure and temperature sensitive paints, 2. edn. Springer
37. Malik MR, Li F, Chang CL (1996) Nonlinear crossflow disturbances and secondary instabilities in swept-wing boundary layers. In: Duck PW, Hall P (eds) IUTAM symposium on nonlinear instability and transition in three-dimensional boundary layers. Springer, Netherlands, pp 257–266
38. Malik MR, Li F, Choudhari MM, Chang CL (1999) Secondary instability of crossflow vortices and swept-wing boundary-layer transition. J Fluid Mech 399:85–115. https://doi.org/10.1017/S0022112099006291
39. Müller B, Bippes H (1989) Experimental study of instability modes in a three-dimensional boundary layer. In: Fluid dynamics of three-dimensional turbulent shear flows and transition, pp 13-1– 13-15
40. Nenni J, Gluyas G (1966) Aerodynamic design and analysis on an LFC surface. Astronaut Aeronaut 4(7):52–57
41. Nitschke-Kowsky P, Bippes H (1988) Instability and transition of a three-dimensional boundary layer on a swept flat plate. Phys Fluids 31:876–795. https://doi.org/10.1063/1.866814
42. Ondrus V, Meier RJ, Klein C, Henne U, Schäferling M, Beifuß U (2015) Europium 1,3-di(thienyl)propane-1,3-diones with outstanding properties for temperature sensing. Sens Actuators A 233:434–441
43. Perraud J, Seraudie A (2000) Effects of steps and gaps on 2D and 3D transition. In: European congress on computational methods in applied sciences and engineering ECCOMAS 2000. ECCOMAS, pp 1–18
44. Radeztsky R, Reibert M, Saric W (1999) Effect of isolated micron-sized roughness on transition in swept-wing flows. AIAA J 37(11):1370–1377. https://doi.org/10.2514/2.635
45. Reneaux J (2004) Overview on drag reduction technologies for civil transport aircraft. In: European congress on computational methods in applied sciences and engineering ECCOMAS 2004
46. Rius-Vidales AF, Kotsonis M (2022) Unsteady interaction of crossflow instability with a forward-facing step. J Fluid Mech 939:A19. https://doi.org/10.1017/jfm.2022.151
47. Saric W, Ruben Carrillo J, Reibert M (1998) Leading-edge roughness as a transition control mechanism. In: 36th AIAA aerospace sciences meeting and exhibit. https://doi.org/10.2514/6.1998-781
48. Saric W, Carpenter A, Reed H (2011) Passive control of transition in three-dimensional boundary layers, with emphasis on discrete roughness elements. Philos Trans R Soc A 369:1352–1364. https://doi.org/10.1098/rsta.2010.0368
49. Saric W, Reed H, White E (2003) Stability and transition of three-dimensional boundary layers. Annu Rev Fluid Mech 35:413–440. https://doi.org/10.1146/annurev.fluid.35.101101.161045
50. Schrauf G (2005) Status and perspectives of laminar flow. Aeronaut J 109:639–644. https://doi.org/10.1017/S000192400000097X
51. Vijgen PMHW, Dodbele SS, Holmes BJ, van Dam CP (1988) Effects of compressibility on design of subsonic fuselages for natural laminar flow. J Aircr 25(9):776–782. https://doi.org/10.2514/3.45659
52. Wassermann P, Kloker M (2002) Mechanisms and passive control of crossflow-vortex-induced transition in a three-dimensional boundary layer. J Fluid Mech 456:49–84. https://doi.org/10.1017/S0022112001007418
53. White EB, Saric WS (2005) Secondary instability of crossflow vortices. J Fluid Mech 525:275–308. https://doi.org/10.1017/S002211200400268X

Improving Source Localization with SODIX for Far-Field Measurements in Static Engine Noise Tests

Sebastian Oertwig and Henri Siller

Abstract The source localization method SODIX is able to determine the contributions of individual sound sources of aircraft engines to the overall radiated noise. Here, the method is applied to new measurements with a short-cowl turbofan engine using a far-field microphone array as it is commonly used in static engine noise testing. The analysis of far-field microphone data is very challenging due to the low number of microphones in these arrays. SODIX is used to evaluate the effect of an intake liner on the sound field radiated from the engine intake by comparing measurements with and without the installation of an intake liner. The source localization results show that SODIX is able to quantify the effect of the intake liner even when the overall sound pressure levels are dominated by other sound sources, e.g. by jet noise. A review of the mathematical formulation of SODIX shows that the point-spread function is relevant for the source localization with SODIX and has to be taken into account in the design of future measurements.

Keywords Aircraft noise · Sound source localization · Microphone array · Static engine noise test · SODIX

1 Introduction

SODIX (SOurce DIrectivity modeling in the cross-spectral matriX) is an advanced source localization technique that is able to determine the individual contributions of the engine components to the overall radiated noise [1, 3]. The method can be used to evaluate acoustic treatments of aero-engines in full-scale and can remove the need to install additional devices like acoustic barriers. In the past, SODIX has mainly been applied to data from full-scale static engine noise tests measured with large

S. Oertwig · H. Siller (✉)
Department of Engine Acoustics (AT-TRA), German Aerospace Center (DLR), Institute of Propulsion Technology, Berlin, Germany
e-mail: Henri.Siller@dlr.de

S. Oertwig
e-mail: Sebastian.Oertwig@DLR.de

F. Boden (ed.), *Numerical Methods and Experimental Means for the Integration of Future Propulsion Systems*, Springer Aerospace Technology,
https://doi.org/10.1007/978-3-031-97725-1_9

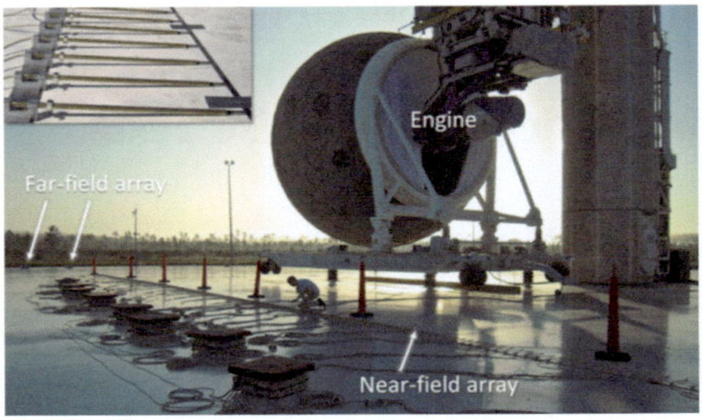

Fig. 1 Experimental set-up for a static noise test with a linear near-field microphone array for source localization measurements and a polar array for far-field measurements of the sound field

linear microphone arrays with up to 250 microphones near the engine (see Fig. 1). Recently, SODIX has also been applied to a sparsely populated far-field array with only 31 microphones. The application to the sparse far-field array promises benefits in terms of, reduced computational time and complexity of the experimental setup. It has been shown that the source localization results with the far-field array are similar to those of a large microphone array despite spurious sound sources that occur due to the low spatial sampling of the microphones in the far-field array. Recent extensions of the original SODIX methods are the added capabilities to analyze fully coherent sound sources [6] and partially coherent sources of sound [5, 7, 9].

This chapter reports the application of the source localization method SODIX to data measured with a polar far-field microphone array during static noise tests of a short-cowl turbofan engine. The measurements have been performed with and without an acoustic liner installed in the intake of the engine. The sound field radiated from the source at the engine intake and the overall sound pressure levels have been calculated using SODIX and compared with the sound pressure levels measured directly by the far-field microphones. The overall sound pressure levels calculated by SODIX agree well with the levels measured in the far-field and the SODIX method is able to quantify the effect of the intake liner even when it is not visible in the overall sound pressure levels because they are dominated by other sound sources such as jet noise.

2 The SODIX Algorithm

The SODIX method determines a distribution of directive sources that fits the data that has been measured with a microphone array. Figure 2 compares the simple monopole model (shown on the left-hand side) that many source localization methods are based on [2] to the SODIX model of directive sources, where one source point can radiate different amplitudes in the direction of every microphone position in the array (shown on the right). SODIX has to solve for source amplitudes from every source position

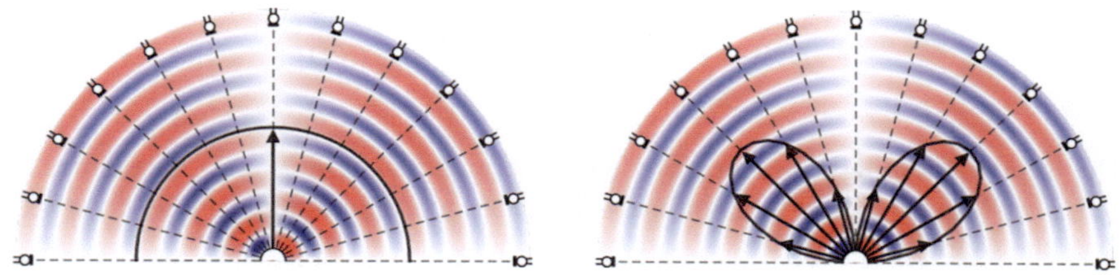

Fig. 2 The sound field of a dipole (shown on both plots) with the source at the bottom center of the semicircle, around which the microphones of the array are arranged. The amplitude and phase of the sound waves are indicated in colour. The monopole source model with uniform directivity in the direction of all microphones (shown on the left) cannot resolve the sound field, but the SODIX model of a directive sound source (shown on the right) can represent different values of the amplitude and the phase in the direction of every microphone

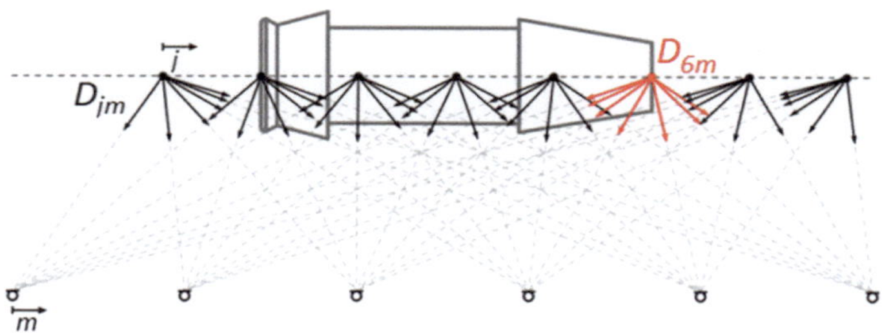

Fig. 3 The SODIX source model that includes different source amplitudes D_{jm} between every source point j and every microphone position m

in the direction of every microphone (see Fig. 3), which significantly increased the computational cost compared to calculations for monopole sources with uniform directivity.

The cross-spectral matrix (CSM) C_{mn} contains the auto- and cross-spectra for every combination of microphones m and n. The CSM from the measured microphone array data is approximated by a model CSM

$$C_{mn}^{\text{mod}} = \sum_{j=1}^{J} g_{jm} D_{jm} D_{jn} g_{jn}^{*} \tag{1}$$

with the amplitudes of the directive point sources D_{jm} and the steering vectors g_{jm}, which describe the propagation of sound between the sources and the microphones. The indices j stand for the positions of the sources, m for those of the the microphones. SODIX solves for the directive source amplitudes D_{jm} by minimizing the difference between the cross spectral matrices C_{mn} of the measured data and the modeled distribution of sources by evaluating the cost function $F(D)$:

$$F(D) = \sum_{m,n=1}^{M} ||C_{mn} - C_{mn}^{\mathrm{mod}}||^2 + \sigma R(D) \ . \tag{2}$$

The term $\sigma R(D)$ describes a regularization R with a weight σ. R(D) can either be an L2 regularization or a function that sets constraints on the shape of the directivity D of the source (see [4]).

For static engine noise tests on free-field test facilities or in indoor tests, linear microphone arrays are used to obtain the input data for the SODIX calculations. The SODIX solution is a distribution of directive sources on the engine axis with different source amplitudes in the direction of every microphone position. The results are usually presented in graphs of the amplitude as a function of the positions of the equivalent sources on the x-axis and the emission direction or emission angle θ on the vertical axis (see for example in Fig. 5).

The directivity of the SODIX equivalent sources can be read from the sound source maps. The directivity of the sum of the amplitudes of all SODIX sources is generally very close to the directivity measured directly from the microphones in the far-field array (unless there are other sources outside of the spatial domain covered by the microphone array). The directivities of the individual source regions show their relative contribution to the total noise radiated in a particular direction.

3 Measurements and Experimental Data

When static engine noise tests are performed to measure data for a SODIX source localization analysis, a large linear microphone array is mounted near the engine and in parallel with its axis. A photograph of a typical experimental set-up is shown in Fig. 1. The engine is mounted on a pylon above the arena and the near-field linear microphone array with 250 microphones is oriented parallel to the engine axis at a lateral distance of about 10 m.

However, the industrial standard for static engine noise tests is to use a permanently installed polar arc array for measurements of the engine noise in the far-field, as shown in Fig. 4. The microphones are installed on a circular arc around the engine at a radial distance of about 46 m (approximately 150 feet) at fixed angular positions with angular increments of 5°. For such tests, without a SODIX analysis, a separation of sources can only be achieved by installing a sound barrier wall that shields the microphones in the forward arc of the engine from sound emitted from the rear of the engine, i.e. from the exhaust nozzle and the jet.

Airbus [8] has provided the experimental data from a static acoustic engine test with a short cowl engine (a turbofan engine with separate core and bypass nozzles as opposed to a long cowl engine where the nacelle extends downstream over the common exhaust nozzle) described in this chapter. The experiments have been performed using only a far-field array. A schematic of the set-up is shown in Fig. 4. The array was sparsely populated with 31 microphones in angular increments of the

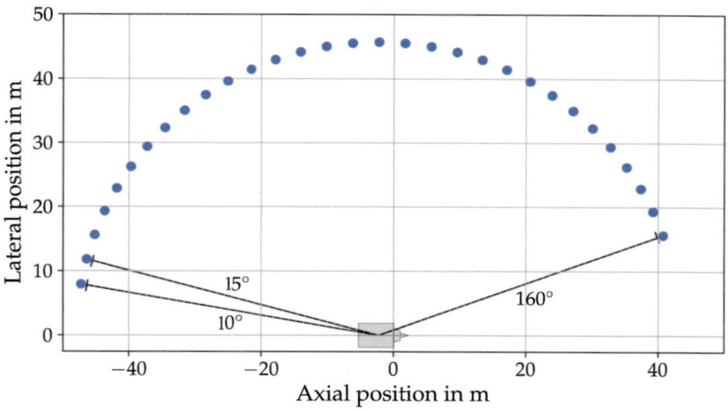

Fig. 4 A schematic of the setup for the static engine noise test with a far-field microphone array. The location of the engine is shown at the origin of the axis system with the engine inlet on the left-hand-side. The direction of flight is oriented towards the negative axial positions

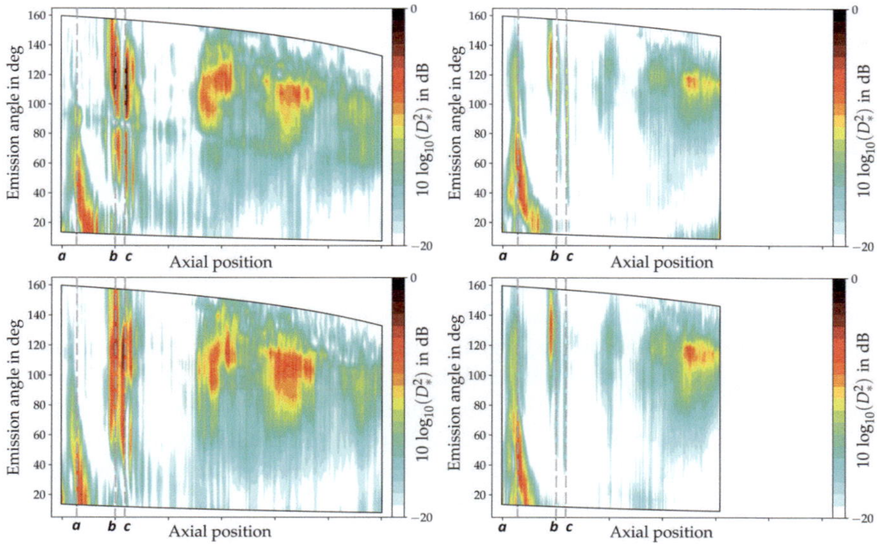

Fig. 5 SODIX sound source directivity maps calculated from the data measured with the polar arc array in the far-field. The sound source amplitudes D_*^2 are plotted as color maps in decibels with arbitrary scaling as a function of the axial position in arbitrary units on the horizontal and the emission angle in degrees on the vertical axis. Left column: for a one-third-octave band in the mid frequency range, right column: for a high frequency one-third-octave band; top row: without an intake liner, bottom row: with an intake liner. The positions of the engine inlet (**a**), the bypass nozzle (**b**) and the core nozzle (**c**) are marked with grey dashed vertical lines

emission angle of 5°, where the emission angle is defined as the angle between the engine axis in the direction of flight and the connecting line from a source point to a microphone (see Fig. 4). Flight coordinates are used, meaning that low values of the emission angle represent radiation in the direction of flight and high values to the rear, in the direction of the jet. Due to concerns of the industrial partners, absolute sound pressure levels, frequencies, and dimensions of the engine cannot be shown here. However, the qualitative values are fully sufficient in order to understand the results of the SODIX analysis.

4 Data Analysis and Results

SODIX has been used to analyze the data of the sparse far-field array. The focus of the analysis has been on the forward radiation from the engine intake in the angular range between 10° and 80°. The benefit of an acoustic liner could be detected in this range with the help of the SODIX analysis, whereas this had not been apparent in the direct analysis of the spectral data of the far-field microphones.

4.1 SODIX Source Maps

Figure 5 shows SODIX source directivity maps calculated from the data of the 31 microphones in the polar arc far-field array. There are four plots altogether, the top row presents the results measured without a liner in the engine intake duct, the bottom row results measured with an intake liner installed. Maps for two one-third-octave bands are presented: those on the left-hand side in the mid frequency range, those on the right-hand side for a high frequency band. Vertical gray dashed lines mark the axial positions of the intake (a), the bypass nozzle (b), and the core nozzle exit (c).

The maps have a dynamic range of 20 dB and the amplitudes are scaled relative to a common peak amplitude. Sound sources are localized at the engine intake, the bypass and the core nozzle exit. As can be expected, the sources located at the engine inlet radiate into the forward arc, in the direction of flight, namely at low emission angles. Sources at the bypass and core nozzles and in the jet radiate in the downstream direction with high emission angles. Sources at the relatively large engine intake and at the bypass nozzle are spread out around their true axial positions due to the parallax effect that occurs when a source distribution that has a significant extension in the radial direction is mapped onto a one-dimensional source grid. The true sources at the engine intake and bypass nozzle radiate at radial positions that are relatively far away from the engine center line—this causes an axial shift of these sources for observer angles other than 90°.

The localization results show that SODIX can be successfully applied to data from a sparse far-field array. Sound sources are localized at the proper positions at the intake, the bypass, and the core and the dynamic range is higher than than 20 dB.

4.2 Source Directivity

Figure 6 compares the polar directivities measured without and with the intake liner. The graphs show the overall sound pressure levels (OASPL) calculated directly from the far-field microphone data with closed circles for the case without the liner and with open circles with the liner. The SODIX solution that has been integrated over all source regions and extrapolated to the positions of the microphones in the far-field array is shown in black (solid lines without and dashed lines with the liner). The directivity of the SODIX solution for the source at the engine intake is shown in magenta.

The results are presented, again, for a one-third-octave band in the mid- and another one in the high frequency range. The range of the emission angles in Fig. 6 is limited to the forward arc, between 10° and 80°, where the effect of the intake liner can be expected to be strongest.

The difference between the sound pressure levels with and without the intake liner is between 1 and 2 dB for both frequency bands shown for both the OASPL calculated directly from the microphone data and for the SODIX total solution integrated over all sources and extrapolated into the far-field. The overall sound pressure level is dominated by contributions from sources located in the downstream direction, at the bypass and core nozzles and the jet, which mask the acoustic benefit of the liner. However, the SODIX calculation for the directivity of the source at the intake (the magenta lines in Fig. 6) shows an acoustic benefit of the liner even in directions where the OASPL in the far-field (black circles) are dominated by sources from the rear of the engine. In the mid-frequency range, the far-field extrapolated directivity of the intake source shows an acoustic benefit of the liner over the full angular range, with greater benefit at higher polar angles. In the high-frequency one-third octave band, the benefit of the liner becomes apparent for polar angles above 30°.

4.3 The Influence of the Point Spread Function on SODIX Results

The maps in Fig. 5 show spurious sound sources located downstream in the jet area, e.g. at the downstream end of the axial range at emission angles between 80° and 130°. These secondary sources cannot be related to actual noise sources and are caused by an aliasing effect due to the low spatial sampling of the sound field by the microphone array.

Figure 7 compares the point spread function of a linear array with 250 microphones to that of a sparse far-field array with only 31 microphones in the same mid range frequency band. These data were measured in a different experiment, where both near-field and far-field array had been installed. The SODIX results from the sparse far-field array (shown on the right) are degraded by grating lobes in the point spread function. These appear, when the distance between microphones is larger than half

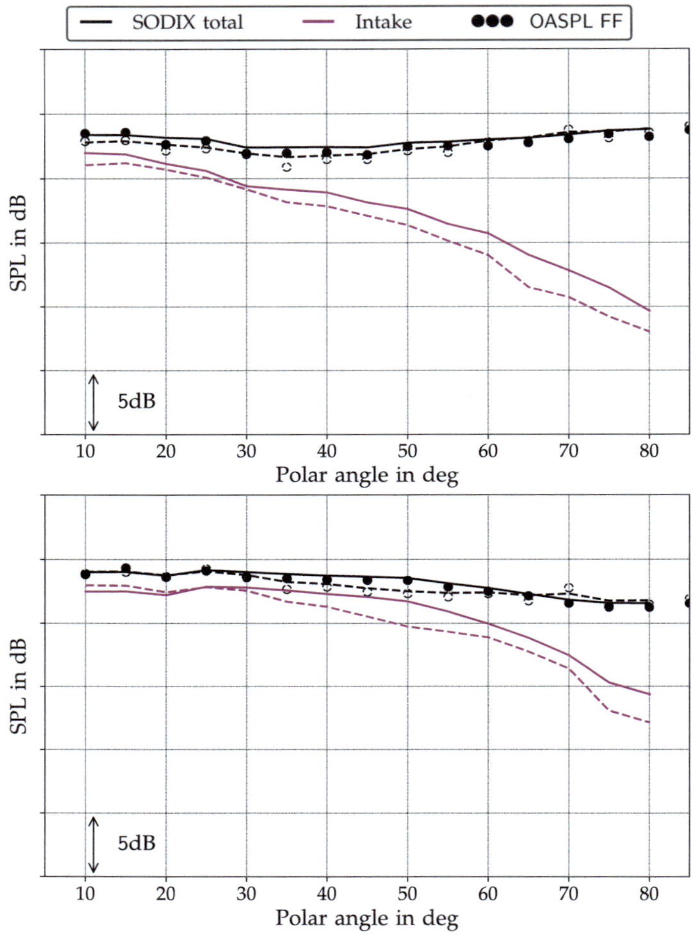

Fig. 6 The directivity of the sound field with the sound pressure levels in dB as a function of the polar emission angle of a short cowl engine in the forward arc measured without (solid lines) and with an intake liner (dashed lines). Top: for a one-third-octave band in the mid frequency range, bottom: in the high frequency range. The total SODIX solution has been integrated over all source positions and is compared to the SODIX result for the source region at the engine intake and to the overall sound pressure levels (OASPL) from the far-field (FF) microphones (closed black circles without, open circles with the liner)

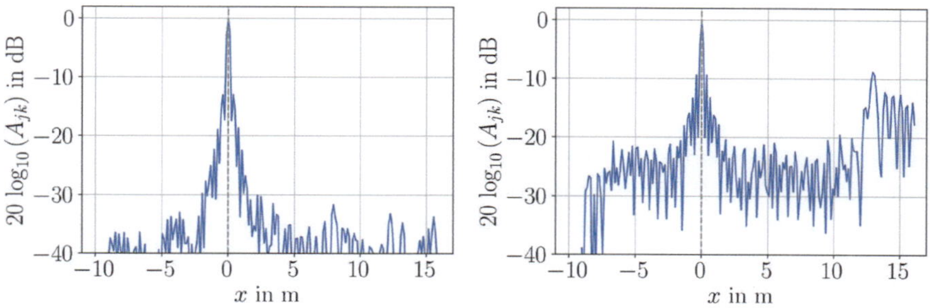

Fig. 7 The point spread functions for a densely populated linear near-field array (left) and a sparse polar arc array in the far-field (right). The x-axis is defined as in Fig. 4. The far-field array has a lower spatial sampling of the sound field which causes spatial aliasing and hence a lower signal to noise ratio

the wavelength of the sound waves. With the far-field array, grating lobes do occur at higher frequencies and are visible in the graph on the right-hand-side of Fig. 7 for $x > 10$ m. In order to avoid grating lobes, the array design for future measurements can be improved by adding more microphones to the far-field array and by optimizing their positions.

4.4 Evaluation of the Results of the SODIX Analysis

The results of the SODIX analysis of the data measured with the sparse far-field array are quite good, considering that the number of microphones is relatively low. Of course, better results can be obtained using a large linear near-field array, which should have about eight times the number of microphones.

However, the near-field array has other benefits compared to a dedicated linear array: it is usually permanently installed in static engine noise test facilities, while a linear array in the near-field of the engine needs to be set-up specifically for one test and stands in the way when work has to be done on the engine or the turbulence control screen. The possibility to use the far-field array data for a SODIX analysis is attractive to analyze available data from previous experiments or to perform simpler and cheaper experiments without having to set up a large near-field linear array with a large number of microphones.

5 Conclusions and Outlook

SODIX is an optimization method that reconstructs a distribution of directive sources that represent the cross-spectral matrix measured with a microphone array. So far,

SODIX has mainly being used to quantify the directive sound radiation from individual engine components in static engine noise tests. SODIX can be used to perform an acoustic source breakdown and a quantification of the contribution of individual engine components to the overall radiation into the acoustic far-field. A source breakdown can be performed to evaluate the acoustic effects of components such as liners that would have only a small impact on the overall sound pressure level in the far-field. The results of the source breakdown can be used to quantify the acoustic effects of different modifications of an engine. As an example, an application of the source localization method SODIX to far-field microphone array data from a static engine noise test provided by Airbus has been discussed in this chapter. SODIX is able to determine the positive effect of an intake liner using the data of the sparse microphone array in the acoustic far-field. The analysis shows good results in the forward arc, but is limited in the rearward arc by the low spatial sampling and by the influence of the point spread function of the microphone array.

Additionally, the understanding of spurious imaging characteristics within SODIX has been improved. It has been found that the point-spread function (PSF) is relevant for the source localization with SODIX. When the spatial resolution of the microphones is too low compared to the acoustic wavelength, aliasing occurs in the form of grating lobes. The appearance of grating lobes is more challenging than in beamforming approaches because the SODIX method has an energy-preserving formulation, which means that the energy in the grating lobes is taken away from the true sound sources, which are consequently underestimated. This has to be taken into account in evaluations of source localization results from data measured with sparse microphone arrays. Also, the design of microphone arrays for future static engine noise tests can be improved by adding more microphones to a standard far-field array. The application of SODIX to data measured with sparse far-field arrays promises substantial benefits in terms of reduced computational time and experimental complexity. A significant benefit is the possibility of analyzing existing data measured in previous static engine noise tests.

Abbreviations

CSM	Cross-spectral matrix
FF	Far-field
OASPL	Overall sound pressure Level in dB
PSF	Point-spread function

Acknowledgements Parts of this work have been funded by the Clean Sky 2 Joint Undertaking under the European Union's Horizon 2020 research and innovation program under grant agreement N° CS2-LPA-GAM-2020-2023-01.

Additional support has come from the EU Clean Aviation program under the grant agreement GA-101140627-COMPANION.

References

1. Funke S, Skorpel A, Michel U (2012) An extended formulation of the SODIX method with application to aeroengine broadband noise. In: 18th AIAA/CEAS aeroacoustics conference, 4-6 June 2012, Colorado Springs, USA. No. AIAA 2012-2276
2. Merino-Martínez R, Sijtsma P, Snellen M, Ahlefeldt T, Antoni J, Bahr CJ, Blacodon D, Ernst D, Finez A, Funke S, Geyer TF, Haxter S, Herold G, Huang X, Humphreys WM, Leclère Q, Malgoezar A, Michel U, Padois T, Pereira A, Picard C, Sarradj E, Siller H, Simons DG, Spehr C (2019) A review of acoustic imaging methods using phased microphone arrays. CEAS Aeronaut J 10(1):197–230
3. Michel U, Funke S (2008) Noise source analysis of an aeroengine with a new inverse method SODIX. In: 14th AIAA/CEAS aeroacoustics conference (29th AIAA aeroacoustics conference), May 5–7, 2008, Vancouver, British Columbia. No. AIAA 2008-2860
4. Oertwig S, Siller H, Funke S (2019) Advancements in the source localization method SODIX and application to short cowl engine data. In: 25th AIAA/CEAS aeroacoustics conference, 20–23 May 2019, Delft, The Netherlands, AIAA 2019-2743. https://doi.org/10.2514/6.2019-2743
5. Oertwig S, Siller H, Funke S (2022) SODIX for fully and partially coherent sound sources. In: 9th Berlin beamforming conference, 8–9 June, 2022, Berlin. https://www.bebec.eu/fileadmin/bebec/downloads/bebec-2022/papers/BeBeC-2022-S10.pdf
6. Oertwig S, Schumacher T, Siller HA, Funke S (2021) Extension of the source localization method SODIX for coherent sound sources. In: AIAA AVIATION 2021 forum. American Institute of Aeronautics and Astronautics. https://doi.org/10.2514/6.2021-2128
7. Oertwig S, Siller H (2023) Validation of the source localization method SODIX for coherent sound sources. In: 10th Forum acusticum, FA 2023, Torino, Italy. https://elib.dlr.de/202189/
8. Oertwig S, Siller H, Julliard E (2020) Quantifying the effect of an acoustic liner from far-field measurements in static engine noise tests. In: eForum Acusticum 2020. Lyon, France, pp 353–357. https://doi.org/10.48465/fa.2020.0719, https://hal.archives-ouvertes.fr/hal-03229458
9. Oertwig S, Siller HA, Schumacher T, Funke S (2022) Extension of the source localization method SODIX for the determination of partially coherent sound sources. In: 28th AIAA/CEAS aeroacoustics conference, June 14–17, 2022, Southampton, UK. https://doi.org/10.2514/6.2022-2812

MEMS Microphones for Wind Tunnel and Flight Tests

Thomas Ahlefeldt, Daniel Ernst, and Carsten Spehr

Abstract The application of micro-electro-mechanical systems (MEMS) microphones in the field of aeroacoustics has gained significant attention in recent years. Traditional measurement techniques often struggle with size, weight, and cost limitations, making MEMS microphones an attractive alternative. Nevertheless, the utilization of MEMS microphones for aeroacoustic measurements reveals some challenges to overcome regarding the field of application (i.e. aircraft, wind tunnel), the data acquisition, the calibration and typical implementation problems in general. The following article will briefly summarize the preparation and calibration of MEMS sensors for inflight aeroacoustic measurements as well as two flight testing campaigns performed on DLR aircraft.

Keywords Microphone array · Turbulent boundary layer · Kapton foil · Flight test · Wave number decomposition · Propeller noise · Spatial resolution · Flexible circuit board

1 Introduction

This paper summarizes the preparation and calibration of MEMS (Micro-Electro-Mechanical-Systems) sensors for aeroacoustic inflight testing [1] which was conducted in the framework of the EU-funded project ADEC and which introduced a new MEMS microphone array for in-flight measurements of the turbulent boundary layer. The pressure fluctuations here are either caused by the turbulent boundary layer (TBL) itself or directly by acoustic waves which can themselves be altered by their transmission through the TBL. The measurements can be used for the experimental observation of fuselage excitation or even in-situ measurements of the sound radiation of aircraft engines. When looking into the near future, new engine features for commercial aircrafts are needed to reduce the climate impact of air traffic.

T. Ahlefeldt (✉) · D. Ernst · C. Spehr
German Aerospace Center (DLR), Institute of Aerodynamics and Flow Technology, Göttingen, Germany
e-mail: Thomas.Ahlefeldt@dlr.de

F. Boden (ed.), *Numerical Methods and Experimental Means for the Integration of Future Propulsion Systems*, Springer Aerospace Technology,
https://doi.org/10.1007/978-3-031-97725-1_10

These new features will have different sound radiation patterns which may cause a different acoustic footprint in the vicinity of the airport or on the fuselage. The aim of the ADEC project was to prepare the measurement technique to meet these new challenges. This task requires pressure sensors which are suitable for the rather harsh conditions on an aircraft fuselage in flight, and they have to be small enough to spatially resolve the fluctuating pressure structures of interest. The dynamic range must resolve both the highest-pressure levels underneath the turbulent boundary layer as well as lower pressure levels of higher wavelength which may also be important to determine the sound radiation from the engine or the fuselage. This paper presents insights into the preparation and evaluation of a flight test with a phased MEMS microphone array, which had been developed in-house. After a short introduction regarding the state of the art, the paper provides the challenges of measuring in a flow using sensors of different sizes (Sect. 1.2). The in-house MEMS pressure sensor array is introduced in Sect. 2.1, followed by the ground tests conducted with and without flow (Sect. 2.2). The setup for the first in—flight measurements with an analog array is presented in Sect. 2.3, and subsequently with a digital MEMS array in Sect. 3, including selected results demonstrating the feasibility and performance of the introduced MEMS microphone array.

1.1 State of the Art

Until 2020 the state-of-the art measurement technology to measure pressure fluctuations was to use piezo-resistive pressure transducers (e.g. Kulite), as in in-flight measurements [3–6, 9, 15] or with thin surface condenser microphones [1]. A new type of microphone, first described in 1983 by Hohm [7] and Royer [11], was the silicon microphone; these were intensively developed in the following years and—due to their small size and low costs—used especially in consumer electronics such as smart phones. To improve the performance in aeroacoustic measurements, Sheplak et al. developed custom MEMS microphones [14]. This customized sensor had an upper frequency limit of 6 kHz with a noise floor of the packaged sensor of 92 dB/Hz^2 and a maximum sound pressure level of 155 dB.

A summary of the most important publications on the developments of custom aeroacoustic MEMS microphones is given in Williams [16] and Reagan [10], where both authors had further developed custom MEMS microphones to meet the requirements of aeroacoustic measurements. In 2022 Erbacher et al. [2] presented a piezoresistive MEMS microphone sensor concept with an upper sound pressure level of 175 dB. However, these sensors are neither available nor in production.

In 2021 Ahlefeldt et al. [1] described first in-flight measurements with MEMS microphones based on commercially available MEMS sensors with analogue output. The MEMS microphones in this study were installed with a Kapton foil cover of 25 μm thickness. This foil attenuates the pressure excitation by approximately 38 dB. Calibration measurements of the MEMS microphone included the influence of this attenuation by the cover foil.

1.2 Sensors in Flow

The size of the sensor in relation to the size of the turbulent structures in the boundary layer governs the output of the measured pressure spectra. Where turbulent structures are small compared to the sensing surface of the sensors, they can easily be overlooked due to the averaging effect of the larger sensing surface. Schewe [13] describes this effect and introduced a dimensionless sensor diameter $d^+ = du_\tau/\nu$ with the wall shear velocity u_τ and the kinematic viscosity ν. He found that the strength of the measured pressure fluctuations decreases with this dimensionless sensor diameter. Furthermore, the measured pressure fluctuations are connected to rare high-power events, indicated by an increasing flatness of the probability density function of the pressure fluctuations with the dimensionless sensor size. Therefore, the small size of the MEMS sensors affects the measurements in the following aspects:

- the low price allows the use of a high number of sensors, increasing the accuracy of wavenumber spectra (by decreasing the noise floor);
- it allows a high spatial resolution due to a smaller averaging sensing surface;
- it allows a high spatial resolution due to smaller spacing between the sensors;
- the decreased damping of higher wavenumbers increases the probability of small scale turbulent structures exceeding the overload point of the sensors;
- the attenuation of surface pressure fluctuations as a function of the pressure-wavelength has to be taken into account.

The attenuation of surface pressure fluctuations can be modeled as a function of frequency at different Mach numbers and sensitive surface radii. Ko [8] gave the effect of a cosine-shaped surface pressure fluctuation over a circular shaped transducer by

$$s(k, r) = \frac{2J_1\,(k \cdot r)}{k \cdot r}, \tag{1}$$

Here, the gain $s(k, r)$ is given as a function of the size ratio $a = k \cdot r$, with sensitive surface radius r, wavenumber k of the surface pressure fluctuations and with $J_1(...)$ being the Bessel function of first kind of order one.

In Fig. 1 the resulting gain $s(k, r)$ from Eq. (1) is shown in dB for a range of frequencies for Mach numbers of M = 0.2 and M = 0.3. The phase propagation wavenumber k_p of the TBL is used for the relation of sensor size and wavenumber in Eq. (1) by using $k_p = f/u_p$, with u_p being the phase velocity and f the frequency. The signal phase on the surface changes by a factor of $u_p \approx 0.7u$, slower than the flow velocity u_∞, defined by the Mach number $\mathrm{M} = \mathrm{u}_\infty/\mathrm{c}_0$, where $c_0 = 340\,\mathrm{m/s}$ is the speed of sound. Figure 1 shows that the gain loss does not exceed 2 dB below 10 kHz even over the largest sensitive surface area ($d = 12.7\,\mathrm{mm} = 1/2$ inch) at a Mach number of M = 0.2.

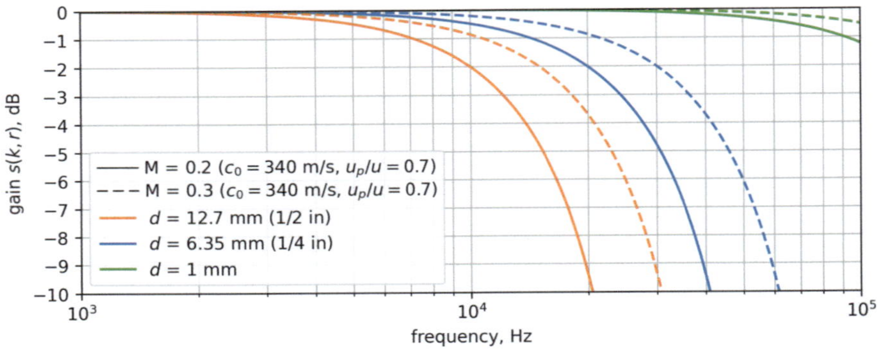

Fig. 1 Modeled attenuation of hydrodynamic surface pressure fluctuations as a function of frequency at different Mach numbers and sensitive surface diameters [1]

2 Analogue MEMS Array

2.1 *MEMS Array Structure*

This section describes the analogue MEMS array, used in the flight test on a Dornier 228-101 aircraft (see Sect. 2.3). This array used MEMS sensors of type ICS-40617 [17] with an analog output. The size of the MEMS microphone is 3.5 mm by 2.65 mm with a thickness of 0.98 mm. The diameter of the sensor port and the diaphragm is 1 mm. The dynamic range of 111 dB is rather high for a MEMS microphone and the data sheet gives an Acoustic Overload Point (AOP) of 129 dB with 10 % THD (total harmonic distortion). The MEMS microphone array is composed of a flexible Printed Circuit Board (PCB) and can therefore be applied onto curved surfaces. It is made up of 8 layers (two outer layers for shielding) with a size of 400 mm × 300 mm as depicted in Fig. 2 (left side). The total thickness of the array is 1.45 mm made up of the thickness of the PCB at 0.45 mm plus that of the MEMS sensor. The MEMS sensors used in the array are of bottom port type, and the sensor ports are aligned with pass-through holes of 1 mm diameter. The rear side of the array was cast with silicone to protect the components and to maintain a uniform height of the entire array of 1.55 mm. The MEMS microphone array on the flexible PCB reacted well to bending (see Fig. 2 (right side)) without any damage occurring to the solder joints on the circuit board.

The array contains 45 MEMS microphones arranged in a spiral, with which the focus was on demonstrating the applicability of MEMS microphones to flight testing. The data were recorded via the data acquisition system GBM Viper-HDR, which is capable of recording 64 channels at sampling rates up to 250 kHz with a bit depth of 24 bit. The data acquisition system was used with an internal second-order high-pass filter with a cutoff frequency of 500 Hz prior to the A/D conversion.

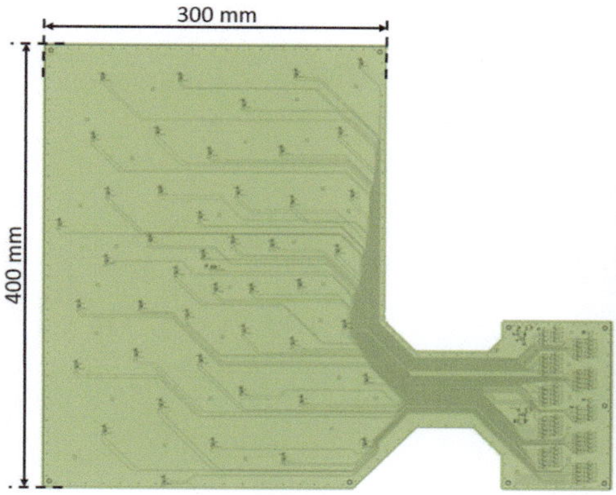

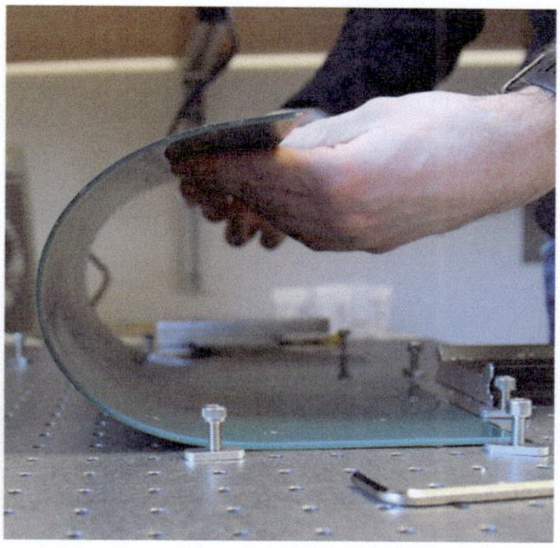

Fig. 2 Left: Schematic of the printed circuit board with dimensions. Right: Bending demonstration of the array [1]

2.2 *Preliminary Tests*

Three different tests were performed in preparation for the flight test to ensure the functionality and to calibrate the frequency response of the MEMS sensors in the array:

- the in-situ frequency response of the individual microphones was determined without flow in an anechoic chamber;
- the influence of the foil on the frequency response was measured and calibrated;
- validation measurements at comparable Mach-number in a transonic wind tunnel were performed to assess the influence of the foil and flow on the MEMS microphone frequency response and compared to a reference sensor.

2.2.1 Frequency Response

The frequency response of the MEMS sensors was measured in the anechoic chamber of the aeroacoustic laboratory of DLR Göttingen and compared to the data sheet [17]. The measurements used two loudspeakers: a calibration speaker (B&K Omnisource 4295) for frequency range up to 6 kHz and a tweeter (ELAC-4Pi-II) for the frequency range from 5 to 100 kHz. The MEMS microphone array was positioned at a 3 m distance to the speakers together with a condenser microphone (B&K type 4944) with known frequency response as reference. The frequency response H was calculated in comparison of the reference microphone spectrum G_Y to the spectra of the MEMS array microphones G_X:

$$H(f) = \frac{G_X(f)}{G_Y(f)} \tag{2}$$

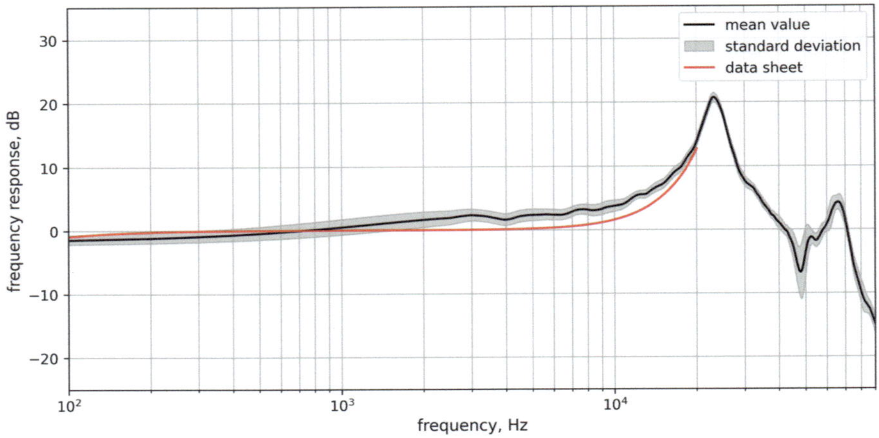

Fig. 3 Mean frequency response of all array sensors with standard deviation in comparison with the data sheet values [1]

The measurement time was 30 s at a sampling frequency of 200 kHz. The data was then processed using an overlap of 50% and a fast Fourier transform block size of 5,000 samples, with a Hann window, yielding 2,399 averages and a narrow-band frequency resolution of 40 Hz. The calculated results were then smoothed with a Savitzky-Golay filter [12] with a window length of 51 and a polynomial of third order. Figure 3 shows the frequency response from 100 to 90 kHz, where the results with the two different loudspeakers have been merged. Between 300 and 10 kHz the frequency response is approximately constant, rising from −1.5 dB to approximately 3 dB. The resonance peak of the MEMS microphones appears at 22 kHz, followed by a deep drop of the frequency response at 44 kHz. The standard deviation of the frequency response of the MEMS microphones is approximately 1 dB. The small differences between measured frequency response and the data from the data sheet are probably due to limited accuracy in the measurement setup in the DLR anechoic chamber.

2.2.2 Surface Treatment: Application of Kapton Foil

For the test under flow conditions the port holes of the MEMS microphones were covered with one-sided adhesive Kapton foil of 25 μm thickness. This has the positive side-effect of protecting the MEMS microphones on the fuselage from humidity and dust particles.

The impact of this foil on frequency response was assessed in the laboratory. A rectangular plate with surface-mounted MEMS microphone array and a reference condenser microphone (B&K type 4944) was used, see Fig. 4. An Adam A3X speaker

Fig. 4 Photo of the measurement setup for the frequency response of the Kapton foil application. The MEMS microphone array is mounted onto a solid plate together with flush-mounted reference microphones [1]

produced sound in the 50 Hz to 60 kHz range. Kapton foil response was determined by comparing measurements with and without foil relative to the reference microphone. The transfer function H_{K} of Kapton foil was calculated by comparing cross-spectra R between MEMS microphones with Kapton (X_{K}) and without Kapton ($X_{\not{\mathrm{K}}}$) using the reference microphone Y.

$$H_{\mathrm{K}} = \frac{R_{X_{\mathrm{K}},Y}(f)}{R_{X_{\not{\mathrm{K}}},Y}(f)} \tag{3}$$

The 24-bit GBM Viper-HDR data acquisition system was used. Measurements lasted 60 s at 200 kHz and were subsequently processed with a fast Fourier transform block size of 5,000, with 50% overlap, a Hann truncation window, yielding 4,799 averages, with a narrowband frequency resolution of 40 Hz. Results were smoothed with a Savitzky-Golay filter [12] (window length: 51, polynomial third order). To examine the effect of means of application, Kapton foil was applied a second time with a felt card as a means of obtaining contact pressure—measurement #2. Measurement #1 used the exact foil as in the final flight test, without removal between applications.

Figure 5 presents the frequency response with confidence intervals for both measurements. Both show an approximately flat attenuation of −4 to −38 dB for measurement #1 and to −34 dB for measurement #2. After a minimum at 23 kHz, attenuation drops to about −10 to −5 dB at 36 kHz. Subsequently, it fluctuates in the −10 to −25 dB range with varying trends up to 60 kHz. This demonstrates that Kapton

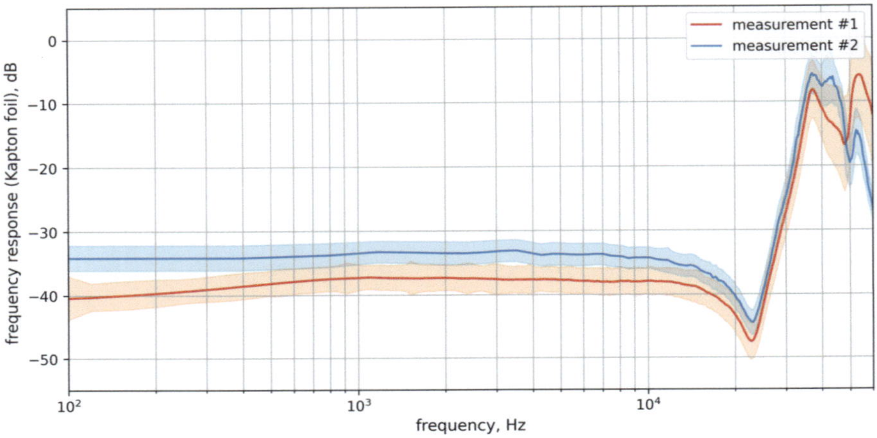

Fig. 5 Mean frequency response of all array sensors treated with Kapton foil with standard deviation for two different applications of the coating with the Kapton foil (measurement #1: exact treatment as used in flight; measurement #2: with felt card—see text). The standard deviations are as given by the respective envelopes [1]

foil application results can differ based on the method used (pressure tool, contact pressure). Therefore, the transfer function should be individually measured for each Kapton foil application.

2.2.3 Validation Measurements (TWG)

Validation measurements at increased Mach numbers were conducted in the Transsonic Wind Tunnel Göttingen (DNW-TWG), a facility at the DLR Göttingen site; DNW-TWG features a 1.0 m by 1.0 m closed test section, operating from M = 0.3 to M = 2.2 under vacuum and pressurized conditions. The Kapton-foil-treated MEMS microphones and a 1/2 in. B&K microphone (type 4134) were compared at Mach 0.3, flush-mounted into the wind tunnel wall (Fig. 6). The 16 bit GBM Viper-48 data acquisition system was used with a 500 Hz high-pass filter, and results were corrected for the filter response.

Figure 7 presents spectra of a Kapton-foil-treated MEMS microphone and a reference microphone. The black line represents the reference microphone spectrum. To illustrate frequency response corrections (explained in Sect. 2.2.1 for the MEMS microphone and Sect. 2.2.2 for Kapton foil), three MEMS spectra are shown: uncorrected (red dotted line), Kapton-foil-corrected (red dashed line), and fully corrected (red solid line). The uncertainty range arising from the combined standard deviations of the correction measurements is depicted by the orange band envelope.

From 100 Hz to 3 kHz, both spectra (MEMS microphone and reference) align well, both with a slight 10 dB downward slope. Peaks at 500 Hz, 1 kHz and a double peak at 2.5 kHz can be seen. Above 3 kHz, the reference microphone spectrum

Fig. 6 Photos of the microphones on the wind tunnel wall. Left: MEMS microphone array with Kapton foil. Right: flush mounted reference 1/2 in. microphone [1]

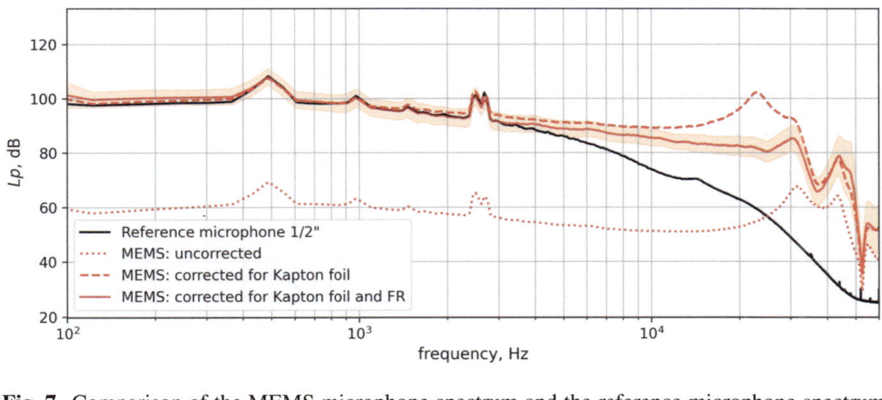

Fig. 7 Comparison of the MEMS microphone spectrum and the reference microphone spectrum for measurements in the TWG at a Mach number of 0.3: a plot of the sound power level L_P versus frequency. The orange envelope illustrates the combined error margin obtained from the Kapton foil and frequency response measurements [1]

gradually decreases up to 500 kHz, flattening out at 25 dB by 60 kHz. In contrast, the MEMS microphone spectrum has a lower downward slope up to 25 kHz, after which it oscillates and decreases, finally reaching 50 dB at 50 kHz. The broadband spectra are primarily dominated by boundary layer noise. Peaks at 500 Hz, 1 kHz, and around 2.5 kHz are assumed to be acoustic in nature. The onset frequency of the strong downward slope is in agreement with modeled attenuation for corresponding sensor sizes. The estimated amplitude attenuation due to surface integration shows less attenuation than observed between MEMS and reference microphone signals; the remaining difference is attributed to the limited coherence length of TBL pressure fluctuations. The Kapton foil correction shifts the spectrum up by approximately 40 dB. With only the foil correction, the MEMS and reference microphone spectra align well below 3 kHz. A peak at 22 kHz is due to the resonance frequency of the MEMS microphones, which has, however, been removed by applying the MEMS calibration curve up to 26 kHz. Beyond 26 kHz, the spectrum exhibits a wavy response, indicating the limited meaningful application of corrections beyond this frequency.

2.3 Application of Analogue MEMS for In-Flight Measurements

Experiments were conducted on a DLR test carrier, the Dornier 228-101 propeller aircraft with the call sign D-CODE (see Fig. 8).One window bank on the port side was fitted with a dummy window to accommodate the cabling for the MEMS array mounted on the fuselage, opposite to the propeller. The fuselage, coated with flight-approved aluminum tape, featured a MEMS microphone array mounted via adhesion between the tape and silicone on the rear side of the PCB, which was also coated with silicone. To attenuate pressure excitation, the array was covered with a Kapton foil as described in Sect. 2.2.2 and secured with flight-approved tape around the edges. A 1/2 in. aerospace surface microphone array (B&K type 4948) was mounted on the fuselage as a reference. All signals were routed to the 24-bit GBM Viper-HDR data acquisition system inside the aircraft, using a 500 Hz high-pass filter. Recorded simultaneously at a 250 kHz sampling frequency, spectra will be presented up to 5 kHz, reducing low-frequency boundary layer noise with the aforementioned 500 Hz high-pass filter. The acquired data were corrected for the filter response and processed using a 50% overlap and a fast Fourier transform block size of 65,536 samples, applying a Hann truncation window for a narrow-band frequency resolution of 3.91 Hz.

The spectral content of signals from both reference and MEMS microphones, presented in an array, provides valuable insights into surface pressure fluctuations. Figure 9 compares the mean spectra of reference and MEMS microphones from a single in-flight measurement. Both spectra exhibit a general broadband decrease over frequency with a dominant peak at 133 Hz, and including its overtones. This peak—likely of acoustic origin—is in agreement with the propeller rotational speed of approximately 1,590 rpm, which results in a Blade Passing Frequency (BPF) of about 133 Hz.

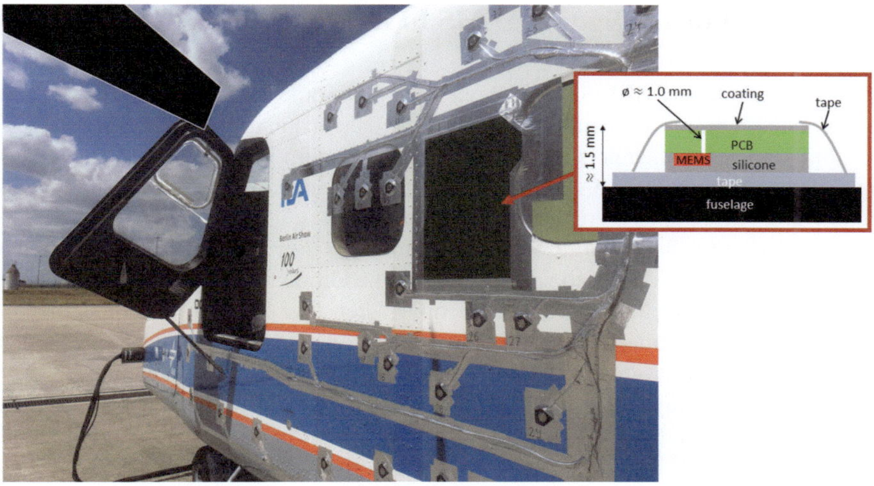

Fig. 8 Photo of the propeller aircraft with the MEMS microphone array mounted on the fuselage and a sketch depicting the MEMS microphone array mounting [1]

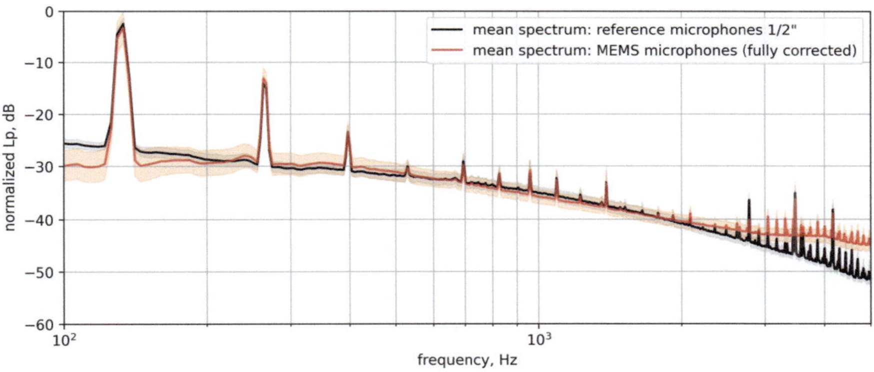

Fig. 9 Comparison of the mean spectra of the MEMS microphones and the reference microphones with standard deviation, in a plot of normalized sound power L_P versus frequency [1]

While the dominant tonal components are similar for both microphones, differences appear in the broadband signal shape. Above 2 kHz, the reference microphone shows a steeper drop in signal amplitude compared to the MEMS microphone. This amplitude reduction at higher frequencies is attributed to surface integration effects, as described for the larger microphone in Sect. 1.2 and as illustrated in Fig. 1.

3 Application of Digital MEMS to In-Flight Measurements

The use of analogue MEMS microphones also has some disadvantages. First, each microphone requires a separate data line to the data acquisition system. In addition, the total number of channels is limited by the maximum number of available channels of the data acquisition system. This can be remedied by using a digital signal path. Here, two options were available: Option 1 lies in using digital MEMS where the A/D converter has been directly integrated into these MEMS. Option 2 is the use of analogue MEMS sensors followed by an additional A/D conversion directly connected to the MEMS board. At the time when this decision had to be made, the commercially available analogue MEMS sensors had better acoustic properties (AOP), which is why a decision was made in favor of option 2. This new type of acoustic MEMS arrays was developed in a collaborative project between gbm gmbH (Software), D-Signt (hardware) and DLR in the framework of the nationally funded LuFo project "FlexMEMS". For this work, these arrays were used in the flight tests. As for the analogue MEMS array, a thin design had been used to carry out measurements in thin fuselage boundary layers, thus enabling relevant flow structures to be recorded and characterized.

The relevant equipment for the flight test consists of the following (see Fig. 10): (1) three MEMS panels; (2) a Digital Signal Processing unit (DSP) connected behind each panel; (3) a concentrator module for collecting the data of the different arrays and synchronizing them; (4) a PC for recording measurement data.

Each array has 32 sensors, which means that a total of 96 sensors were used for these measurements. The data from all three arrays was digitised by an A/D conversion included on the PCB, after which the DSP unit then passed on the data

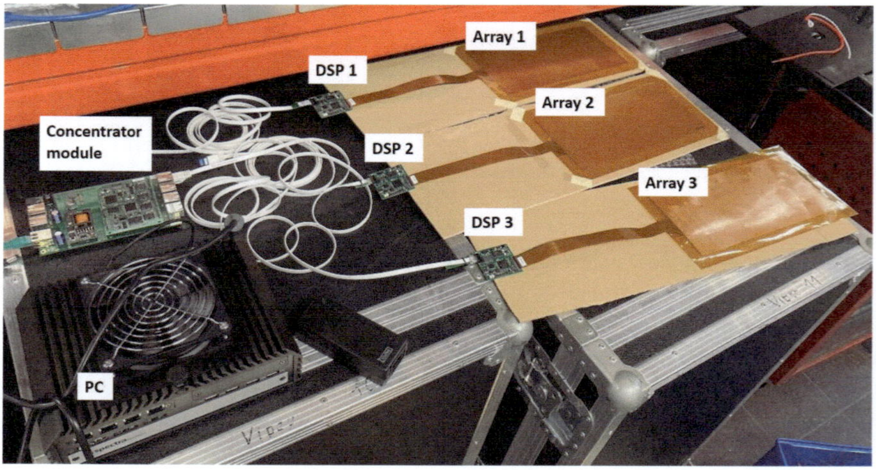

Fig. 10 Laboratory setup of the MEMS system with 3 microphone arrays, the concentrator module and the measurement PC

Fig. 11 Photo of the ISTAR airplane at take-off with the 3 microphone arrays mounted on the fuselage: in front of the first window and above the cockpit

over the network to the concentrator module, before being finally stored on the data acquisition PC with a sampling rate of 20 kHz. For the flight test, these three MEMS array foils are to be adhered to the fuselage of the ISTAR Dassault Falcon 2000LX with the call sign D-BDLR (see Fig. 11). The setup work encompassed: (1) a flow panelling using silicone encapsulation of the array and tape; (2) an EMC panelling of the DSP and the concentrator module; (3) a preceding function test in a low temperature chamber at −60 °C; (4) the cable routing. The PC for the data acquisition was integrated into a flight-test rack inside the cabin.

The signals were analysed for the basic functional test and a simple wave number evaluation was carried out independently of the flight status. The analysis of the signals showed in general good results, but nevertheless also regular signal failures, especially during longer measurement times and at high speeds. In the hope of finding the cause for these, a future metadata analysis will be carried out to correlate these failures with the flight data. Figure 12 shows a representative result from a wave number analysis, with a clearly identifiable convective ridge at $k_x/k_0 \approx 2.5$.

4 Summary

This chapter outlines the preparation of MEMS microphones in a flexible circuit board array for measuring unsteady pressure fluctuations on an aircraft fuselage

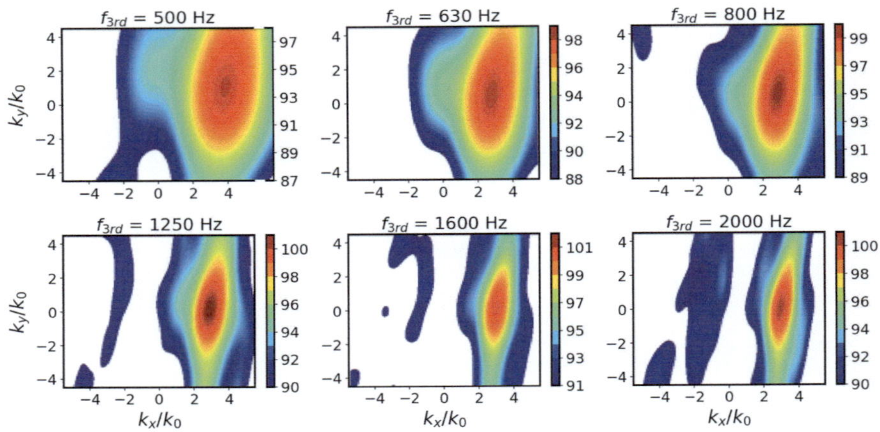

Fig. 12 Wave number spectra of one array in front of the windows at selected third-octave frequency bands with a dynamic range of 10 dB

during flight tests. In low-speed wind tunnel tests, the small sensitive surface area of the MEMS sensors captured high-energy small-scale structures within the turbulent boundary layer. The measured high amplitudes, resulting from the matched size ratio of turbulent boundary layer pressure fluctuations to the sensor sensitive surface area, can induce nonlinear behavior which exceeds the dynamic range. To address these limitations, Kapton foil of 25 μm thickness was applied to decrease the sensitivity of the MEMS array. The Kapton-covered MEMS were tested in a high-speed wind tunnel, demonstrating good agreement with reference signals after applying frequency response corrections. In-flight tests on a DLR Dornier 228 propeller aircraft showed good signal agreement with the 1/2 in. reference microphones, especially after having applied the frequency corrections. To overcome the limitations of analogue MEMS—such as the total number of channels being limited by the maximum number of channels available in the data acquisition system—a digital MEMS setup has also also been developed and tested in flight. Here the whole setup showed good functionality in general, but also regular signal failures related to the flight conditions. As with the analogue MEMS array, the Kapton film effectively damped pressure fluctuations, but nevertheless could also preserve critical phase information for wave number analysis. MEMS microphones, with their small size, close spacing and flush-mounted arrangement, offer advantages for spatially high-resolution measurements during challenging flight tests. The wavenumber analysis from flight test measurements highlights the presence of acoustic and hydrodynamic pressure fluctuations, showcasing MEMS microphones as a cost-effective alternative for such experimental conditions.

Abbreviations

A/D	Analogue/Digital
AOP	Acoustic Overload Point
A/D	Analogue/Digital
D-CODE	DLR research aircraft (Dornier DO 228-101)
DLR	German Aerospace Center
DNW	German-Dutch Wind Tunnels
EMC	Electromagnetic compatibility
EU	European Union
ISTAR	DLR research aircraft (Dassault Falcon 2000LX)
MEMS	Micro-Electro-Mechanical Systems
PCB	Printed Circuit Board
PC	Personal Computer
TBL	Turbulent Boundary Layer
THD	Total Harmonic Distortion
TWG	Transonic Wind Tunnel Göttingen

Variables

ν	Kinematic viscosity
c_0	Speed of sound
d	Diameter
d^+	Dimensionless sensor diameter
f	Frequency
H	Transfer function frequency response
J_1	Bessel function of first kind of order one
k	Wave number
k_p	Phase propagation wavenumber
L_P	Sound power level
M	Mach number
R	Cross-spectra
r	Radius
s	Gain
u	Velocity
u_∞	Free stream velocity
u_P	Velocity at surface
u_τ	Wall shear velocity
x, y, z	Spatial coordinates
X, Y, G	Spectra

Subscripts

K	With Kapton foil
~~K~~	Without Kapton foil
x, y, z	Spatial coordinates

Acknowledgements This project has received funding from the Clean Sky 2 Joint Undertaking under the European Union's Horizon 2020 research and innovation program under grant agreement N° CS2-LPA-GAM-2020-2023-01.

References

1. Ahlefeldt T, Haxter S, Spehr C, Ernst D, Kleindienst T (2021) Road to acquisition: preparing a mems microphone array for measurement of fuselage surface pressure fluctuations. Micromachines 12(8):961. https://doi.org/10.3390/mi12080961
2. Erbacher KJ, Ngo HD, Mackowiak P, Wu L, Julliard E, Spehr C (2022) Design and modeling of a novel piezoresistive microphone for aero acoustic measurements in laminar boundary layers using FEM and LEM. https://doi.org/10.2514/6.2022-1023
3. Haxter S (2016) Extended version: improving the DAMAS 2 results for wavenumber-space beamforming. J Sound Vib
4. Haxter S, Spehr C (2017) Comparison of model predictions for coherence length to in-flight measurements at cruise conditions. J Sound Vib 390:86–117. https://doi.org/10.1016/j.jsv.2016.10.038
5. Haxter S, Spehr C (2017) Determining flow propagation direction from in-flight array surface pressure fluctuation data. https://doi.org/10.2514/6.2017-3205
6. Haxter S, Spehr C (2021) Wavenumber characterization of surface pressure fluctuations on the fuselage during cruise flight. Springer International Publishing, pp 157–180. https://doi.org/10.1007/978-3-030-64807-7_8
7. Hohm D, Sessler GM (1983) An integrated silicon-electret-condenser microphone. In: Proc. 11th Congr. Acoust., Paris, vol 6, pp 29–32
8. Ko SH (1993) Performance of various shapes of hydrophones in the reduction of turbulent flow noise. J Acoust Soc Am 93(3):1293–1299. https://doi.org/10.1121/1.405414
9. Palumbo D (2012) Determining correlation and coherence lengths in turbulent boundary layer flight data. J Sound Vib 331(16):3721–3737. https://doi.org/10.1016/j.jsv.2012.03.015
10. Reagan T (2017) MEMS on a plane: a flush-Mount MEMS piezoelectric microphone for aircraft fuselage arrays. PhD thesis, University of Florida Gainesville, FL, USA
11. Royer M, Holmen JO, Wurm MA, Aadland OS, Glenn M (1983) ZnO on SI integrated acoustic sensor. Sens Actuators 4:357–362. https://doi.org/10.1016/0250-6874(83)85044-6
12. Savitzky A, Golay MJE (1964) Smoothing and differentiation of data by simplified least squares procedures. Anal Chem 36(8):1627–1639. https://doi.org/10.1021/ac60214a047
13. Schewe G (1983) On the structure and resolution of wall-pressure fluctuations associated with turbulent boundary-layer flow. J Fluid Mech 134(-1):311. https://doi.org/10.1017/s0022112083003389
14. Sheplak M, Seiner J, Breuer K, Schmidt M (1999) A mems microphone for aeroacoustics measurements. https://doi.org/10.2514/6.1999-606

15. Spehr C, Hennings H, Buchholz H, Bouhaj M, Haxter S, Hebler A (2012) In-flight sound measurements. A first overview. https://doi.org/10.2514/6.2012-2208
16. Williams MD (2011) Development of a MEMS piezoelectric microphone for aeroacoustic applications. PhD thesis, University of Florida
17. ICS-40618 datasheet. https://invensense.tdk.com/wp-content/uploads/2016/02/DS-000044-ICS-40618-v1.0.pdf. Accessed 24 Feb 2021

Experimental Vibroacoustic Characterization of an A320 Wing Utilizing a Controlled Shaker System for Artificial Engine-Vibration Excitation

Marco Norambuena, Sebastian F. Zettel, Ray D. Dewald, and René Winter

Abstract The wing is the main transfer path for engine-induced vibrations into the fuselage. Therefore, investigating the high-frequency structural behavior of wings is important for cabin acoustics. In cooperation with Airbus, DLR has conducted a vibroacoustic experiment on an actual A320 wing. The goals were to create an updated numerical model of the wing and to quantify the vibrational power input into the wing suspension utilizing structural intensity calculated from measurement data. Additionally, a novel multi-axial excitation system, capable of performing an excitation about multiple axes simultaneously, has been utilized. With this system a realistic wing response due to engine-induced vibrations during in-flight operation has been generated.

Keywords Vibroacoustic · Wing structure · Experimental · Multi-axial excitation system

1 Introduction

In recent years [3, 17], vibroacoustic topics have drawn considerable attention in the aeronautic industry. This is because emphasis has evolved from safety and reliability topics of previous decades of research to cabin quality and passenger comfort. This current trend has pushed for the development of new tools and methodologies in

All authors have equally contributed to this work.

M. Norambuena (✉) · S. F. Zettel · R. D. Dewald · R. Winter
German Aerospace Center (DLR), Institute of Aeroelasticity, Göttingen, Germany
e-mail: marco.norambuena@dlr.de

S. F. Zettel
e-mail: sebastian.zettel@dlr.de

R. D. Dewald
e-mail: ray.dewald@dlr.de

R. Winter
e-mail: rene.winter@dlr.de

F. Boden (ed.), *Numerical Methods and Experimental Means for the Integration of Future Propulsion Systems*, Springer Aerospace Technology,
https://doi.org/10.1007/978-3-031-97725-1_11

order to tackle these new challenges [26]. A set of refocused topics has emerged from the interest in understanding how vibroacoustics play a crucial role in cabin noise.

One of the problems commonly faced in numerical modeling of vibroacoustics is the limited availability of suitable models; for example, with sufficiently high resolution to treat high frequency applications, or with accurate models which have been sufficiently validated by measurements. The advantage of having such high-quality models opens the full potential of digital design as well as evaluation of components and systems interfaces. For example, if a precise model of an aircraft wing exists, this will enable the study of new attachment interfaces between pylon-wing or wing-fuselage in order to minimize vibrational energy that subsequently transforms into radiated noise into the cabin.

A fundamental requirement to obtain any kind of validated model is to perform measurements of a structure that captures their behavior in the most realistic way. One could easily argue, for instance, that flight test data provides the best type of excitation since it is in fact an operational condition; however, the elevated costs associated with flight tests make this option unfeasible and unaffordable in many cases. Moreover, this type of operational excitation, although ideal for many applications, is not fully known and difficult to quantify since there are multiple sources acting simultaneously on the fuselage during flight. The unavailability of such excitation reference will limit the types of processing that can be performed.

If one neglects for the moment the effect of the Turbulent Boundary Layer during flight, the major source responsible for the vibrations on the fuselage at low and middle frequencies is the loads generated by the rotating components of the engines [17]. Given the fact that the starting point of almost all vibroacoustic phenomena is structure-borne propagation, it becomes crucial to have complete control and knowledge of the excitation characteristics in order to perform any meaningful vibroacoustic and cabin noise research.

In order to achieve this, it would be most desirable to have a suitable structural excitation system capable of reproducing a real engine vibration signature. Such a characteristic will then enable the replication, for instance, of the vibrational operational conditions of an entire aircraft, its subcomponents, such as a wing, or demonstrator platforms. From the vibroacoustic point of view, a system with these characteristics could eventually replace flight tests by laboratory testing.

This work describes the measurement campaign performed on an A320 wing and the subsequent analysis of the acquired data. The focus of this project was set on three key elements that play an important role in vibroacoustic research: on the numerical side, the modal characterization and the assessment of power flow through the structure by means of structural intensity (STI); on the experimental side, the application of a novel multi-axial excitation system (MAES) to a real scale environment.

Within the framework of Clean Sky 2, Airbus Operation in Bremen provided access to an A320 wing that had been decommissioned and was currently used for testing of hydraulic systems and actuators. From the structural point of view, the wing was in perfect condition for the proposed project; the only further prerequisite was having to build a suspension system to keep the wing in a position similar to its normal operation on an aircraft in flight.

The wing of an aircraft can be considered as the only transfer path of vibrational energy between the engines and the fuselage. The engines induce vibrations on the wing structure and a significant part of these vibrations travel into the fuselage structure where they eventually lead to cabin noise.

In summary, the measurement campaign was carried out as follows with three objectives: first, the wing needed to be characterized by its modal parameters. This will allow one to perform the second objective, which is to create an updated Finite Element model (FE) of the wing for further investigations. For vibroacoustic purposes, the FE model must be updated for both low and also for high frequency ranges.

The third objective was to estimate the power flow through the wing. In order to do this, generic shaker excitation was used to perform the experimental modal analysis. Additionally, the MAES was used to excite the wing with a vibrational spectrum which mimics the excitation forces of a simulated engine.

Section 2 will present the measurement setup including an overview of the experimental structure, the used sensor grids and the excitation. In Sect. 3 a description of the FE model representing the wing and the subsequent model updating is presented. In Sect. 4 the power flow analysis is performed. The fundamentals of the STI are covered and a power flow analysis for the generic excitation of the wing in z-direction (lift direction) is compared to the power flow resulting from an excitation with the MAES. A detailed description of the MAES is presented in Sect. 5.

2 Measurement Setup

This section covers a detailed overview of the structure measured during the experiment. The structural components will be shown in order to gain an appreciation of the full setup. Additionally, the sensor grid on the wing and the adopted excitation setup will be explained.

2.1 Experimental Structure

The experimental structure consists of a starboard A320 wing attached to a support frame, see Fig. 1. The wing has a span of 18 m and chord length of 6.5 m at the root and 0.4 m at the tip. The whole wing weighs approximately 4.5 tons (without fuel and without pylon and engine attached).

The suspension structure (blue frame on the right side of Fig. 1) was designed to hold the wing in a similar manner as it would be installed on an aircraft. This suspension system has been specifically designed, constructed and installed for this measurement campaign. The wing is fully equipped and, apart from the main struc-

Fig. 1 Overview of support frame with attached wing

tural components such as wing box and aerodynamic covers, all other attachment parts have been installed (flaps, ailerons, fairings, slats). Fuel tanks were empty and the pylon as well as an engine were not attached to the wing.

Close to the root of the wing there is an external structure installed which is originally not part of the wing structure. This additional structure is used by Airbus to demonstrate and test the inboard flap movement. While this structure is not part of the original wing it had to stay installed on the wing during the measurement campaign. The size (2 m length) and mass (approximately 250 kg) of this beam-like structure is negligible in comparison with the rest of the wing, and is therefore not considered in the analyses.

2.2 Accelerometer Grids

Main responses of the wing structure are measured using accelerometers attached to the structure. Subsequently, accelerometer data are converted to other suitable quantities such as displacements when needed for other parts of the analyses. Two setups were used: First, one to capture the modal behavior of the wing structure, and secondly, to perform an analysis using structural intensity estimations to calculate balances of power flows in the structure. This second grid-layout requires a high-density placement of sensors in certain areas of the wing, while for the grid-layout for the modal analysis, a coarse distribution along the full wing is sufficient.

2.2.1 Modal Grid

In Fig. 2 the grid-layout for the modal analysis is indicated by the black markers distributed along the wing bottom surface. A total of 64 sensors were installed along the leading and trailing edges and additionally along a center line of the wing. A previous numerical study with an FE model showed that the chosen sensor distribution is sufficient to capture all necessary modes. At every sensor location the accelerometers are installed normal to the surface. At some locations a second sensor pointing in x-direction (flight direction) was installed in order to capture in-plane modes.

2.2.2 Structural Intensity Grid

A total of 328 sensors were installed in the layout for the structural intensity analysis. In Fig. 2 the sensor locations (red markers) are indicated in two sections on the top and bottom surfaces of the wing (in the Root and Pylon areas). A photo of the actual installed measurement grid for the experiment is shown in Fig. 3. There are areas on the top and the bottom surfaces of the wing near to the root and around the pylon. The areas near the root of the wing were chosen to capture the power flow which enters the wing suspension in this experiment, and which would be entering the fuselage

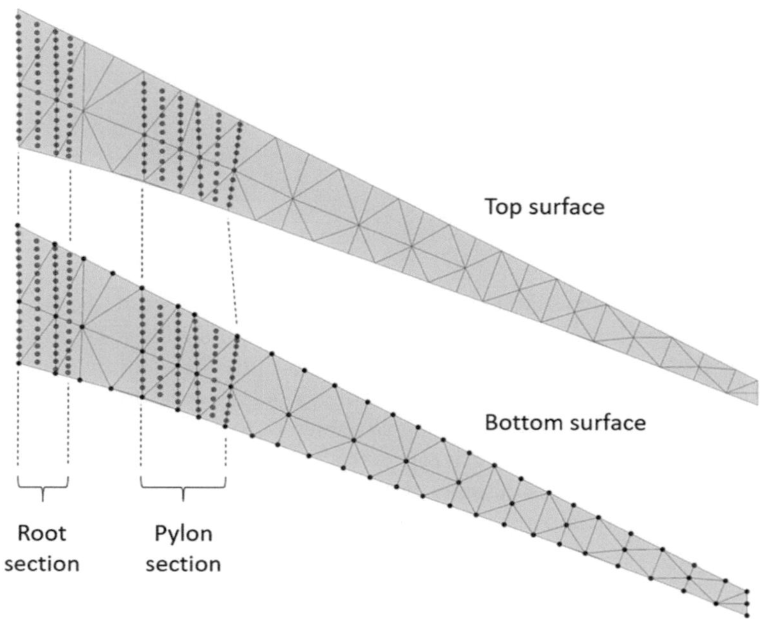

Fig. 2 Combined sensor grid layout, black markers: modal analysis grid; red markers: structural intensity grids

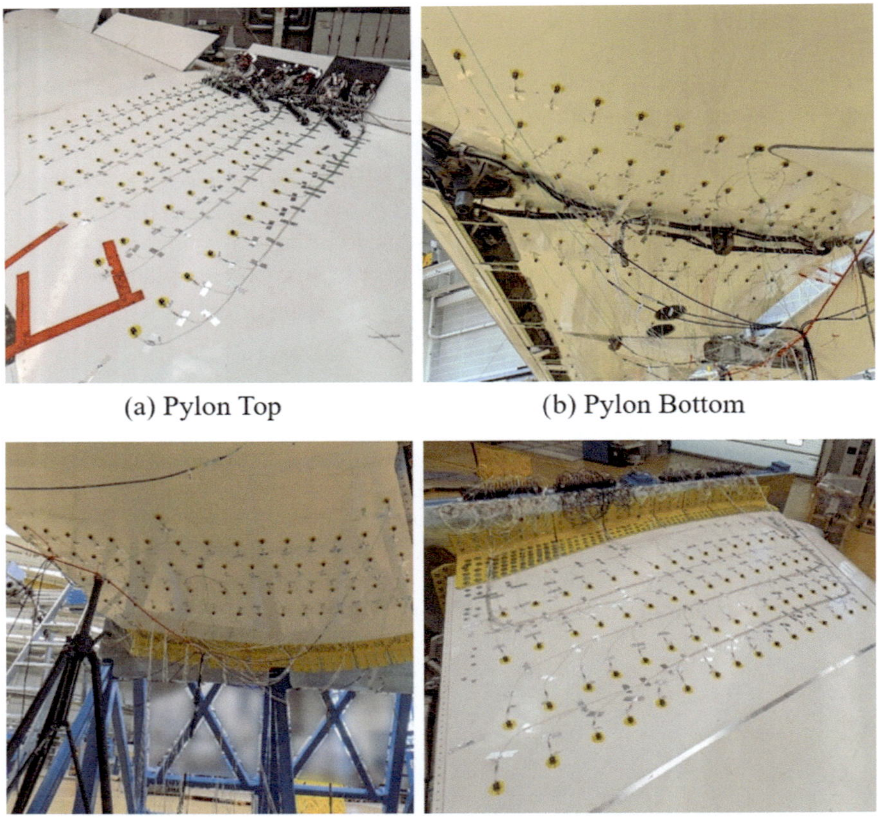

(a) Pylon Top

(b) Pylon Bottom

(c) Wing Root Bottom

(d) Wing Root Top

Fig. 3 STI grid installed on wing

in the real case. The areas around the pylon were chosen to capture the power flow which results from the excitation on the pylon location along the wing. In this way, the power flowing from the excitation location towards the wing tip and towards the root of the wing is quantifiable. All these sensors measure the response of the structure normal to the surface. All sensor cables were taped to the wing surface, and patch panels were installed on a part of the wing where there would be no direct influence on the vibrations of the relevant wing surfaces. This minimizes any negative influences of the structural response in the area of interest.

2.3 Excitation

The excitation of the wing was performed with electrodynamic shakers in a frequency range up 400 Hz. For the Experimental Modal Analysis (EMA), the excitation was performed at two locations. The first excitation point was located half way along the wing span, and the second close to the wing tip. At both locations the wing was excited in z-direction (lift direction) to capture bending modes. Additionally, an excitation at the wing tip along the x-axis (towards the tail) was performed to capture in-plane modes.

For the STI analysis, the excitation was performed at the pylon attachment points. Both points were excited in z-direction (lift direction) and y-direction (root to wing tip direction)—see Fig. 5 for coordinate system.

A single shaker excitation and the MAES shaker setup were used for excitation. For this case, four shakers in a controlled system are used to reproduce a given force/moment profile. The input profile used was obtained by a simulation of a pylon-engine vibration model. MAES will be covered in more detail in Sect. 5.

3 Experimental Modal Analysis and Model Updating

This chapter covers an overview of the validation and updating of the Finite Element Model (FEM) of the primary structure of the A320 Wing. The lower frequency range up to 14 Hz will be analyzed using the methods of Experimental Modal Analysis (EMA) in order to identify modal parameters that will subsequently be used to perform FE model updating. Furthermore, by means of the Energy Correlation Criterion (ECC), the updated FE model will be validated by comparing spatial energy distributions for different frequency bands.

3.1 FE Model

Figure 4 shows the FE model of the suspension with the attached A320 wing for the calculation program Nastran. The wing structure grid, which is marked in yellow, was generated using the model generator ModGen [11] developed by DLR. The element resolution can be defined parametrically. The suspension, which is marked red, is coupled to the wing structure via rigid body elements (RBE3). The model structure consists of beam elements (CBAR), used in the suspension of the wing and as stiffeners on the top and bottom surfaces of the wing along the wing span. Additionally, shell elements (CQUAD4, CTRIA3) were used for the wing attachment area of the suspension, the top and bottom surface of the wing as well as its spars and ribs. In total, the FE model has 5,359 CBAR elements, 4,034 CQUAD4 elements and 31 CTRIA3 elements.

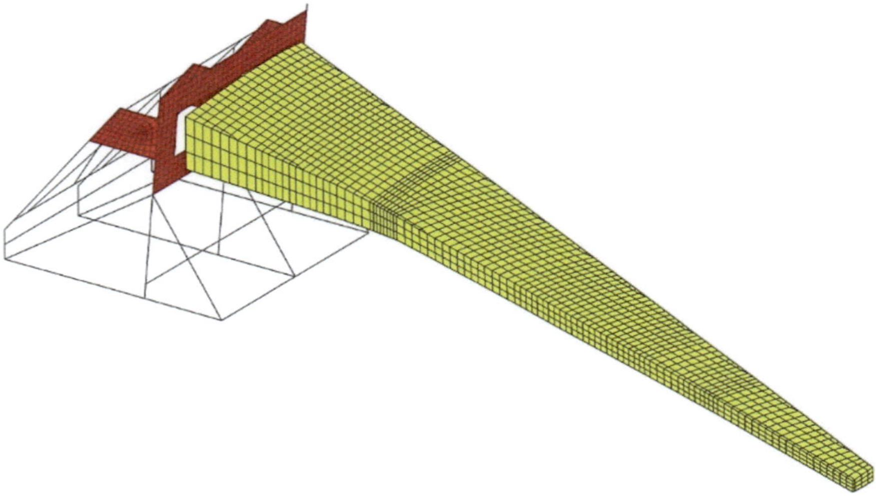

Fig. 4 FE model of A320 wing structure (yellow) including suspension (red)

3.2 Experimental Modal Analysis

In order to optimize and validate the FE model, the modal properties of the real structure were investigated by means of experimental modal analysis in the low-frequency range up to 14 Hz. The modal parameters (eigenfrequency, normal mode, damping) were identified using the POLYMAX frequency-domain method [22]. The modal analysis was based on frequency data with a frequency resolution of 0.1 Hz; they result from a Fourier transformation of the measured accelerometer time data (approx. 5 min) with a sampling rate of 3,200 Hz.

Figure 5 shows the first 5 experimentally identified modes. Vertical (z-direction) and lateral (x-direction) deflections are indicated by blue and red arrows, respectively. The modes Fig. 5a, c and e show the first three bending modes of the wing structure. In contrast, mode Fig. 5b shows the first in-plane mode. The mode Fig. 5d depicted here shows an opposite movement of wing root and wing tip in x-direction. The relatively large displacement at the wing root indicates a mode of the wing suspension structure and therefore adds a global superposed mode to the wing mode.

These modes shown here served as the basis for the optimization of the FE model, which will be discussed in more detail in the following section.

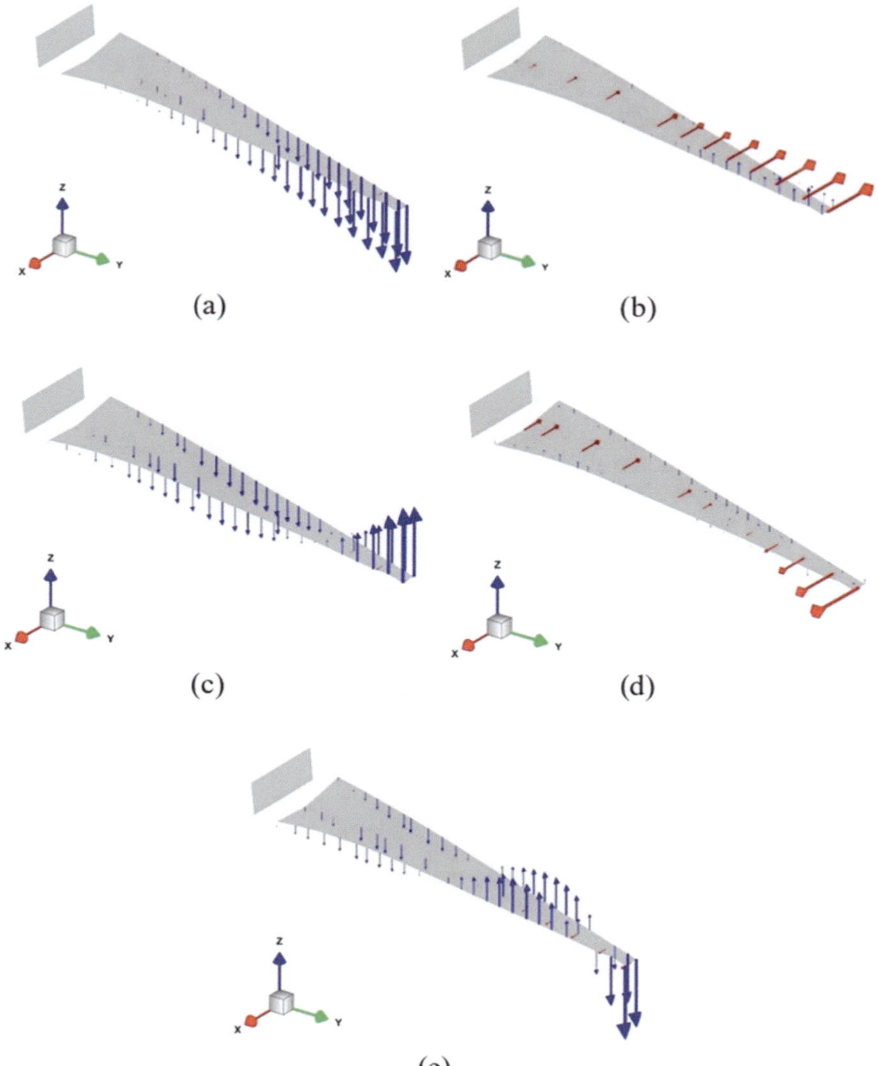

Figure panel labels: (a), (b), (c), (d), (e)

Fig. 5 Visual representation of the experimentally identified modes up to 14 Hz showing first three bending (**a**, **c**, **e**) and first in-plane (**b**, **d**) modes

3.3 Model Updating

In order to reproduce the real vibration behavior of the wing structure as accurately as possible, model optimization is necessary. As part of the model optimization, the eigenfrequencies of the FE model were updated to the experimentally identified

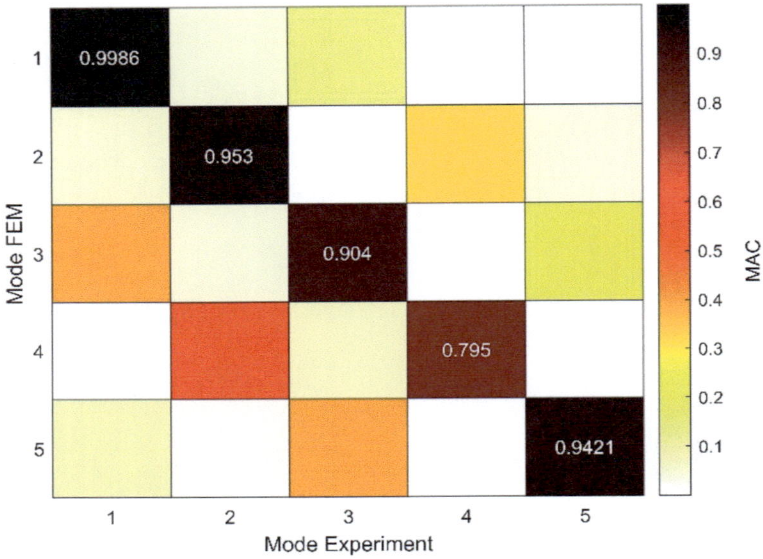

Fig. 6 MAC evaluation

eigenfrequencies. The correct assignment of the structure-specific eigenfrequencies between experiment and FEM was carried out via the Modal Assurance Criterion (MAC) [1]:

$$MAC_{k,l} = \frac{|\{\phi_k\}^H\{\phi_l\}|^2}{\{\phi_k\}^H\{\phi_k\}\{\phi_l\}^H\{\phi_l\}} \tag{1}$$

where ϕ are the eigenvectors of the datasets to be compared and H indicates a Hermitian transposition. The MAC criterion checks the linear dependence of two eigenvectors ϕ. Nearly matching mode shapes can be indicated by a high MAC value close to 1. Due to different resolutions adopted for the measurement and the finite element grids, it is necessary for the calculation of the MAC value to narrow the eigenvectors down to corresponding nodes. The result of the MAC evaluation between FEM and experiment is shown in Fig. 6. From the high MAC values obtained here, it can be seen that the mode shapes shown in Fig. 5 could also have been determined numerically by performing an eigenvalue analysis.

The corresponding modes from the FE model determined by the MAC analysis are shown in Fig. 7 for illustration. The color bar scaling from −1 to 1 shows the relative displacement of the structure in x- and z-directions. As expected, due to the high MAC results in Fig. 6, the numerical mode shapes depicted in Fig. 7 are very similar to those in Fig. 5 obtained by EMA.

Figure 8 shows the frequency deviation between experiment and FEM before and after optimization. The frequency values are only disclosed qualitatively to show the effect of the updating procedure. Before optimization, the eigenfrequencies of

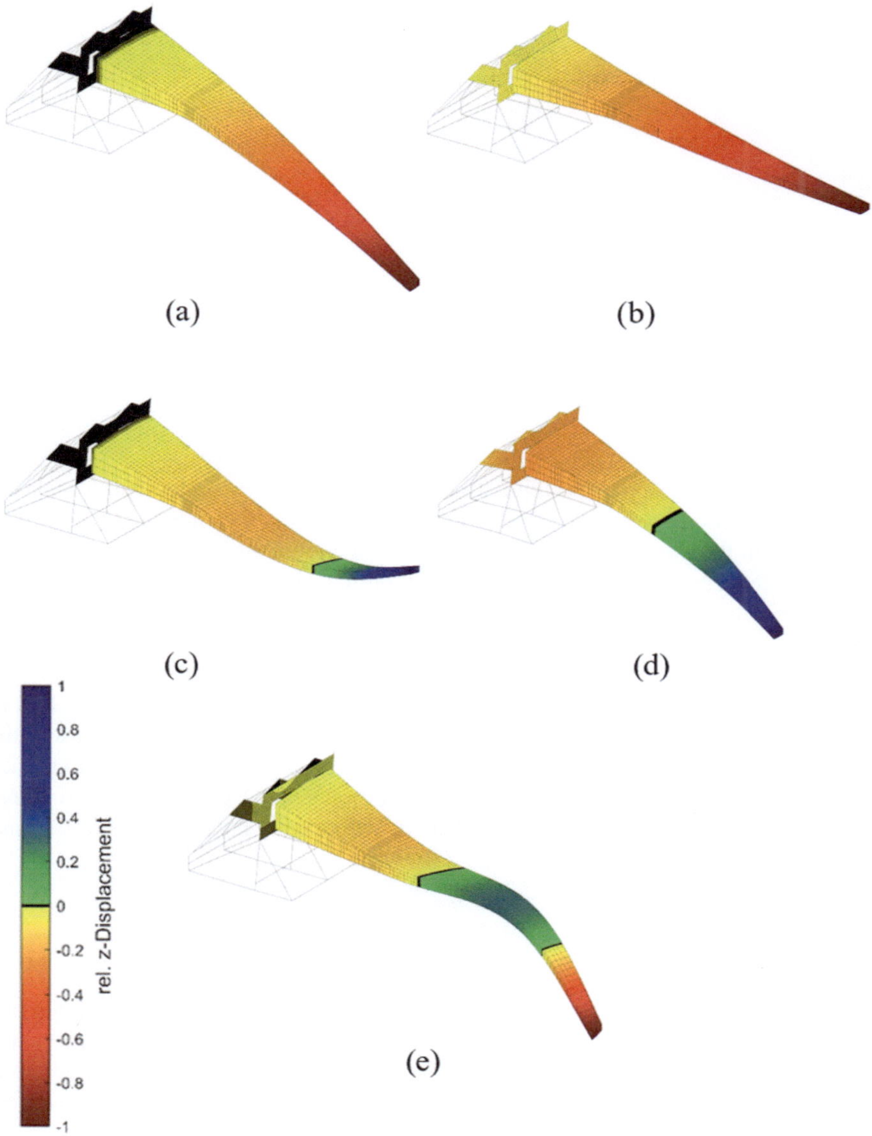

Fig. 7 Visual representation of correlating numerical modes calculated for the five cases shown in Fig. 6

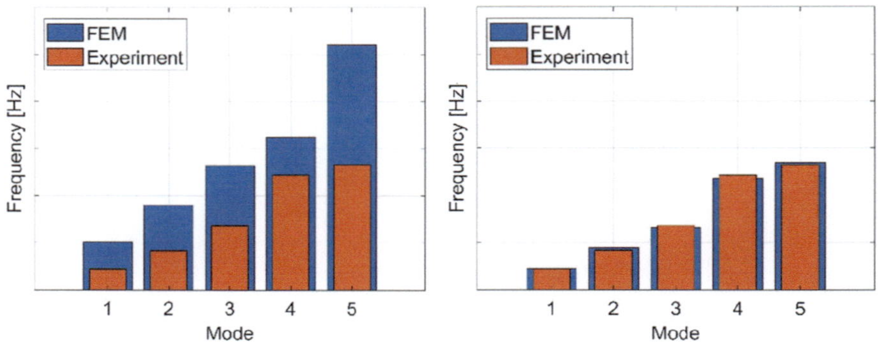

Fig. 8 Frequency differences between experiment and FEM before (left) and after (right) optimization

the FE model were clearly too high. To explain this, it was assumed that there had been a mass deficit in the FE model, so that the mass was then adjusted within the framework of the model optimization by means of a gradient-based method in such a way that the frequency deviation becomes minimal. Here, individual point masses (CONM2), which were added to the FE model, were updated. After the optimization, a clear improvement of the frequency deviation can be seen for all modes.

In order to validate the optimized model for a larger frequency range, an ECC [4] evaluation between FEM and experiment was performed. The ECC is suitable for checking whether a model is able to provide sufficiently accurate predictions with regard to the vibration behavior. The ECC is a method that utilizes the total kinetic energy of a determined number of spatial subcomponents of the analyzed structure for different frequency bands. The ECC(n) for frequency band n is calculated as follows:

$$ECC(n) = \frac{\left(\{E_{kin}^{exp}\}_n^T \{E_{kin}^{num}\}_n\right)^2}{\{E_{kin}^{exp}\}_n^T \{E_{kin}^{exp}\}_n \{E_{kin}^{num}\}_n^T \{E_{kin}^{num}\}_n} \tag{2}$$

with $\{E_{kin}^{exp}\}_n$ and $\{E_{kin}^{num}\}_n$ describing, respectively, the experimental and numerical energy distributions for the frequency band n. Similar to the MAC analysis, a value close to 1 indicates high linearity between the energy distributions. For the validation the spatial segmentation of the wing structure was set to 5 segments, as shown in Fig. 9, and the spectral segmentation was carried out using 1/3-octave bands.

Figure 10 shows the result of the ECC evaluation for an excitation of the wing in z-direction (vertically). The center frequencies of the respective frequency bands are plotted on the x-axis. As can be seen, the ECC values up to the frequency band 125 Hz are all at least 80%, which indicates an acceptable similarity between the experimental and numerical energy distributions. In contrast, the ECC values 125 Hz drop significantly. This means that analyses in this frequency range are unusable. However, it must be considered that the spatial distribution and number of the sensors of the modal grid had been primarily designed for the modal analysis. As a

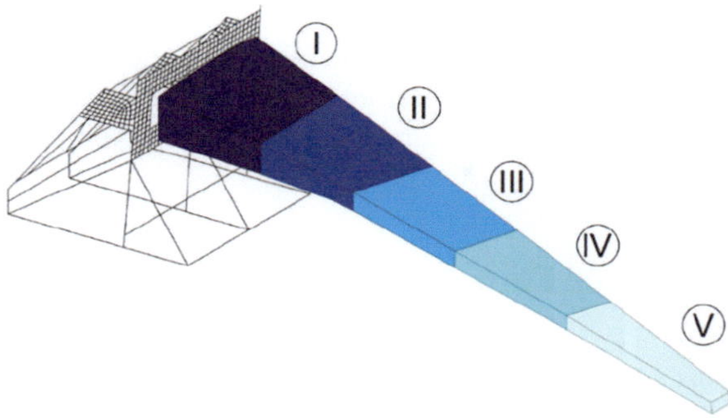

Fig. 9 Spatial segmentation of the wing structure for ECC

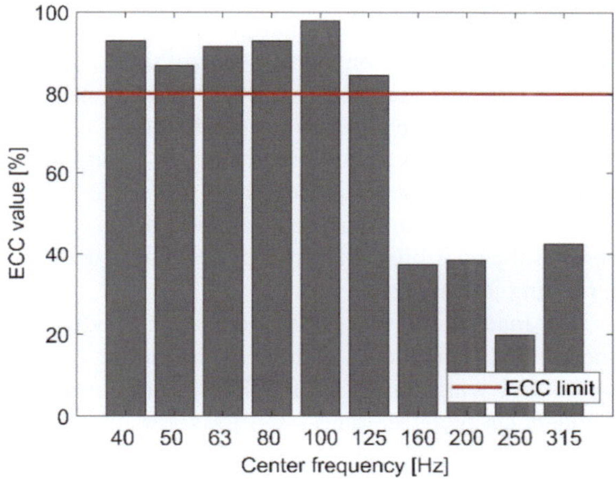

Fig. 10 ECC evaluation of optimized FE model (excitation in z-direction)

consequence, only a very limited number of sensors were present in the relevant spatial segments for a meaningful ECC evaluation compared to the corresponding grid used from the FEM model. This is because the ECC requires a certain amount of vibration data (viz. number of sensors) for spatial averaging in order to be robust against uncertainties in measurement and modeling, which is especially important for an evaluation in the higher frequency ranges.

On the other hand, the significance decreases if the number of spatial segments is too small. For this reason, the results of the ECC evaluation shown here should be treated with caution. For a particularly meaningful ECC analysis, however, a sufficiently high sensor density on the entire wing surface is needed, which would allow a higher spatial discretization both along and lateral to the wing direction.

One way to reduce the potential weaknesses of the ECC evaluation described here is the use of the sensor layout for the structural intensity instead of the modal grid layout, see Fig. 2. The spatially accurately measured area around the wing root and pylon allow higher spatial discretization for the characterization of the energy distribution, which eventually leads to more reliable results when comparing numerical and experimental data. Additionally, in order to properly simulate the vibration behavior of the highest engine frequency, the FE model needs to be updated up to at least 200 Hz band. Therefore, the ECC will be included into an updating process in combination with the already performed modal updating approach.

4 Power Flow Analysis

This analysis utilizes the concept of structural intensity to achieve three goals: (1) to visualize the power flow induced by vibrations throughout the wing; (2) to quantify the power transfer from the wing structure into the suspension; (3) to visually compare and quantify the generic excitation versus the MAES excitation in order to show the impact of the MAES on the response of the structure.

The wing was excited using two different shaker setups. The first setup was a single shaker excitation in z-direction (lift direction). A generic pseudo-random excitation with frequency band limitation between 22 and 400 Hz and with a uniform energy distribution over the frequency range was used. The second setup was MAES, which will be described in more detail in Sect. 5. The key advantage of the MAES shaker setup is the possibility of exciting the wing structure with the realistic load profile of a running jet engine.

The load profile was generated using a previously developed jet engine vibration model [28]. This model permits an estimation of the loads at the interface between pylon and wing during cruise-flight condition. It includes the two main rotation frequencies of jet engines—the low-pressure shaft speed also called N1 and the high-pressure shaft speed N2. The rotation speeds of N1 and N2 range between 70–90% of the respective maximum rotation speeds during cruise flight. Additionally, the model includes one sub-harmonic ($0.5 \times$ N1) and one higher-harmonic ($2 \times$ N1) of the low-pressure shaft. The power flow analysis will be performed for both of the excitation setups to allow a comparison between them.

The four areas on which the STI grid was installed, see Sect. 2, will be used to perform the power flow analysis. These assessment areas are here referred to as "Pylon Bottom", "Pylon Top", "Root Bottom" and "Root Top"; these areas on the wing geometry are shown in Fig. 11.

The light-grey area represents the subset of elements which will be used to quantify the power transmission from wing to suspension. The suspension itself is not shown here, but can be seen in Figs. 1 and 4. Also, the thickness of the wing has been exaggerated here for illustration purposes.

Section 4.1 presents the fundamental theory of structural intensity and the necessary equations to quantify the power flow, while Sect. 4.2 gives the aforementioned comparison between generic and MAES excitation.

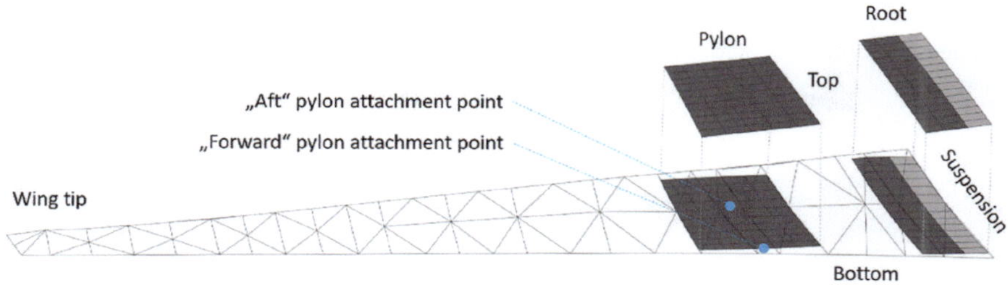

Fig. 11 Lower wing surface shown with STI assessment areas (grey) and pylon attachment points (blue)

4.1 Fundamentals of Structural Intensity and Power Flow Assessment

The power flow calculation is performed by utilizing structural intensity estimation from measurement data. A hybrid method is used, with which the acquired measurement data of the wing structure can be projected onto an FE mesh. The projected measurement data is used to estimate missing values of element forces and moments, as well as of rotational degrees of freedom (DOF) [2]. As an example, the equations for calculating power flow for shell elements $\vec{\vec{\mathrm{P}}}_{\mathrm{Shell}}$ in the frequency-domain can be expressed as [8]:

$$\vec{\vec{P}}_{Shell} = -\frac{1}{2} \cdot Re\left\{\begin{bmatrix} N_x & N_{xy} & Q_{xz} \\ N_{yx} & N_y & Q_{yz} \\ 0 & 0 & 0 \end{bmatrix}\begin{bmatrix} v_x \\ v_y \\ v_z \end{bmatrix} + \begin{bmatrix} M_{xy} & M_{xx} & 0 \\ M_{yy} & M_{yx} & 0 \\ 0 & 0 & 0 \end{bmatrix}\begin{bmatrix} \omega_x \\ \omega_y \\ \omega_z \end{bmatrix}\right\} \tag{3}$$

where N are the normal and shear forces, Q are transversal forces, v is velocity, M are moments and ω is angular velocity.

Structural intensity and power flow have units of [W/m^2] and [W], respectively. In the following evaluations, power rather than intensity will be considered.

In order to perform assessments of power transfer over certain boundaries of the structure, the open path-integral:

$$\hat{\mathbf{P}}_{path} = \int_0^S \mathbf{P} \cdot \mathbf{n}\, \mathrm{d}S \cong \sum_1^i \mathbf{P}_i(x, y, z) \cdot \mathbf{n}_i(x, y, z) \cdot \frac{S}{i} \tag{4}$$

is applied, where P is the power flow in vector form, S the path length and n the normal vector indicating the outflow direction. The outflow normal $\mathrm{n}_{\mathrm{flow}}$ is orthonormal to the plane formed from the path direction vector $\mathrm{n}_{\mathrm{path}}$ and the surface normal $\mathrm{n}_{\mathrm{surf}}$, see Fig. 12.

The absolute power input $\mathbf{P}(\mathrm{f})_{\mathrm{input}}$ from the shaker can be calculated by using the velocity $\mathbf{v}(\mathrm{f})$ at the excitation point as well as the force signal $\mathbf{F}(\mathrm{f})$ measured by

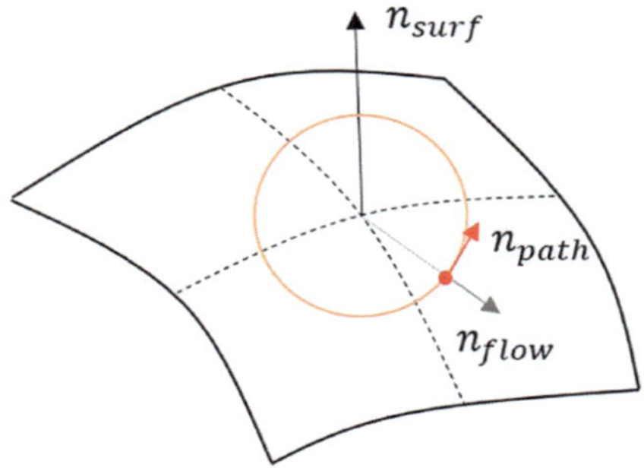

Fig. 12 Visualization of path integral normals; orange circle: path

the force cell. The force cell sits at the end of the shaker-structure-connection and is oriented towards the structure. The power input for discrete frequencies in the frequency-domain can then be calculated by:

$$\mathbf{P}(f)_{input} = \frac{1}{2}\Re\{\mathbf{F}(f) \cdot \mathbf{v}(f)^*\} \tag{5}$$

where f is frequency and * indicates complex conjugation ($\Re$ is the real part).

4.2 Power Flow Assessment for Single and MAES Shaker Excitation

4.2.1 Power Flow Visualization and MAES Excitation Impact

In Fig. 13 the power flow vector fields for the generic and MAES excitations at the discrete engine frequency N1 are shown. For generic excitation the vector fields are presented in subplots a) and b) with the excitation locations marked by red dots. MAES excitation utilizes the two pylon attachment points simultaneously, and are labelled as "Aft" and "Forward" pylon attachment points, respectively, as shown previously in Fig. 11.

Looking at subplot Fig. 13(c) it is clearly visible that emerging from the excitation locations the power is flowing towards tip and root of the wing along the bottom and top surface. In the root section of the bottom surface, the power flow is higher than on the top which means most of the power flowing into the suspension is transferred along the bottom surface. As the excitation in both cases is performed on the bottom of the wing, this finding may seem obvious. However, due to the box design of wings, it is also possible that injected power which reaches the top of the wing is transferred back to the bottom part during its way throughout the structure. This assumption is supported by the power flow emerging from the middle line of the pylon section on top of the wing because it seems to be transferred through the rib located along the line of the excitation points. The ribs themselves were not included in the measurements

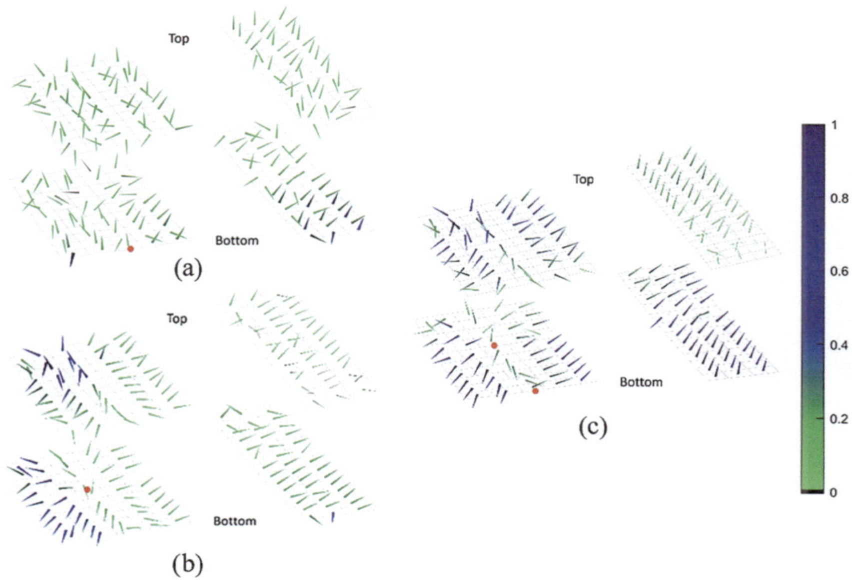

Fig. 13 Vector fields for generic excitation: at leading (**a**) and trailing (**b**) edges, compared to MAES excitation (**c**) at engine frequency N1; color scale: normalized to maximum power in respective fields

during this campaign and thus this possible mechanism will remain unresolved until a more detailed evaluation of the updated FE model can be carried out.

The comparison of subplots Fig. 13(a) and (b) with subplot Fig. 13(c) shows that the MAES excitation results in a vector field with different flow characteristics compared to a generic excitation. The vector field in subplot Fig. 13(b) shows a similar general characteristic, only that the major portion of the power flow is now directed towards the wing tip. In subplot Fig. 13(a) the vector field shows only a rather random behavior. A reason for this might be a too small phase-difference between sensor points at this frequency, due to wavelengths spanning more than one of the rib-separated wing box segments, which would be unfavorable for the power flow determination. Also, the excitation point itself could be the reason. It is directly located at the lower edge of the front spar which might transfer the main part of the power flow at this frequency. The spars were not equipped with accelerometers and can therefore not be evaluated regarding their power flow.

In Fig. 14 the same comparison of excitations as in Fig. 13 is presented, only now for the engine frequency N2. Comparing subplots Fig. 14(b) and (c), it becomes apparent that the vector fields look very similar, both in flow characteristics and normalized magnitudes. On the other hand, subplot Fig. 14(a) does not seem to have much similarity with the resulting vector field of the MAES excitation in subplot Fig. 14(c). This indicates, in contrast to the vector field comparison in Fig. 13, that for certain frequencies the MAES excitation does not necessarily have a significantly

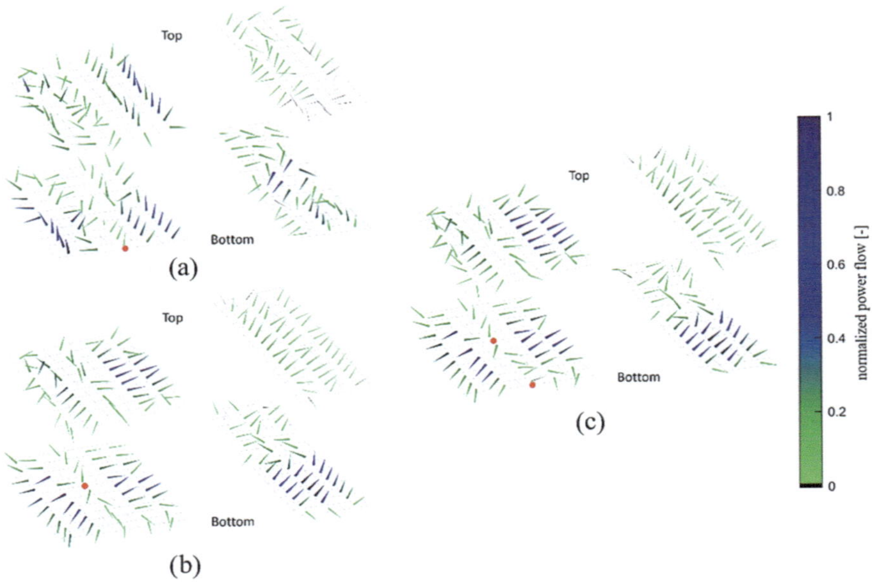

Fig. 14 Vector fields for generic excitation at leading (**a**) and trailing edges (**b**) compared to MAES excitation (**c**) at engine frequency N2; color scale: normalized to maximum power in respective fields

different impact compared to the generic excitation. This might be because there is a low level of excitation for this particular frequency of the load profile used with the MAES, see Sect. 5.5. Alternatively, the respective force/moment combinations at this frequency, which result in an insignificant change in structural response compared to a generic excitation, might be the reason.

In conclusion, the visualization of the power flows, resulting from the MAES excitation was successful and allows to clearly see the flow directions and regions of highest power. This is the first step to allow assessments of vibrational-power transmissions in aircraft structures. However, only with the respective quantification of the power flows, relevant comparisons can be made. Additionally, the comparison between power flow fields based on generic and MAES excitation showed a significant difference. This indicates that the more realistic excitation profile of the MAES shows the desired effect.

4.2.2 Quantitative Power Flow Evaluation

The wing can be considered as the major transfer path for vibrations induced by the jet engine transferring to the fuselage structure. Besides the visualization of the power flow the quantification of the transferred power from the wing into the fuselage needs to be calculated for a full power balance for the fuselage.

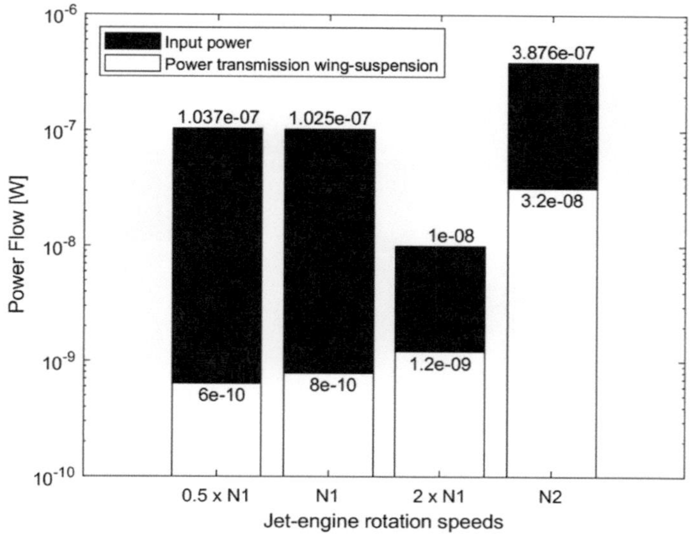

Fig. 15 Power input and transmission for generic excitation (forward attachment point)

In the following, the power transmission from wing into the suspension as well as the injected power at the excitation locations are calculated. The power transmission will be determined using Eq. (4) along the elements at the top and bottom root section of the wing which are closest to the suspension, see light-grey elements to the right in Fig. 11. The power injected from the shaker setups will be determined using Eq. (5); in case of MAES excitation the four single powers will be added together.

Both, the input power and the power transmission, are shown for the generic excitation at the four load profile frequencies at forward and aft pylon attachment points, in Figs. 15 and 16, respectively.

The distribution of power input and power transmission for both excitation locations is different. For the forward attachment point the transmitted power from wing into suspension increases with frequency. However, comparing the transmitted to the input power, there are more than two orders of magnitude in difference for $0.5 \times$ N1 and N1. For N2 the difference is about one order of magnitude and for $2 \times$ N1 less than one. This shows that almost no energy is transmitted into the suspension at the first two frequencies while about 10% is transmitted for $2 \times$ N1 and N2. In total (over all frequencies) a power of $P_{fwd} = 3.46 \cdot 10^{-8}$W is transmitted. In case of the aft attachment point, the transmitted power increases from $0.5 \times$ N1 to $2 \times$ N1 but for N2 it drops to the level of N1. In general, the transmitted powers are higher than for the forward attachment point. Also, in relation to the respective input powers the aft attachment point transmits more power per input into the wing as the forward point. Only for $0.5 \times$ N1 the difference between input and transmission is higher than one order or magnitude. In total the aft point accounts for $P_{aft} = 8.02 \cdot 10^{-8}$W in transmitted power, so more than double than for the forward point. An explana-

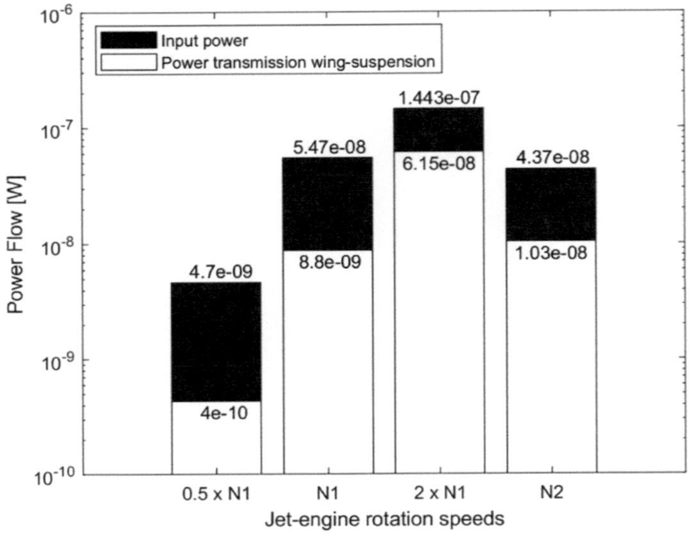

Fig. 16 Power input and transmission for generic excitation (aft attachment point)

tion for the difference in transmitted power between the two points might be that the major portion of the transmitted power for the excitation on the forward pylon attachment point flows via the wing spars into the suspension. This component of the wing was not equipped with sensors during this campaign and cannot be assessed.

In comparison, Fig. 17 shows the power input and power transmission from wing into suspension for the MAES excitation. For the main engine frequencies N1 and N2 the transmitted power is higher than for the remaining frequencies. Also, ratios between input power and power transmission lie less than one order of magnitude apart for N1 and N2. For the case of the sub- and higher-harmonics of the engine frequencies (0.5 $\times$ N1, 2 $\times$ N1), the values are separated by more than one order of magnitude. This implies that the power transmission into the suspension per Watt input power is higher at the fundamental engine rotation speeds than for the harmonics. This is a reasonable assumption and indicates again that the MAES method results in a more realistic structural response than with the generic excitation. An open question is how the spectrum of power transmission would change if also the power flow along the front spar would be part of the assessment as was already mentioned in the prior paragraph.

A note regarding the power values for input and transmission in Figs. 15, 16 and 17: Intuitively, the shown power values seem incredibly small. However, it must be remembered that these values were calculated for single specific frequencies and not for frequency bands. Additionally, not the full absolute input energy was considered in these plots. In accordance with the calculation of the power flow, utilizing the structural intensity approach, only the real-part also called the active part of the vibrational power flow was calculated. The active part is the ratio of the power which

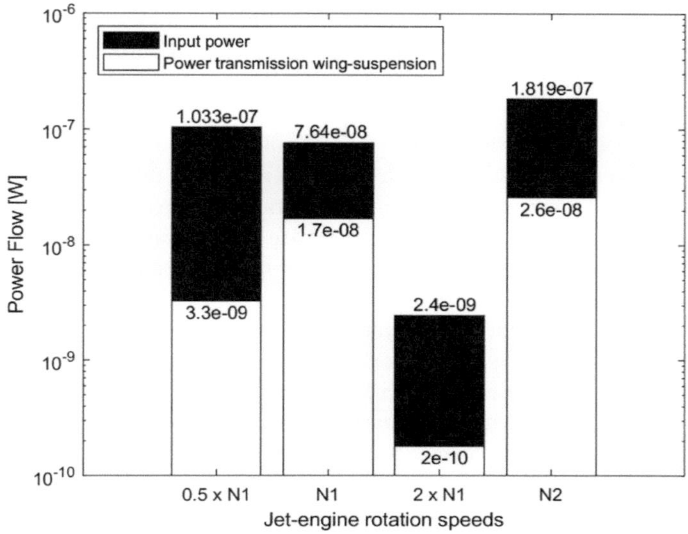

Fig. 17 Power input and transmission for MAES excitation

is actively transferred in the structure while the imaginary part can be compared with blind power from electrical engineering theory [12] and is usually left out in STI evaluations. However, this part still carries power. Furthermore, the publications of Hambric et al. [8, 9] show power values in similar ranges for numerical and experimental evaluations on beam structures.

5 Multi-axial Excitation System

Through simulation, it is possible to accurately calculate the loads generated at connection points between engines and fuselage. However, a key problem arises at the moment when one tries to reproduce those loads on the structure, since there is no way to compensate for the modifications introduced by the system created between amplifier, shaker and structure. Such an unaccounted system will certainly change the predefined reference loads. Nonetheless, if a reference load signal is available, it can be used to design an active excitation system able to generate and inject into the structure an exact copy of the simulated loads. As will be shown in the following sections, such a real-time excitation system is based on an adaptive controller. As part of the development process, an initial single-input single-output (SISO) controller had been investigated and implemented; these finding have been presented previously [5, 6, 18–21]. The fundamentals of this developed system served later as a predecessor for a more complex multi-input multi-output (MIMO) controller. Such a multichannel approach is part of a long-term goal to create a system able to reproduce forces in three spatial directions, including also their associated moments. Previous research had already established the capabilities of the controller to perform MIMO force control, so in this work focus was placed on the scaling and application of such system to a real aircraft structure.

Therefore, if one could implement a system able to excite a structure in a realistic manner, mimicking true loads, such a system will enable the study of the vibroacoustic behavior of a structure in the most accurate way possible. This will help in the better understanding of structure-borne noise mechanisms on different parts of a fuselage and will assist in the determination of relevant energy transmission paths between sources and receivers.

As mentioned earlier, the development of the system had been carried out for several years already and its proper operation has been validated countless times through laboratory experiments. Hence, the natural next step is to place the system in a relevant test scenario, such as, in this case, with a real aircraft structure, to thence assess proper scaling and performance operation.

5.1 Anatomy of the System

The hardware implementation of the controller evolved together with that of the control algorithm software, from a simple one channel SISO to a multi-channel MIMO configuration. The development cycle consisted, for the most part, of prototyping and validating in Matlab, implementation and validation in Simulink, deployment in the hardware and experimental evaluation.

The hardware used during the main part of the project was an ADwin-Pro II real-time controller; this system offers a real-time processing unit with a set of interfaces for analog inputs and outputs that allow one to connect all sensors and actuators required for the adaptive excitation system. The ADwin-Pro II was able to import and execute code generated directly from Simulink, so the deployment of developed algorithms was quite direct, just small adjustments were required on the ADwin-Pro II software in order to execute the controller.

As actuators of the system, multiple electromagnetic shakers were used, depending on the configuration; they are responsible for injecting into the structure the desired vibrational profile while the system generates the control signals. The system uses multiple force cells as sensors to drive the controller. For every shaker used, a force cell is used. In order to measure the injected forces, each force cell is placed at the attachment points between the shaker and the structure. Between the analog output of the real-time controller (ADwin-Pro II) and the amplifiers, an analog reconstruction filter is inserted; this is usually a requirement in order to eliminate aliasing effects of digital-to-analog converters (DAC) that do not have any post-output signal conditioning. Basically, reconstruction filters smooth a discretized signal, so that all artificial artefacts created by the quantization process of the DAC are removed. A diagram of the whole interconnected system can be seen in Fig. 18. There, the real-time controller has N number of shakers and force cells connected to its outputs and

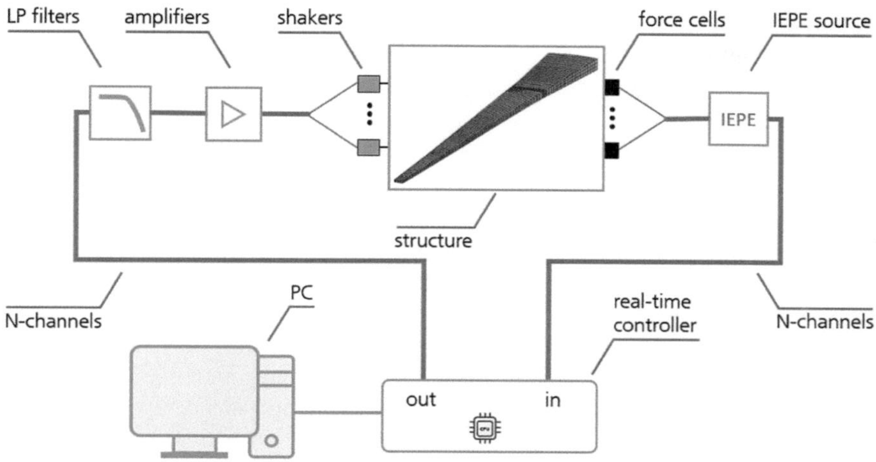

Fig. 18 Interconnection diagram of components of the Multi-Axial Excitation System (MAES)

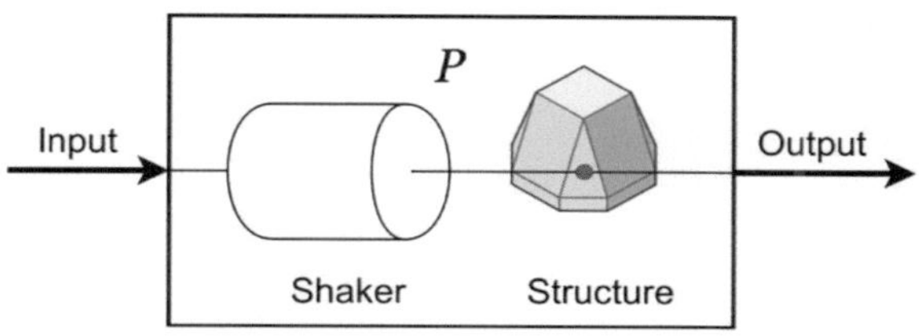

Fig. 19 Schematic of the plant, which consists of a shaker, a structure and a force cell (denoted by the red dot) which measures the injected force into the structure

inputs, respectively, where the number N depends on the configuration under test and can be as large as needed.

5.2 Control Algorithm

The type of problem presented in a typical application, as mentioned before, where it is required to inject on a structure a copy of an arbitrary reference force signal, seems to be a perfect candidate for an adaptive control solution. The two main reasons for this are the availability of a reference signal, in this case the load forces, and the need to compensate a relatively stable plant, which is defined as the path created between amplifier, electrodynamic shaker, the structure and a force cell used to quantify the injected force; Fig. 19 shows the configuration of the plant.

The type of configuration described above is known as inverse control, since the task of the adaptive process is to generate an inverse system $\hat{C}$ of the plant P. If this has been successfully accomplished, the output of the plant $\hat{y}(k)$ will be an unmodified copy of the reference force signal r(k). The typical schematic of an idealized inverse controller is shown in Fig. 20, where $\hat{C}$ represents the compensation filter of the plant

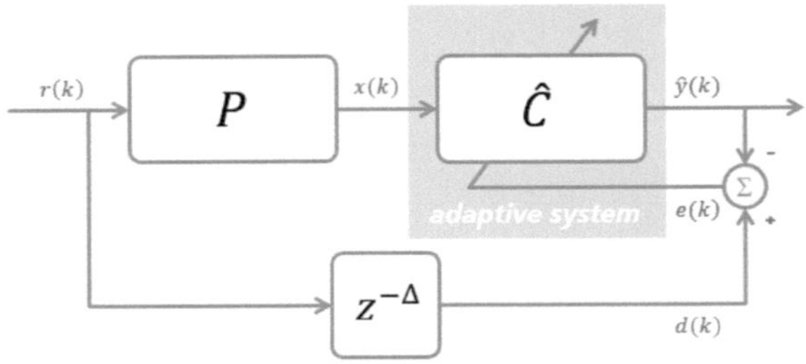

Fig. 20 Block diagram of an idealized adaptive inverse controller

P, driven by the adaptive procedure. The error signal represents the deviation of the output of the control filter $\widehat{C}$ in comparison with the delayed reference signal d(k) at the input of the plant. And Δ is a small time delay, so d(k) is a delayed version of r(k).

There are multiple options available to implement the adaptive algorithm. From the most simple and reliable Least Mean Square (LMS) variations to the most sophisticated and demanding Recursive Least Square (RLS) variations.

Considering the requirements for this type of application, the following algorithms were tested in order to find the best tradeoff between speed of convergence and overall quality of output signal: Normalized-LMS (NLMS), Correlation-LMS, Discrete Cosine Transform-LMS and RLS. The comparisons showed that all algorithms performed equally well. After this comparison, and given that there is no requirement for high-speed tracking of the reference signal, it was decided that at this stage the NLMS was the appropriate choice. Having the simplicity of an NLMS algorithm will make it easier to debug any problem that could potentially appear during the implementation on hardware.

The idealized system shown in Fig. 20 is the most basic configuration for an inverse controller. However, this simplicity helps to illustrate how such a configuration works. Unfortunately, such an idealized controller does not work properly in a real application since it would require conditions that are almost impossible to attain; for example, that there is no internal disturbance at the output of P. One of the options suggested in the literature [25] that can be used for an actual implementation has a configuration similar to what is called the filtered-x LMS (Fx-LMS) algorithm. This configuration is well known and widely used in Active Noise Control applications.

One of the characteristics of this configuration is that the control filter Ĉ is not applied directly to the plant P but rather to a modeled version P̂ of the plant. The Fx-NLMS configuration ensures stable inverse control of the plant even for the case where there are errors in the modeling of P̂, and independent of how P̂ is modeled, online or offline. In this schema it is assumed that on a previous process the plant P̂ had been identified and now a copy is connected at the input of the control filter Ĉ. Under certain conditions it might be desirable that the plant does not directly track the Reference signal but rather a delayed or modified version of it. The introduction of a Reference model M serves this purpose.

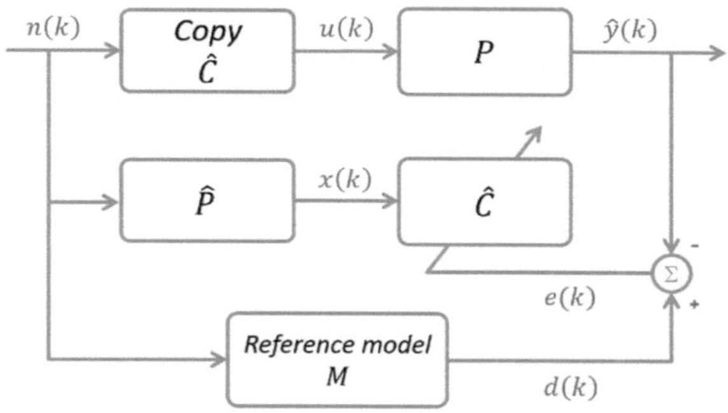

Fig. 21 Block diagram of an inverse controller with added reference model

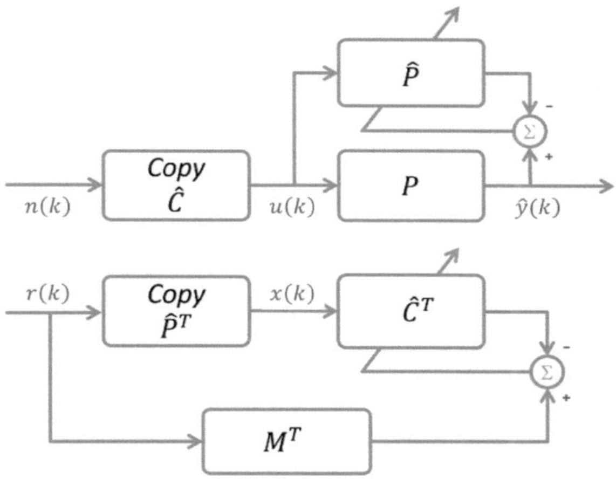

Fig. 22 Block diagram of MIMO hybrid controller

The aforementioned implementations rely on the SISO system being commutative, i.e. P(z) C(z) = C(z) P(z). However, such a property is in general not true for MIMO systems, as pointed out by Plett [23]. Special care needs to be taken in how the transfer function matrix of the controller C is manipulated. Plett also introduces a highly efficient controller by using a Recursive Least Square update strategy. It is well-known that RLS algorithms, although fast, are computationally demanding.

During the development phase it became clear that our hardware system was not capable of running the controller at the required sampling rate with an RLS algorithm. Therefore, a simplified version was implemented with an LMS algorithm, in a similar manner as for the SISO controller. The penalty incurred here is the somewhat slower convergence; however, for our application it is not crucial if we have to wait a little longer before the control coefficients converge and an optimal solution is reached. The scheme of the MIMO controller can be seen in Fig. 22. This is a more sophisticated online version of the Fx-LMS controller of Fig. 21.

5.3 Inherent Phase Control

As part of an ideal replication of a vibrational signature, it is not only important to generate the same magnitudes of forces defined by the *Reference Model M*, but it is also crucial to maintain the same phase relationship between the excitation channels. This means that if the reference model defines a particular phase at a given frequency among the force channels, the control system will try to inject into the structure the forces that also keep intact the same phase relationship.

In order to illustrate this behavior, a simulation was created where the *Reference Model M* has been defined with three force components F_x, F_y, F_z and one moment component M_x. An arbitrary spectrum profile with three peaks at 73, 150 and 180 Hz was used for each of the four components. Although the phase relationships can be defined in different ways, in this case they were defined such that F_x has an arbitrary but fixed phase relationship with the input reference signal r_k, as shown in Fig. 22. Therefore, the phase relationship of F_y, F_z and M_x can be defined with respect to F_x. Thus, the target phase between r_k and F_x is $[0, 0, 0]^\circ$ for each of the three aforementioned frequency peaks. Now, we define a target phase ($\angle$) for each component with respect to F_x, as shown in Table 1, where the phases are given in degrees.

These phase values were arbitrarily defined, the only purpose being to provide an easy visualization on the evaluation shown in Fig. 23; similar results would be obtained for any other chosen values of the phase. In Fig. 23, one can see the *Response Phase* (blue continuous line) generated by the control system and the *Target Phase* (red cross) which had been set as the objective for each frequency component. The phases were calculated by a Frequency Response Function (FRF) between r_k and each injected force on the structure. The good agreement between *Response* and *Target* phases verifies the correct behavior of the control algorithm. A specific experimental validation of the phase control characteristic will be shown later; however, it has always been assumed that the phase adaptation was taking place; if this were not the case, the compensation generated by $\hat{C}$ would not be fully correct with only magnitude adaptation occurring.

Table 1 Target phase specified for simulated evaluation

	Frequency [Hz]		
	73	150	180
$\angle F_y$	−90°	−90°	−90°
$\angle F_z$	−45°	−135°	−45°
$\angle M_x$	−135°	−45°	−90°

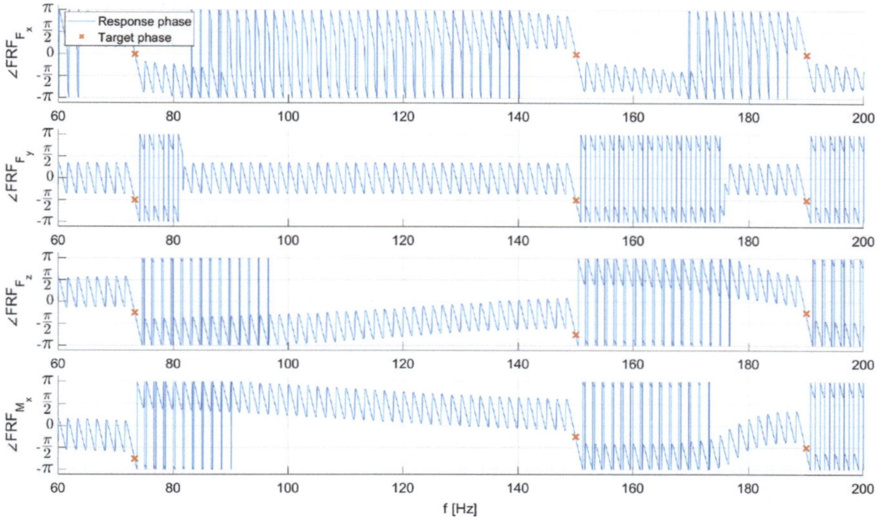

Fig. 23 Simulated control system with arbitrary target phase definition and generated response phase

5.4 Laboratory Validation of Operation

In order to evaluate the proper operation of the adaptive excitation system, a set of laboratory experiments was designed and performed. One of the laboratory structures that is commonly used by the DLR-AE Vibroacoustics group for research purposes is a single block aluminum plate of size $1.0 \times 0.8 \times 0.03\,\text{m}^3$ with an asymmetrically milled surface pattern that resembles the skin fields, stringers and frames typically found on a fuselage. The thicknesses of the skin fields, stringers and frames are 1, 5 and 10 mm, respectively. The integration of all these sub-structures into a single structure with no additional coupling elements provides ideal conditions for multiple types of vibroacoustic research. For each experiment the plate was suspended with a bungee system in order to generate free displacement boundary conditions, as shown on the top side of the plate. A photo of this test structure and the setup can be seen in Fig. 24.

In the following example configuration presented, a comparison is made between the reference vibrational profile that is the target, i.e. the desired injected forces and moments on the structure, that we call *Reference*, and the measured forces and moments that the adaptive system is actually injecting, that we call *Control ON*.

The reference targets used throughout the project were chosen based on the assumption that the adaptive excitation system will mainly be used to replicate forces and moments originating from turbo engine vibrational profiles. As such, it was considered reasonable to assume that these profiles are the result of rotational loads of the engine fans. Therefore, fundamental and harmonic frequencies were used to

Fig. 24 Combined force and moment control in a 4 DOF MIMO configuration, F_x, F_y, F_z and M_y

create these references. Profiles of different number of peaks and bandwidth per peak were defined arbitrarily. These profiles were then set into the controller as the so-called *Reference model M*. Here, we will focus only on showing the result of the latest MIMO controller version; previous results that show the evolution from an early SISO to a mature MIMO implementation can be found in the literature [5, 6, 18–21]. It should be mentioned here that there were other partners in this project that were responsible for accurately modelling and calculating the vibrational response of the engine and for providing this data as input for our controller; however, these data had not become available in time for these experiments, so that the *Reference Model M* made use of the aforementioned design parameters to create an approximate but still valid model of an engine.

Earlier systems focused on the control of either injected force or moment of force. The following example shows the merging of both strategies in order to control simultaneously forces and moments. This example presents a 4 DOF system configuration with a control of three forces and one moment, where 5 input/output (I/O) channels are used. This is due to the fact that 2 I/O channels are needed to control one moment DOF. In this configuration, the forces are defined for each direction of space F_x, F_y and F_z; and the moment is defined as M_y, i.e. moment around the y-axis.

The evaluation result is shown in Fig. 25, where the good agreement between the *Reference* (full lines) and the *Control ON* (dashed lines) conditions is immediately

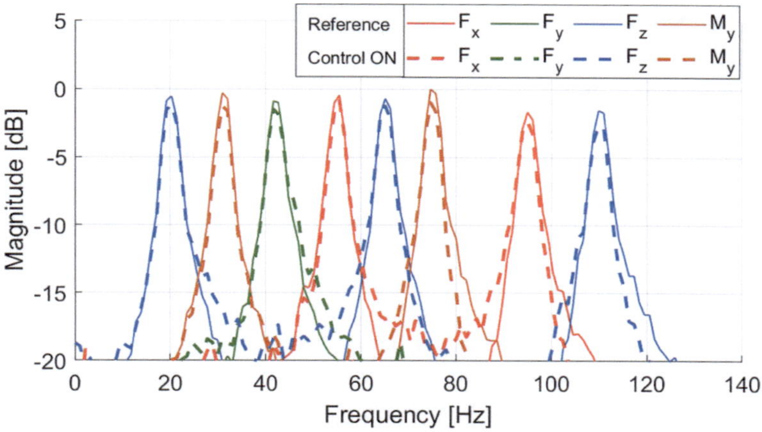

Fig. 25 Spectral comparison of *References* and *Control ON* condition for the 4 DOF MIMO combined force and moment control. Ordinate is expressed in dB

evident. Such agreement between both conditions confirms that the adaptive structural excitation system correctly generates the desired profiles on the test structure.

As mentioned earlier, the reference profiles had been arbitrarily assigned; they were chosen, however, so that they still maintained their relationship with the intended final application of this controller as an artificial excitation system. In this configuration, as well as the previous one, the bandwidth of each frequency peak was reduced in order to minimize overlapping of frequencies: this increased their visibility in the plot, and was additionally relevant since real spectra of rotating machinery tend to have this very narrow bandwidth feature. This modification in the profiles uncovers an important characteristic of this type of adaptive system. Figure 25 shows that as peaks decay laterally from their center frequency, the agreement between *Reference* and *Control ON* becomes worse, and also the signal-to-noise ratio (SNR) starts to decrease rapidly, indicating that below -10 and -15 dB control would become difficult. For example, from a reference value of 0 dB, a decay of -10 to -15 dB implies that a value has becomes between 10 and 30 times smaller. At these levels, SNR is so low that it becomes extremely difficult to perform any control. Thus, we can infer from Fig. 25 that the controller would perform well in the range of approximately 0 to -15 dB. A comparison of this range with studies of other active systems found in the literature shows that this range of performance is quite common amongst many adaptive controllers, see for example Misol [13], Zimicik [29] or Xiao [27].

5.5 Real Scale Validation

As mentioned before, the availability of the A320 wing presented an ideal opportunity to evaluate the MAES in a relevant environment of operation. At the beginning of the project it had been decided that an engine model should become available to serve as input for the MAES; however, the development of this model had been delayed for reasons beyond our control. However, it was fortunate that at this stage a model generated from a simulated engine-pylon design had been available. This is the main reason why all previously reported steps of the developed system used artificial references to evaluate and validate proper operation. With the newly available model, a more realistic evaluation could be performed on the A320 wing. The results of the simulated model showed that engine vibration had two main contributions in Y and Z directions (using the global aircraft coordinates system convention, x axis: from nose to tail, y axis: starboard direction, z axis: from the center of the fuselage upwards). Both components belong to the rotation plane of the engine blades. Since simulations did not consider vibrations due to thrust forces, the X direction contribution has been neglected: A consideration of the X direction contribution here would have increased the complexity of the model design and, more importantly, the necessary data for such a study is extremely limited. It was therefore felt that for a realistic approximation of an engine model, a consideration of just the Y and Z components would be adequate. More precisely, both components were calculated for the engine-pylon connection points to the wing, resulting in two Y and two Z components; for the forward and aft attachment interfaces, namely Y_{fwd}, Y_{aft}, Z_{fwd} and Z_{aft}. System and attachment points are shown in Fig. 26. A diagram of the attachment points is shown in Fig. 27.

In the same way as before, a comparison is made between two conditions: The target *Reference* (full lines), which represents the desired force profile defined by the engine model, and the *Control ON* condition (dashed lines), which represent the actual injected forces of the MAES. For both conditions, the magnitude and phase are considered. The phase for both conditions is referenced to an internal signal of the control algorithm that is shared amongst all channels, so that this signal can be used to calculate a frequency response function in order to verify correct phase output, either for the *Reference* profile or the *Control ON* injected forces.

The resulting magnitudes and phases are depicted in Figs. 28 and 29, showing a good agreement between both profiles. This agreement confirms that the MAES is able to properly generate and inject the forces specified by the reference model, both in magnitude and phase. Peaks are located at $0.5 \times N1$, N1, $2 \times N1$ and N2 of their respective shaft rotating speed. It is important to notice the large disparity between the peak magnitudes of the first (N1) and last two ($2 \times N1$ and N2) frequencies: All peaks of the last two high frequencies are below -10 dB of the maximum peaks, which means that they are strongly masked by the lowest frequency. This is one of the reasons that explains the poorer matching between *Reference* and *Control ON* for these two high frequency peaks. Additionally, the N2 peak is the weakest one with three of the signals being barely above the background noise level. The low signal-to-noise ratio in this region is almost certainly the reason why the phase matching at this

Fig. 26 General view of the Multi Axial Excitation System (MAES) attached under the A320 wing

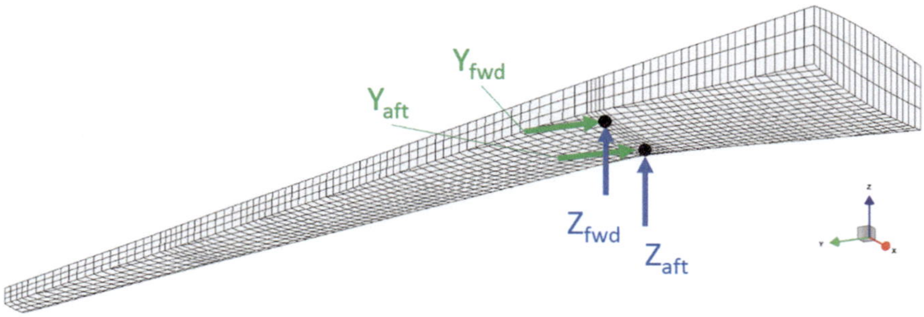

Fig. 27 Diagram of forward and aft attachment points for each shaker direction

frequency is not as good as for the rest of the profile. The large magnitude differences for the peaks arises from the particular simulated engine model; a different design would lead to a different profile, so that these results should just be considered as one example of proper operation.

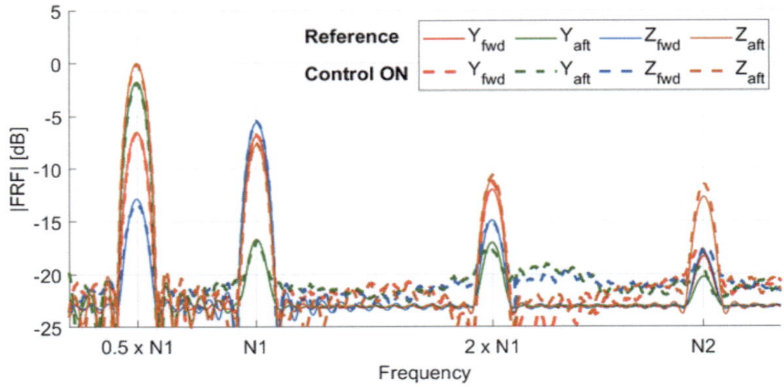

Fig. 28 Spectral comparison between *Reference* and *Control ON* conditions for the injected engine model profile: FRF magnitude

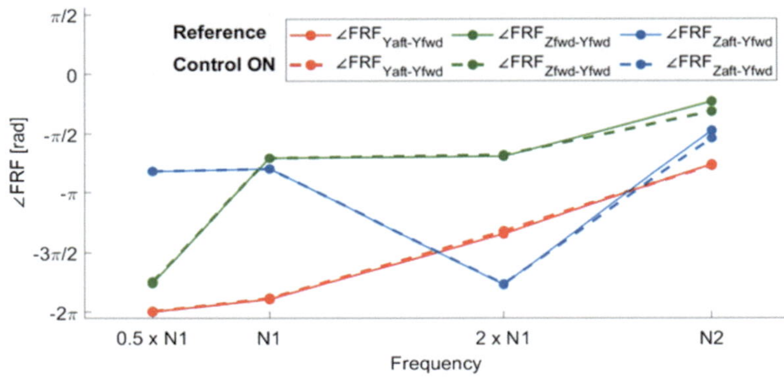

Fig. 29 Spectral comparison between *Reference* and *Control ON* conditions for the injected engine model profile: FRF phase

5.6 Relation to Other Testing Strategies

MAES originated from the need to replace a real engine excitation by a system that is easier to operate in a laboratory environment, with which the capabilities of vibroacoustic research can be extended. A real engine in operation would require a large infrastructure, which would impose severe limitations on the experimental setup and would also complicate the measurements. MAES is seen as serving as a simple replacement for a real engine, at least from the point of view of carrying out vibration measurements. The end result of such a concept would be to obtain realistic mechanical responses on the rest of the fuselage or substructures in order to support further research related to, for example, reduction of cabin noise, or to a better understanding of energy propagation processes between source and receiver. Since the main result from the application of MAES is operational responses, the system has occasionally been categorized by the testing community as a MIMO environmental testing method. However, there are fundamental differences that set it apart from this field.

Environmental testing focuses on directly generating operational responses over a whole test sample in order to establish, for example, structural integrity, safety and proper functioning during its operational life. In the ideal case, the key concept would be the reproduction of responses of the structure which are as realistic as possible. The required input to achieve this would be one or many previously acquired operational responses, obtained either from measurements or by simulation. The most widely adopted environmental testing methods today [7, 10, 14] rely on the measurement of a set of control responses at different positions on the structure, which should help to ensure proper generation of desired responses in the whole structure. However, one of the known issues with this strategy is that the response at other locations of the structure which are far from the available references is not necessarily that of the desired operational response.

The concept of MAES is the replication of the source responsible for the generation of operational responses. The underlying assumption is that if the source is characterized, and more importantly reproduced, as accurately and faithfully as possible, then one can be confident that the subsequently measured responses will be equally accurate. MAES strives to emulate the actual vibrational source behavior, so that it uses as inputs to the control system only the injected forces (via force cells) at the shaker positions in order to drive the controller. Thus, all responses generated at any other location of the structure are just assumed to be the natural result of a real operational condition generated by the source (in the MAES case, the set of attached shakers to the structure).

One of the main problems that arise from MIMO environmental testing is the need to solve the inverse FRF matrix of the created system and all the numerical complications that come attached to this operation. Extensive research focuses on this issue—see, for example [15, 24]. MAES adopts a different approach: by being a purely time-domain algorithm, the FRF inversion is replaced by a real-time adaptive process that should converge to the desired solution, namely the *Reference* profile, as exemplified in the previously shown evaluation results. Although arguably a time-domain algorithm could be considered more computationally demanding from the implementation point of view, it is still seen as a simple approach to tackling this problem. Similar approaches have also been explored recently by others, see [16]. This is the reason why this strategy had been selected years ago when this project had first been conceived [5, 6, 18–21].

6 Conclusions

6.1 FE Model and Model Updating

The data obtained during the measurement campaign was used to update and validate a finite element model of the A320 wing. For this purpose, the low-frequency range was analyzed by using methods of Experimental Modal Analysis to identify the modal properties up 14 Hz, which were subsequently used to update the finite element model. Here, the focus was on the first five elastic modes. Through adjusting the mass properties of the FE model, the numerical and experimental eigenfrequencies could be successfully matched.

In addition, a further comparison between experimental and numerical data was performed using an energy-based approach to allow an updating up to higher frequency ranges. The frequency range considered here was up to 350 Hz. Within this range, it was shown that the numerically and experimentally determined statistical energy distributions were very similar up to 125 Hz. However, the results need to be treated with caution as the sensor grids installed on the wing are not optimized for an ECC analysis. The ECC methodology benefits from a high spatial sensor distribution. While this was given for the grid of the FEM model, the sensor grid used for the measurement data was actually design to only capture the first significant wing modes (experimental modal analysis). Therefore, the corresponding number of sensors in the given wing segments, see Fig. 9, for FEM and measurement data, differ highly. One idea to improve the ECC results would be to use the STI sensor grid in comparison to the FEM data in the respective wing segments to utilize a higher number of sensor locations. Although, this would mean that not the response over the full length of wing geometry is included into the analysis as the STI grid only exists in local regions on the wing.

6.2 Power Flow Analysis

A power flow analysis was performed for two sets of measurement data. The first set was measured using a generic shaker excitation and the second set using MAES. The goal was to evaluate these datasets regarding its power flow vector fields and the quantification of power transmission from wing into suspension. The results of the two datasets were compared at four specific single frequencies (engine rotation speeds: $0.5 \times$ N1, N1, $2 \times$ N1 and N2) which the MAES used for the structural excitation.

The visualization of the power flows, resulting from the MAES excitation was successful and allows to clearly see the flow directions and regions of highest vibrational power. Additionally, the comparison between power flow fields based on generic and MAES excitation showed a significant difference, indicating that the MAES operated as expected. During the evaluation of the power flows originating from the generic excitation it was identified that the front spar of the wing might be a component which transfers a considerable amount of vibrational power. However, this component was not equipped during the measurement campaign and leaves this question unanswered for now. A numerical study with the updated wing model might bring further insights.

The quantification of power transmissions, based on power flows approximated from measurement data, was successful and resulted in physically reasonable values. The power flow evaluations of the MAES excitation showed that the transmitted power flow into the suspension is highest for the main engine rotation speeds (N1, N2) and also that the ratio between transmitted power and input is the highest for them. This was not the case for the generic excitations while it is reasonable to assume that the main rotation speeds are significant for the power flow transfer throughout the wing. Again, this shows also that the MAES allows to create more realistic structural responses.

Additionally, the successful attempt of evaluating the power flow based on measured structural response of a complex mechanical structure such as this aircraft wing has to be seen as a result itself. To the authors knowledge a comparable study has not been done so far on a large scale like this. The insights acquired during this campaign allow to further develop and improve the currently used methods.

6.3 Multi-axial Excitation System

As described in detail, an adaptive excitation system has been proposed, developed and tested. Such a system provides a significant advance in the vibracoustic research field by enabling the replication of vibrational signatures of aircraft engines and thus simplifying the requirements for this type of research.

The developed system is able to replicate on an arbitrary structure several different configurations of forces and moments of force, from a simple single channel of force injection to a multi-channel force configuration with excitation in all directions of space. It is also possible to combine those forces with the generation of moments of force in multiple directions as well. As such, the system offers great flexibility in terms of possible configurations.

As part of the development, the control algorithm of the system evolved through several iterations in order to improve performance, but more importantly in order to scale up the controller to be able to manage Multi-Input Multi-Output configurations of different numbers of sensors and actuators.

Previous laboratory evaluations were carried out by selecting a set of reference vibrational profiles and then asking the adaptive excitation system to try to replicate these profiles on an arbitrary structure. The results of all laboratory configurations tested demonstrated correct behavior of the system.

The real scale evaluation performed on the A320 wing showed that the system was able to scale up appropriately and correctly. The result of the evaluation demonstrated the correct replication of the pre-defined Reference forces profile on the structure, whereby this good agreement was able to confirm the potential of MEAS as a valuable tool for vibroacoustic research.

Abbreviations

$\angle$	Phase
Δ	Time Delay
$\hat{y}(k)$	Output Signal
ω	Angular Velocity
ϕ	Eigenvectors of the Datasets
$\mathbf{F}(f)$	Force Signal
$\mathbf{v}(f)$	Velocity
$-_{aft}$	Aft Attachment
$-_{fwd}$	Forward Attachment
$\widehat{C}$	Control filter
d(k)	Delayed Reference Force Signal
DAC	Digital to Analog Converter
DLR	Deutsches Zentrum für Luft- und Raumfahrt (German Aerospace Center)
DLR AE	DLR Institute of Aeroelasticity
DOF	Degree of Freedom
$E_{kin}^{exp}, E_{kin}^{num}$	Experimental and Numeric Kinetic Energy Distributions
ECC	Energy Correlation Criterion
EMA	Experimental Modal Analysis
f	Frequency
F_x, F_y, F_z	Forces in x-, y- and z-direction
FE	Finite Element
FEM	Finite Element Method
FRF	Frequency Response Function
Fx-LMS	Filtered-x Least Mean Square
I/O	Input and Output
IEPE	Integrated Electronics Piezo Electric
LMS	Least Mean Square
M	Moment
M_x, M_y	Moment around x- and y-direction
MAC	Modal Assurance Criterion
MAES	Multi-Axial Excitation System
MIMO	Multiple Input Multiple Output
n, **n**	Frequency Band, Normal Vector
N_x, N_y	Normal Forces
N_{xy}, N_{yx}	Shear Forces
N1	Low-pressure Shaft Speed
N2	High-pressure Shaft Speed

NLMS	Normalized Least Mean Square
P	Power Flow or System Plant
Q	Transversal Forces
r(k)	Reference Force Signal
RLS	Recursive Least Square
S	Path Length
SISO	Single Input Single Output
SNR	Signal to Noise Ratio
STI	Structural Intensity
$.^T$	Transposition
$.^H$	Hermitian Transposition
x, X, y, Y, z, Z	Cartesian Co-ordinates

Acknowledgements This project has received funding from the Clean Sky 2 Joint Undertaking under the European Union's Horizon 2020 research and innovation program under grant agreement N° CS2-LPA-GAM-2020-2023-01.

References

1. Allemang RJ (2002) The modal assurance criterion—twenty years of use and abuse. In: European air and space conference (CEAS) 2009. Los Angeles, CA, USA
2. Biedermann J, Winter R, Norambuena M, Böswald M (2018) A hybrid numerical and experimental approach for structural intensity analysis of stiffened lightweight structures. In: 28th International conference on noise and vibration engineering. ISMA 2018. Belgium, Leuven, pp 4101–4115
3. Biedermann J, Winter R, Norambuena Gonzalez M, Böswald M (2017) Classification of the mid-frequency range based on spatial Fourier decomposition of operational deflection shapes. In: 24th International congress on sound and vibration (ICSV24). London, UK
4. Biedermann J, Winter R, Wandel M, Böswald M (2017) Energy based correlation criteria in the mid-frequency range. J Sound Vib 400:457–480. https://doi.org/10.1016/j.jsv.2017.04.024
5. Boden F, Ahlefeldt T, Geisler R, Meier zu Ummeln R, Moreau A, Norambuena Gonzalez M, Schröder A, Spehr C, Stürmer A (2022) Current advancements of numerical methods and experimental means for the integration of future propulsion systems. In: Deutscher Luft- und Raumfahrt Kongress DLRK 2022. Dresden, Germany
6. Boden F, Barth HP, Hein S, Heitmann D, Kirmse T, Klein C, Krüger WR, Lemarechal J, Norambuena M, Schäfer D, Schnell R, Sinha K, Theiß A, Weiss A (2021) Recent developments of common numerical methods and common experimental means within the framework of the large passenger aircraft programme. In: DGLR (ed) Deutscher Luft- und Raumfahrtkongress 2021. Bremen, Germany and virtual
7. Daborn PM, Ind PR, Ewins DJ (2013) Replicating aerodynamic excitation in the laboratory. In: Topics in modal analysis, vol 7: Proceedings of the 31st IMAC conference on structural dynamics. Garden Grove, CA, USA

8. Hambric SA (1990) Power flow and mechanical intensity calculations in structural finite element analysis. J Vib Acoust 112(4):542—549. https://doi.org/10.1115/1.2930140
9. Hambric SA, Taylor PD (1994) Comparison of experimental and finite element structure-borne flexural power measurements for a straight beam. J Sound Vib 170(5):595–605. https://doi.org/10.1006/jsvi.1994.1089
10. Hunter N, Cross KR, Nelson G (2019) The cross spectrum in multiple input multiple response vibration testing. In: Topics in modal analysis & testing, vol 9
11. Klimmek T (2009) Parameterization of topology and geometry for the multidisciplinary optimization of wing structures. In: European air and space conference (CEAS) 2009. Manchester, UK
12. Maysenhölder W (1994) Körperschallenergie. Hirzel, Stuttgart/Leipzig, Germany
13. Misol M (2018) Experiments on noise reduction in aircraft with active sidewall panels. In: 25th International congress on sound and vibration. Hiroshima, Japan
14. Musella U, D'Elia G, Manzato S, Peeters B, Marulo F, Mucchi E, Guillaume P (2017) Tackling the target matrix definition in MIMO random vibration control testing. In: Aerospace testing seminar. Los Angeles, CA, USA
15. Musella U, D'Elia G, Manzato S, Peeters B, Guillaume P, Marulo F (2017) Analyses of target definition processes for MIMO random vibration control tests, pp 135–148. https://doi.org/10.1007/978-3-319-53841-9_12
16. Musella U, Peeters B, Marulo F, Guillaume P (2020) Multi-input multi-output swept sine control: a steepest descent solution for a challenging problem. In: Special topics in structural dynamics & experimental techniques, vol 5. https://doi.org/10.1007/978-3-030-12243-0_13
17. Norambuena M, Böswald M, Govers Y Vibro-acoustic analysis of flight test data comprising fuselage vibrations, external pressure and interior cabin noise measurements. San Diego, CA, USA. https://doi.org/10.2514/6.2016-1013
18. Norambuena M, Winter R (2019) Active structural excitation as known source for structure-borne noise research. In: 48th International congress and exhibition on noise control engineering, INTER-NOISE 2019 MADRID. Madrid, Spain
19. Norambuena M, Winter R (2019) An adaptive structural excitation system as a tool for structure-borne noise research. In: ICA 2019—23rd International congress on acoustics. Aachen, Germany
20. Norambuena M, Winter R, Biedermann J (2018) Accurate structural excitation through adaptive control. In: 28th International conference on noise and vibration engineering. ISMA 2018. Belgium, Leuven, pp 1503–1513
21. Norambuena M, Winter R, Biedermann J (2018) Inverse control applied to structural excitation systems. In: DAGA 2018—44. Jahrestagung für Akustik. Munich, Germany
22. Peeters B, Van der Auweraer H, Guillaume P, Leuridan J (2004) The PolyMAX frequency-domain method: a new standard for modal parameter estimation? Shock Vib 11:395–409. https://doi.org/10.1155/2004/523692
23. Plett GL (2001) Efficient linear MIMO adaptive inverse control. In: IFAC proceedings volumes, vol 34, issue 14, pp 89–94. https://doi.org/10.1016/S1474-6670(17)41604-1
24. Schultz RA (2020) A demonstration of force estimation and regularization methods for multi-shaker testing. In: Sensors and instrumentation, aircraft/aerospace, energy harvesting & dynamic environments testing, vol 7
25. Widrow B, Walach E (1995) Adaptive inverse control. Prentice-Hall Inc.
26. Winter R, Biedermann J, Böswald M, Wandel M (2016) Dynamic characterization of the A400M acoustics fuselage demonstrator. In: Inter-Noise 2016. Hamburg, Germany, pp 3296–3306
27. Xiao T, Qiu X, Halkon B (2020) Ultra-broadband local active noise control with remote acoustic sensing. Sci Rep 10. https://doi.org/10.1038/s41598-020-77614-w
28. Zettel S, Böswald M, Winter R (2023) Jet-engine vibration model for the estimation of pylon-wing interface loads. In: DAGA 2023. Hamburg, Germany
29. Zimcik DG (2004) Active control of aircraft cabin noise. In: NATO. Ottawa, Canada

Full-Scale Particle Image Velocimetry in Aircraft Related Flows Under Open-Air Conditions

Reinhard Geisler, Janos Agocs, Fritz Boden, Oguzhan Erdogdu, Selina Ewald, Carsten Fuchs, Heiko von Geyr, Tania Kirmse, Tobias Kleindienst, Lars Koop, Florian Philipp, and Andreas Schröder

Abstract Particle Image Velocimetry (PIV) is a well-established tool for flow field measurements in laboratory experiments as well as in industrial wind tunnel environments. However, real-scale aircraft investigations are still a challenging task, especially when performed under open-air conditions. In this publication the main problems and possible solutions are highlighted in short reviews. Moreover, an engine intake flow field measurement in an aircraft ground test is presented in more detail.

Keywords PIV · Aircraft ground test · Aircraft in-flight test · Full-scale · Open-air

Coming from the reference case of a laboratory measurement, the application of Particle Image Velocimetry (PIV) faces increasing challenges when the measurements have to be performed in industrial wind tunnel environments or, in one step further, under real-scale open-air conditions. In wind tunnels, high velocities up to the transonic range are common and in general can be handled using PIV. More demanding difficulties arise, for example, from vibrations or model deformation, as well as from limited optical access. Real-scale aircraft measurements encounter similar restrictions, but typically have additional restrictions such as large (optical) distances and large fields of view. Open-air metrology in turn needs additional measures to protect the sensitive equipment from environmental influences, and, conversely, to protect the environment from the application of class 4 lasers [2, 6, 9, 13, 17]. One of the biggest open-air problems is the distribution of tracer particles.

In the first part of the present chapter, selected measurement campaigns are used to give a glimpse into this topic. The focus here is on specific features; for a more

R. Geisler (✉) · J. Agocs · F. Boden · O. Erdogdu · S. Ewald · C. Fuchs · T. Kirmse · T. Kleindienst · L. Koop · F. Philipp · A. Schröder
German Aerospace Center (DLR), Institute of Aerodynamics and Flow Technology, Experimental Methods, Göttingen, Germany
e-mail: Reinhard.Geisler@dlr.de

H. von Geyr
German Aerospace Center (DLR), Institute of Aerodynamics and Flow Technology, Transport Aircraft, Braunschweig, Germany

F. Boden (ed.), *Numerical Methods and Experimental Means for the Integration of Future Propulsion Systems*, Springer Aerospace Technology,
https://doi.org/10.1007/978-3-031-97725-1_12

complete insight the references should be consulted. The second part gives a description of the latest test campaign co-funded by the Clean Sky 2 joint undertaking Large Passenger Aircraft (LPA) Platform 1.

1 Selected Review: Full-Scale Open-Air PIV

1.1 Airbus A320 'ATRA' Ground Test 2013

Within the scope of the DLR Project SAMURAI (**S**ynergy of **A**dvanced **M**easurement Techniques for **U**nsteady and high **R**eynolds Number **A**erodynamic **I**nvestigations), the DLR Airbus A320 ATRA (**A**dvanced **T**echnology **R**esearch **A**ircraft) has been used to perform a ground test campaign. The test was designed to apply different measurement techniques, partly in parallel, to gain information on aerodynamic and aeroacoustic properties. Using data fusion methods, the final outcome can be more complete than the cumulated results of the individual measurement techniques [1, 4, 5, 7, 8, 11, 12, 14, 15, 18].

The test took place in the noise-attenuating hangar of Lufthansa Technik (LHT) at Hamburg Airport, which is a semi-open anechoic chamber designed to protect residents from the noise of aircraft maintenance engine tests (Fig. 1a). The air flow coming from the front side of the building is stratified by a number of guide vanes integrated into the front gates (Fig. 1b), passing through the hangar and the test aircraft and leaving at the open exit in the back.

One of the measurement techniques applied in the one-week test campaign was Particle Image Velocimetry (PIV) within the shear layer between free flow and bypass flow of the IAE V2527 engine [11]. The incoming air is seeded with sub-micrometer-sized Di-Ethyl-Hexyl-Sebacate (DEHS) tracer particles by six Laskin nozzle aerosol generators combined with blowers for turbulent diffusion. These seeding systems are located in the guide channels between the vanes of the front gate. The seeded air is drawn in and accelerated by the engine fan and illuminated by a laser light sheet at the

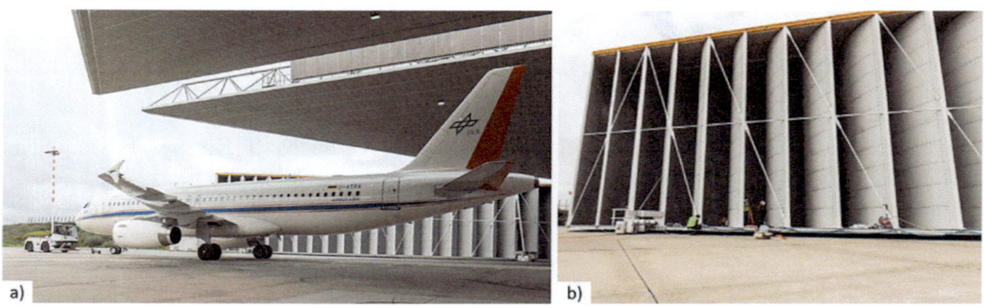

Fig. 1 **a** A320 'ATRA' is pushed into the noise-attenuating hangar at Hamburg airport; **b** Hangar gate with guide vanes

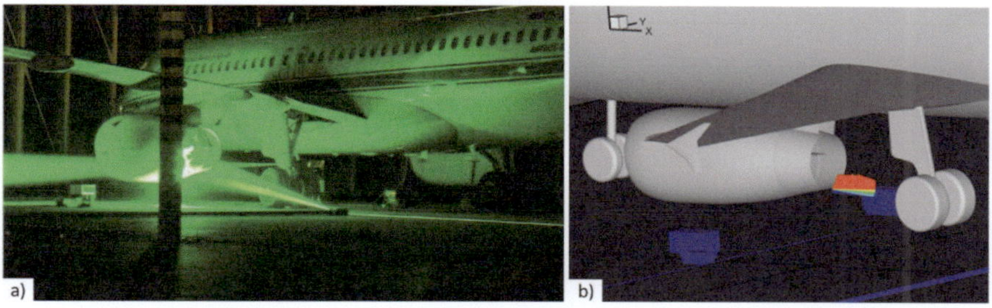

Fig. 2 **a** Laser light sheet during stereoscopic PIV-measurements; **b** field of view in shear layer with sample evaluation

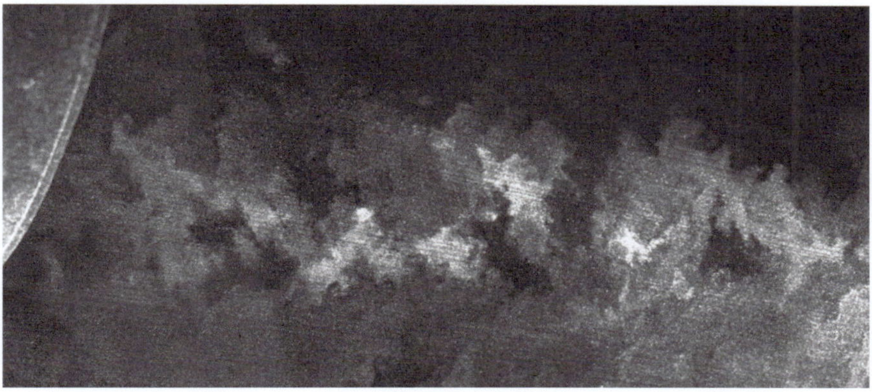

Fig. 3 Raw particle images combined from two camera fields of view. Engine nozzle exhaust to the left

engine nozzle exit—see Fig. 2. Part of the air passes through the combustion chamber where the aerosol is burnt. This unseeded hot core flow is mixed with the bypass flow by a shear layer at the nozzle exit which leads to additional inhomogeneities and schlieren in the corresponding areas of the field of view.

Despite the inhomogeneous seeding distribution visible in the particle images (Fig. 3), it was possible to gain well resolved quantitative results from the cross-correlation evaluation throughout the complete field of view. The only exceptions were two spots with a strong image background which inhibited a proper evaluation (seen as gaps in the 3C velocity vector fields, see Fig. 4).

1.2 Airbus A320 'ATRA' In-Flight PIV 2015

The first PIV measurement on board a civil aircraft took place using the above-mentioned Airbus A320 ATRA as part of the German Luftfahrtforschungsprogramm

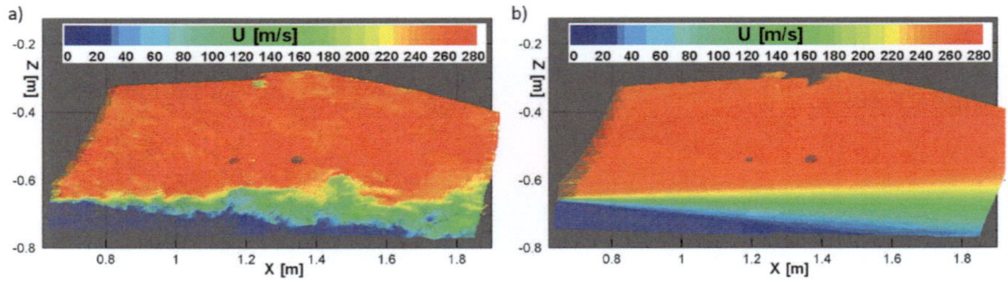

Fig. 4 **a** Instantaneous and **b** averaged 3C velocity vector fields measured by stereoscopic PIV. Colour code shows flow velocity component in X-direction

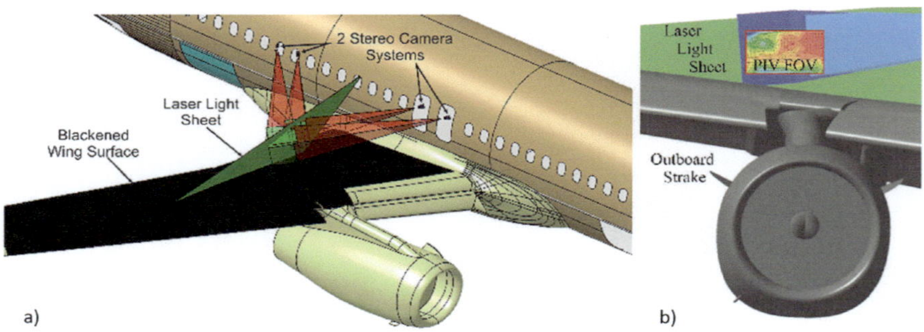

Fig. 5 Sketch of in-flight stereo-PIV [3]: **a** Location of cameras and light sheet; **b** field of view

(Aerospace research program) project HINVA (**H**igh lift **IN**flight **VA**lidation). The aim of the test was the investigation of a high-lift configuration [3, 10].

To compensate for negative effects of engine integration on the high-lift performance of a wing, a vortex generator vane ('strake') can be attached to the nacelle. It is arranged in flow direction during cruise, thus having then only negligible influence on the flow. In contrast, with increasing angle of attack (off cruise mode) it induces an increasingly strong vortex. The A320 ATRA has strakes attached to both sides of the nacelle, inboards and outboards.

For each strake of the starboard nacelle a stereo camera system is used to image the area of the vortex on the suction side of the wing, as shown in Fig. 5. A laser light sheet perpendicular to the flow direction and the wing surface illuminates both fields of view. Laser as well as cameras are installed inside the cabin and optical access is provided through the cabin windows, which results in a backward scatter illumination direction (Fig. 5). The wing surface is blackened with a self-adhesive foil to darken the image background and terminate the laser beam (Figs. 5a and 6).

Since the possibility of providing artificial tracer particles in the air is quite limited for free flight tests, it was decided to use natural seeding in the form of different genera of clouds. This has the disadvantage that the measurements depend on the presence and discovery of clouds with a suitable particle density. Furthermore, with typical cloud particle diameters in the range from 5 to 30 μm, at strong accelerations a slip

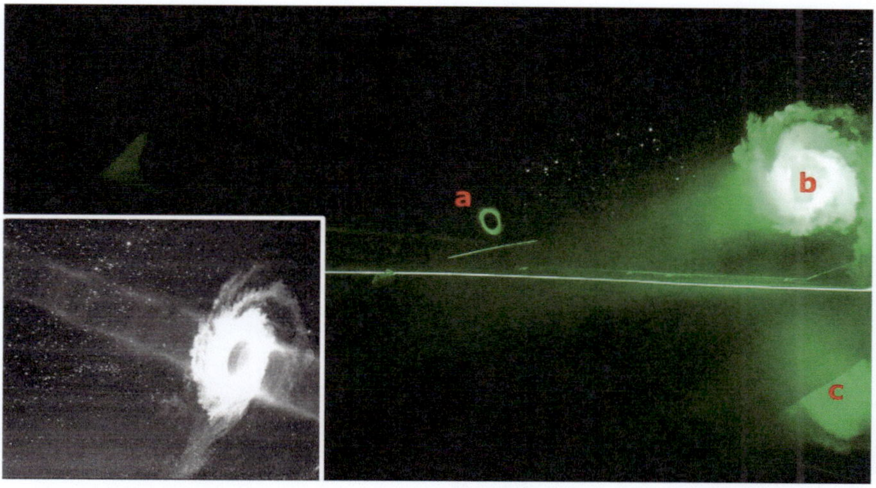

Fig. 6 Visualisation of outboard (**a**) and inboard (**b**) strake vortex illuminated by the laser light sheet on the wing. The inner limit of the wing surface blackening and resulting reflection of inboard strake vortex (**c**) are also seen. A raw image from a measurement camera with in-plane particles and condensation is shown inset to the left, where the vortex tube in the fore- and background illuminated by scattered light can also be seen

between flow and particles will bias the measurements [2]. The cores of sufficiently strong vortices are free from particles due to the temporal integration of the centripetal forces along the vortex tube, which locally hinders all measurements apart from the core localisation itself. Additionally, condensation in the vortices can bias the velocity measurement or even saturate the camera images due to strong laser light scattering (Fig. 6).

In conclusion, most recorded particle images were far from optimal for PIV analysis. The evaluation of single snapshots was only partially successful and hence lead to sparse vector fields. For this reason, the PIV results were averaged over 2000 single snapshots. Selected average results show the strake vortex presence and position as well as its vorticity increasing with the angle of attack (Fig. 7).

1.3 Rolls Royce Engine Test at NASA Stennis Space Centre 2019

A further outdoor test comprising longer optical distances has been performed in the framework of the German Luftfahrtforschungsprogramm (Aerospace research program) project PRESTIGE. It took place at the Rolls-Royce engine test site in the NASA Stennis Space Center in Mississippi. The objective was to investigate

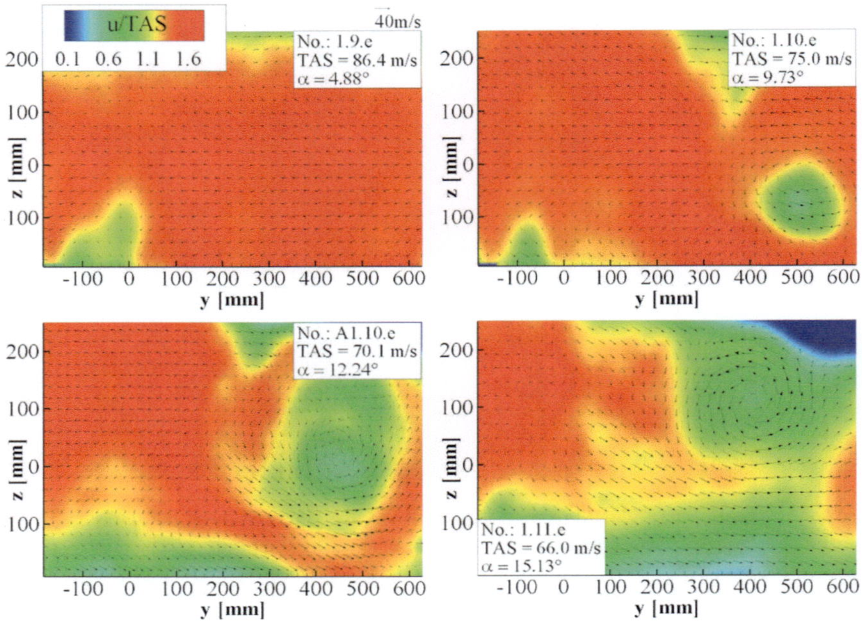

Fig. 7 Averaged velocity vector field of outboard strake vortex measured by PIV at different angles of attack. In-plane flow velocity components are represented by arrows, the out-of-plane component relative to the true air speed (TAS) by the colour-coded contour plot. The measurement area has been limited by the extent of the laser light sheet (blue, for example in the upper right corner)

the inflow properties of an engine under cross-wind conditions and gain data for validation of computational fluid dynamics (CFD) [16].

For this purpose, the engine under test was equipped with windows in the nacelle. Through these windows, laser light sheets can illuminate an observation plane within the nacelle intake. The downside with these windows is that the laser light sheets will be partly reflected from them, and since the engine is mounted at a height of about 7 m from the ground, the incoming laser light is directed upwards and its reflections may enter the observed airspace. To avoid this under all circumstances, different measures have been taken, including the installation of shielding baffles at the upper limit of the windows (see Fig. 8, both sides of the nacelle). As a result, all reflections could be restricted to the controlled laser safety area. The influence of these baffles obviously had to be considered in the CFD calculations.

A set of cameras in stereo geometry at a distance of about 15 m is used to image the observation plane with the laser-illuminated seeding particles. To achieve the best possible seeding conditions, a seeding rake was installed at the intake of the crosswind-blower. The latter is an open jet flow generator similar to an Eiffel-type wind tunnel, which is mobile and thus can be placed at different positions with respect to the engine to simulate different inflow conditions (Fig. 8, right). Due to the

Fig. 8 Test-setup for stereoscopic PIV intake flow measurements, Rolls-Royce test site at NASA Stennis: a part of the cross wind blower nozzle (**a**) is seen at the right; the baffles (**b**) at both sides of the nacelle limit laser reflections (*Photo* J. Agocs)

combination of particle distribution by the seeding rake and turbulent mixture from the crosswind blower, a sufficiently high homogeneity of the seeding fog could be achieved.

Stennis Space Center is located in a swamp area with a plethora of wild animals such as alligators, deer, poisonous snakes and spiders; it is exposed to extreme variable weather conditions ranging from sunshine to heavy rain and thunderstorms. Hence all components of the measurement system had to be housed not only to be protected from contamination by the weather, but also from wild-life such as spiders in the beam path of the lasers (Fig. 9).

A selected result and the corresponding CFD simulation are depicted in Fig. 10.

2 Dassault Falcon 2000LX 'ISTAR' Ground Test

In 2020, the Dassault Falcon 2000LX 'ISTAR' (**I**n-flight **S**ystems & **T**echnology **A**irborne **R**esearch) has joined DLR's research fleet (Fig. 11). Its first test missions were accomplished within the DLR project HighFly (High Speed Inflight Validation). This project covers the characterisation of aerodynamic, aeroelastic, structural dynamic and propulsion properties. In parallel, the aircraft's basic instrumentation

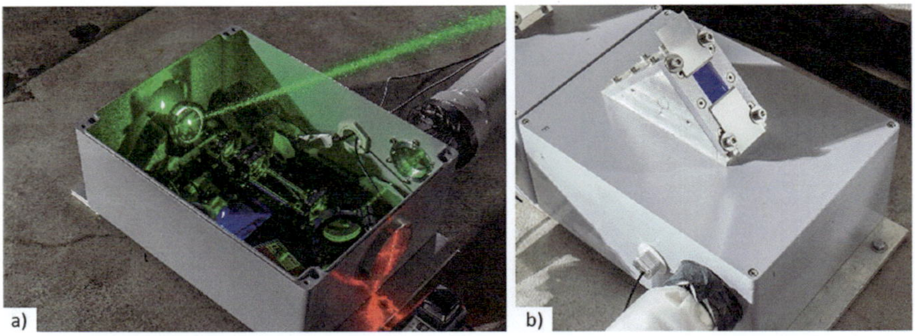

Fig. 9 **a** Beam shaping optics inside protection box; **b** closed protection box with exit window and light sheet limiting apertures

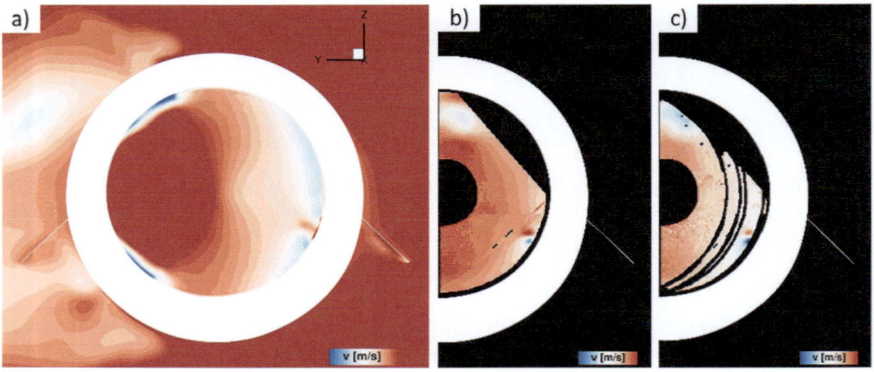

Fig. 10 Comparison of the averaged velocity component in Y-direction (contour plot, cross-wind direction from right to left) between uRANS simulation (**a**), 2C PIV result (**b**) and 3C PIV result (**c**); images from Ref. [16]

has been completed. One of the scheduled tests was a ground test measurement of the intake flow of a rear mounted engine.

2.1 Planning and Preparation

The two Pratt & Whitney PW 300C engines of the ISTAR are mounted on both sides of the rear fuselage, a configuration common to many business jets. The intake flow of the airplane at rest—or at slow velocity—is hence influenced by the close proximity of the fuselage to the engines. An intake vortex is formed which can have an undesirable influence on the fan blades. The present ground test was proposed to measure the properties of the intake flow and in particular of this vortex, with the aim of generating a data base for validation and tuning of CFD codes. To gain a sufficiently

Fig. 11 Dassault Falcon 2000LX 'ISTAR' (In-flight Systems & Technology Airborne Research)

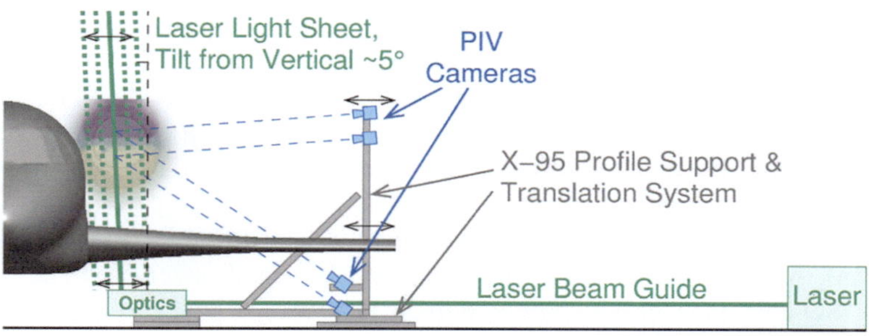

Fig. 12 Sketch of setup for intake flow measurement. The laser light sheet is parallel to the fuselage and slightly off the vertical, and can be placed at different distances to the fuselage, as shown by the dashed green lines

comprehensive insight into the flow features, the measurement is designed to cover the complete intake cross-section. Since PIV can achieve the highest dynamic range for in-plane motions, the two-dimensional measurement field of view is oriented vertically and in main flow direction (i.e. parallel to the aircraft's longitudinal axis). A translation system is foreseen to shift the complete measurement system in spanwise direction and thus capture the complete intake slice by slice without the need of a recalibration. Care was taken to keep all installations at a distance of at least 2 m from the engine axis and downstream of or at least close to the intake plane. A slight tilt of the measurement plane towards the fuselage is chosen to reach areas close to the latter (Fig. 12).

The one-week test campaign was scheduled to take place in the Noise Attenuating Hangar of Lufthansa Technik (LHT) at Hamburg Airport, the facility already used

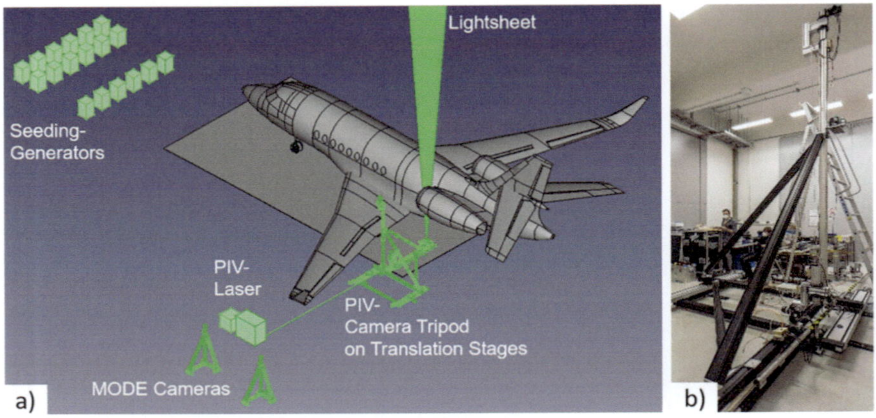

Fig. 13 **a** Total system in the CAD plan; **b** mock-up in lab

for the A320 ground test (Sect. 1.1). Due to the availability of hangar and aircraft the only possible timeslot was in October 2021, a time of the year unfavourable for an outdoor test since unpredictable weather conditions within the limited test time can easily spoil a complete measurement campaign. In addition to the PIV system, a MODE (Marker-Oriented Deformation Estimation) system was foreseen to capture the exact aircraft position which depends, for example, on thrust settings and fuel content. All system components were first arranged with the help of CAD software (FreeCAD, Fig. 13a) and finally implemented and tested in a laboratory mock-up (Fig. 13b). After the proof of reliability, the system was disassembled and packed into an ISO-container for transportation to the test site.

Apart from the technical setup aspects, a variety of administrative preparation steps had to be considered. Open-air application of class 4 lasers, especially near the apron of an operational airport, requires an extensive safety concept. Technical safety measures and shut off laser safety areas have to be established, laser safety distances have to be calculated and the whole operation has to be coordinated with the local authorities. A ground test request has to be submitted to document both the intended installations and especially all aircraft modifications. Additional safety issues have to be identified and addressed, such as:

- Modified ventilation to protect the onboard crew from the seeding-fog (Fig. 14)
- Manufacturer clearance for engine operation in seeding-fog
- Procedure to rule out any influence on the engine prior to the next flight (engine wash, inspection)
- Proof of stability for installations in the flow field close to the aircraft.

Last but not least, the PIV staff needs a clearance for access to the airport apron restricted area and has to be trained in airport safety measures.

Fig. 14 Aircraft with seeding system (**a**) and cabin ventilation duct (**b**) and camera tripods arranged around the port wing (**c**), also showing engine fan (**d**) illuminated for MODE-based localisation of engine axis

2.2 Test Campaign

After arrival at the Hamburg airport apron, the test equipment was unpacked and reassembled. Since the test site itself provides only some basic infrastructure such as electricity and a small test cabin, additional infrastructure had to be installed. A mobile air compressor provides pressured air for the Laskin nozzle-type seeding system, and ISO-containers are used to serve as a test cabin containing the acquisition and control computers and a meeting room.

Prior to the arrival of the aircraft, a preliminary coordinate system had to be defined to enable the reassembly and placement of the measurement system (Fig. 15a). The ISTAR arrived on the second day and was pushed with an accuracy in the centimetre range to a predefined location, after which the final coordinate system was referenced to the actual aircraft position. All components in the vicinity of the aircraft were anchored to the ground. A standard tachymeter with reflectorless distance measurement capability was used to determine their positions. Black sticky foil was attached to the fuselage to darken the background of the PIV images (Fig. 15b). For the dynamic aircraft position measurement with MODE, sticky checkerboard markers were distributed on different parts of the aircraft surface, including the nacelle, and were subsequently located with the tachymeter. Markers on the fan were located by MODE at different fan rotation angles to determine the engine axis position.

To limit operation of class 4 lasers and the resulting safety restrictions to the absolute minimum, eye-safe laser pointers were used for the initial alignment of

Fig. 15 **a** Setup and preliminary placement of tripod on translation system; **b** Application of the black sticky foil onto the fuselage for a darkened image background

Fig. 16 Measurement in progress: MODE camera tripods (**a**) and PIV laser system (**b**) are visible in the foreground, PIV camera tripod and light sheet (**c**) directly behind the port wing

the laser beam path and the optical systems. Only for the final alignment and for the measurements themselves were the main lasers then operated, but only after consultation with and approval from the airport tower (Fig. 16). For the camera-based measurement systems the positions of the cameras were determined with the tachymeter. The imaging systems themselves were calibrated with corresponding calibration targets (Fig. 17). Inaccuracies in the alignment of the PIV calibration target as well as the influence of translation system and light sheet thickness were compensated for in the final evaluation process (disparity correction).

Contrary to the ATRA test 2013 (Sect. 1.1), the seeding system was placed closer to the field of view at about 5 m in front of the aircraft (Fig. 14). Each seeding

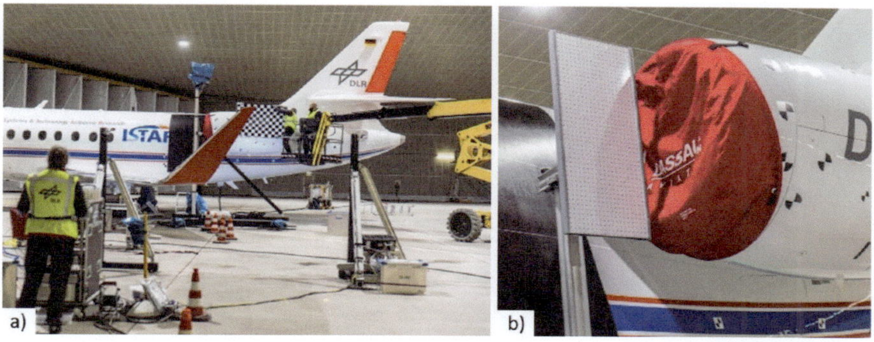

Fig. 17 **a** Calibration of MODE cameras with checkerboard pattern target in different positions; **b** Calibration of PIV cameras with dot pattern target in central measurement plane

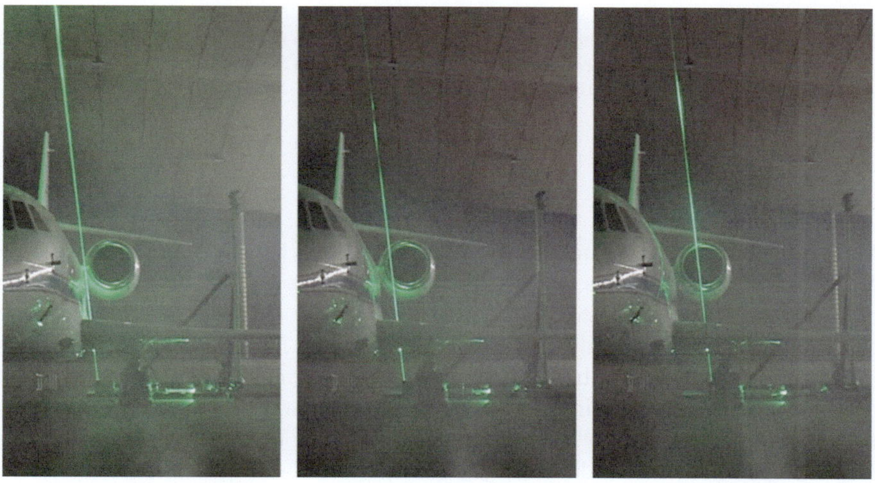

Fig. 18 (Nearly) vertical laser light sheet in seeding fog and camera tripod in different translation positions

generator is supplemented by a blower to direct and distribute the seeded air towards the engine intake. Despite the fact that the sound-attenuating hangar appears to be an area protected from environmental influences, the airflow inside still strongly depends on the outside weather. Squally wind can pass almost unattenuated through the intake vanes of the hangar and thus affect the direction of the seeded airflow. Hence the final adjustment of the seeding system took place with the aircraft engines in operation under test conditions in order to obtain the best possible seeding distribution in the field of view (Fig. 18).

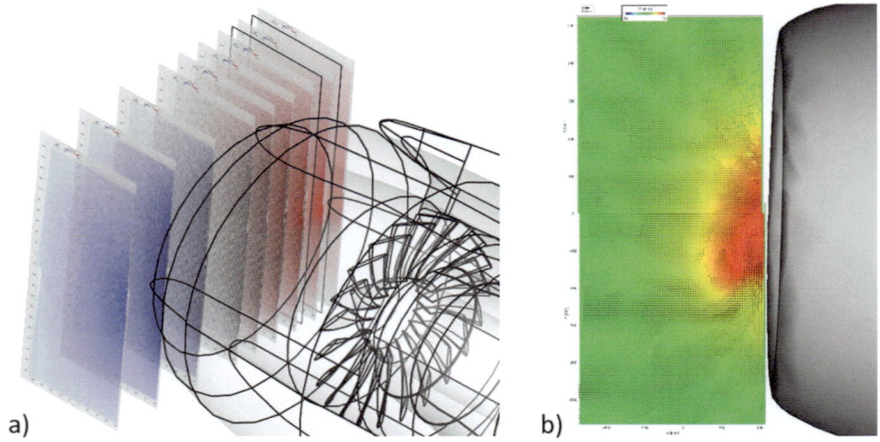

Fig. 19 **a** Sketch of measurement plane positions with respect to the engine; **b** a selected result in a plane close to the fuselage. The colour-coded contour plots in **b** depict the averaged out-of-plane velocity component at thrust setting 23% N1 (idle)

2.3 Results

Measurements were performed at thrust settings 23% N1 (‘idle’), 50% N1 and 80% N1. Since the focus of this test was the characterisation of the intake vortex, the inboard measurement planes were chosen to be more closely spaced than in the centre and towards outboard (4 × 50 mm, 3 × 100 mm, 3 × 200 mm, Fig. 19a).

Despite a careful adjustment of the seeding system, the particle distribution remained inhomogeneous in parts of the field of view and prone to instabilities which were made worse by the squally winds outside. However, almost all image parts—even where the seeding density was quite low—in general provided sufficient signal-to-noise ratio for a good correlation analysis of most instantaneous PIV images.

An exception is the vortex core, which is free from seeding particles due to the centripetal forces. Thus, the flow velocities in the central part of the vortex cannot be determined. Moreover, the wet weather conditions in October had led to condensation in the areas around the vortex core at higher thrust settings. In these areas an evaluation was either not possible or biased due to the high particle density and the scattering brightness. Nevertheless, the core position could be located; results revealed that the vortex core position is highly unsteady (Fig. 20).

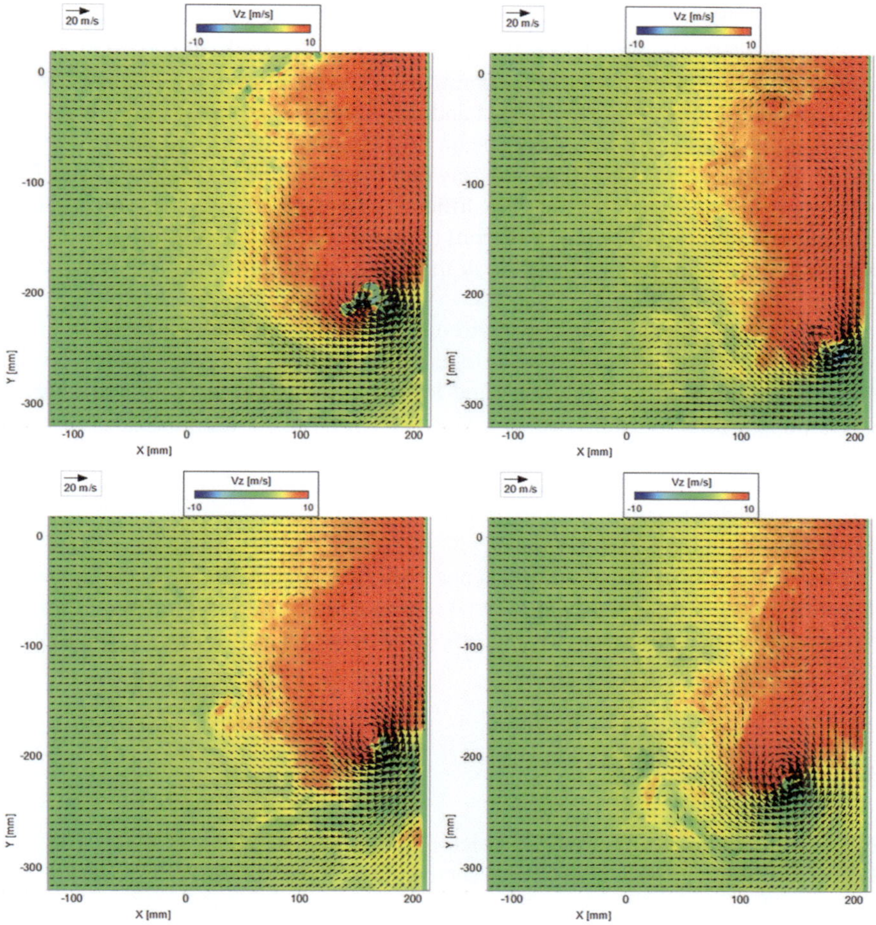

Fig. 20 Selected instantaneous snapshots of unsteady intake vortex near fuselage at thrust level 23% N1. The contour plots depict the out-of-plane velocity component. The four plots show measurements at arbitrary different times

3 Conclusion

A business jet engine intake flow measurement using stereoscopic Particle Image Velocimetry has been performed in a ground test at Hamburg Airport. Three-component velocity vector fields have been obtained in a field of view of 45 cm width in aircraft axis direction and height of 100 cm in vertical direction. This observation area was shifted in span wise direction by use of a translation stage system, thus covering slice by slice the complete intake volume in front of the engine.

The main challenge in open-air PIV is to provide a sufficiently homogeneous distribution of tracer particles in the field of view. Since this is very sensitive to varying wind conditions, seeding generation should be placed as close as possible to the observation area. On the other hand, the seeding distribution itself has to be designed to have a negligible influence on flow conditions, which requires a certain minimum distance. One solution is to rely on natural seeding such as clouds or fog, but since exact weather conditions are difficult to forecast, this will be difficult given the tight scheduling and available aircraft access times. Moreover, one has to accept that natural particle sizes may not be in the optimum range to faithfully follow the flow, especially at higher accelerations.

Further requirements arise from environmental influences such as rain and high air humidity, but also from the local wildlife. Long distances from the aircraft complicate the optical setups, limit resolutions and lead to increased sensitivity with respect to vibrations. Last but not least, the laser safety concepts are more complicated than in closed buildings and require additional measures to limit all direct, reflected and diffuse worst-case laser light emissions to the pre-defined laser safety area. All emissions outside that area have to be shown to stay below the corresponding safety thresholds.

Despite the challenging conditions and identified limitations, it could be shown that with carefully designed experimental setups, real-scale aircraft-related flows can be investigated using stereoscopic PIV, even under various harsh open-air and inflight conditions.

Abbreviations

2C	Two-component (velocity vector)
3C	Three-component (velocity vector)
ATRA	Advanced technology research aircraft (Airbus A320 research aircraft)
DEHS	Di-ethyl-hexyl-sebacate
DLR	Deutsches Zentrum für Luft- und Raumfahrt (German Aerospace Center)
CFD	Computational fluid dynamics
ISTAR	In-flight Systems & Technology Airborne Research (Dassault Falcon 2000LX research aircraft)
MODE	Marker-oriented deformation estimation
N1	Rotational speed of engine fan and low-pressure compressor; percentage of the engine maximum standard operating value
NASA	National Aeronautics and Space Administration
PIV	Particle image velocimetry
TAS	True air speed

Acknowledgements Parts of this work were funded by the Clean Sky 2 Joint Undertaking under the European Union's Horizon 2020 research and innovation program under grant agreement N° CS2-LPA-GAM-2020-2023-01.

Two of the test campaigns in the brief review were funded by the German Luftfahrtforschungsprogramm (aerospace research program) projects HINVA and PRESTIGE. The authors would also like to acknowledge the support of the DLR Flight Facility Braunschweig.

References

1. Bassetti A, Berkefeld T, Schroeder A, Siller HA, Wrede B (2016) Correlation analysis between the jet exhaust velocity and microphone-array acoustic measurements for a turbo-fan engine. In: AIAA SciTech 2016—54th AIAA aerospace sciences meeting. Sandiego, CA, USA. https://doi.org/10.2514/6.2016-0112
2. Dunker C (2020) An in-flight investigation of a turbulent boundary layer at Reynolds numbers up to $Re_\theta = 49{,}000$. Exp Fluids 62(4):1–12. https://doi.org/10.1007/s00348-020-03072-9
3. Dunker C, Geisler R (2018) Full-scale in-flight flow investigation of a high-lift vortex system by means of particle image velocimetry. In: New results in numerical and experimental fluid mechanics XI. Notes on numerical fluid mechanics and multidisciplinary design, vol 136. Springer, Cham, pp 523–532. https://doi.org/10.1007/978-3-319-64519-3_47
4. Ebel PB, Schnell R, Schönweitz D, Kirmse T (2016) Structural modelling and validation of a V2500 honeycomb core fan blade. In: AIAA SciTech 2016—54th AIAA aerospace sciences meeting. Sandiego, CA, USA. https://doi.org/10.2514/6.2016-0387
5. Geisler R, Schröder A, Schanz D, Agocs J, Siller H (2016) Large-scale density gradient visualization of the V2527 engine jet flow at ground operation using background oriented Schlieren (BOS). In: AIAA SciTech 2016—54th AIAA aerospace sciences meeting. Sandiego, CA, USA. https://doi.org/10.2514/6.2016-0109
6. Henning A, Richard H, Kowalski T, Gries T, Huntgeburth S, Loose S (2016) Full scaled high speed train PIV measurement. In: 18th International symposium on applications of laser techniques to fluid mechanics. Lisbon, Portugal
7. Keller D (2016) Numerical simulation of a fully integrated engine ground test. In: AIAA SciTech 2016—54th AIAA aerospace sciences meeting. Sandiego, CA, USA. https://doi.org/10.2514/6.2016-0389
8. Kirmse T, Ebel PB, Schroeder A (2016) Image base fan blade deformation measurements on an Airbus A320 V2500 engine in ground operation. In: AIAA SciTech 2016—54th AIAA aerospace sciences meeting. Sandiego, CA, USA. https://doi.org/10.2514/6.2016-0386
9. Raffel M, Richard H, Agocs J, Otter D, Mattner H, Göhmann U (2001) Experimental aspects of PIV applied to a Bo 105 helicopter in Hover-flight condition. In: 4th International symposium on particle image velocimetry. Göttingen, Germany
10. Rudnik R, Schwetzler D (2016) High lift INflight VAlidation (HINVA)—overview about the 2nd flight test campaign. In: AIAA SciTech 2016—54th AIAA aerospace sciences meeting. Sandiego, CA, USA. https://doi.org/10.2514/6.2016-0041
11. Schröder A, Geisler R, Schanz D, Wrede B, Agocs J (2016) Flow field investigations in the free bypass jet flow of a V2527 engine at ground operation using SPIV. In: AIAA SciTech 2016—54th AIAA aerospace sciences meeting. Sandiego, CA, USA. https://doi.org/10.2514/6.2016-0108
12. Schröder A, Geisler R, Schanz D, Agocs J, Pallek D, Schroll M, Klinner J, Beversdorff M, Voges M, Willert C (2014) Application of image based measurement techniques for the investigation of aeroengine performance on a commercial aircraft in ground operation. In: 17th International symposium on applications of laser techniques to fluid mechanics. Lisbon, Portugal

13. Schwarz C, Bauknecht A, Wolf CC, Coyle A, Raffel M (2020) A full-scale rotor-wake investigation of a free-flying helicopter in ground effect using BOS and PIV. J Am Helic Soc 65:1–20. https://doi.org/10.4050/JAHS.65.032007
14. Schönweitz D, Becker R, Ebel PB, Schnell R, Schroll M (2016) Aerodynamic performance characteristics of the installed V2527 fan at ground operation. In: AIAA SciTech 2016—54th AIAA aerospace sciences meeting. Sandiego, CA, USA. https://doi.org/10.2514/6.2016-0111
15. Siller HA, Bassetti A, Funke S (2016) SAMURAI—jet noise source analysis of a V2500 engine. In: AIAA SciTech 2016—54th AIAA aerospace sciences meeting. Sandiego, CA, USA. https://doi.org/10.2514/6.2016-0110
16. Stuermer A, Geisler R (2023) CFD modeling of turbofan intake aerodynamics at crosswind conditions. In: AIAA AVIATION 2023 forum. San Diego, CA, USA. https://doi.org/10.2514/6.2023-3306
17. Toloui M, Riley S, Hong J, Howard K, Chamorro LP, Guala M, Tucker J (2014) Measurement of atmospheric boundary layer based on super-large-scale particle image velocimetry using natural snowfall. Exp Fluids 55:1737. https://doi.org/10.1007/s00348-014-1737-1
18. Wolters F, Becker R, Schnell R, Ebel PB (2016) Engine performance simulation of the integrated V2527—engine fan. In: AIAA SciTech 2016—54th AIAA aerospace sciences meeting. Sandiego, CA, USA. https://doi.org/10.2514/6.2016-0388

3D Lagrangian Particle Tracking with Shake-The-Box Applied to Large-Volume Measurements

Daniel Schanz, Andreas Schröder, Johannes Bosbach, Tobias Strübing, Tobias Jahn, and Janos Agocs

Abstract The developments over the last decade have transformed 3D Lagrangian Particle Tracking from a tool to measure averaged particle properties in small volumes (a few cubic centimeters) to a fully integrated environment, capable of delivering time-series of spatially highly resolved snapshots within scalable volumes up to several cubic meters. The combination of advances both on the software side (with the introduction of the Shake-The-Box method and data assimilation techniques) and the hardware side (LED illumination, Helium-filled soap bubble seeding, better cameras) enabled a rapid expanse in measurement capabilities. Here we outline the most relevant developments in evaluation methods and demonstrate their capabilities by application to experimental data of the flow around a Quadcopter drone.

Keywords Lagrangian particle tracking · Volumetric velocimetry · Large-scale flow · Quadcopter

1 Introduction and Motivation

The past decades have seen a tremendous increase in the amount of information that can be gained from experimental investigations by applying image-based measurement techniques (complementary or alternative to existing probe techniques). For experiments in unsteady and turbulent flows non-intrusive volumetric and time-resolved measurement techniques determining all three components of the velocity- (and acceleration) vectors at many points inside the investigated volume instantaneously are highly desired [7, 41]. Consequently, particle image-based velocimetry measurement techniques such as Particle Image Velocimetry (PIV) [31] have been extended in recent years from 2D to 3D by tomographic methods (tomo-PIV; [8,

D. Schanz (✉) · A. Schröder · J. Bosbach · T. Strübing · T. Jahn · J. Agocs
German Aerospace Center (DLR), Institute of Aerodynamics and Flow Technology, Göttingen, Germany
e-mail: Daniel.Schanz@dlr.de

A. Schröder
Institute of Transport Technology, Brandenburg University of Technology, Cottbus, Germany

F. Boden (ed.), *Numerical Methods and Experimental Means for the Integration of Future Propulsion Systems*, Springer Aerospace Technology,
https://doi.org/10.1007/978-3-031-97725-1_13

32]) and from snapshot modes to temporal resolution. Tomo PIV enables flow field estimations on regular 3D velocity vector grids with relatively high spatial resolution based on particle image densities around ~0.05 particles per pixel (ppp). For all 3D reconstruction techniques, the particle image density N_I is typically expressed in terms of 'particles per pixel' (ppp) on single camera projections and determines the possible spatial resolution inside the investigated flow. Particle tracking methods, on the other hand, have evolved from relatively sparse 3D PTV schemes (e.g. [6, 21, 22]) at particle image densities of ~0.005 to 0.01 ppp to Lagrangian particle tracking techniques reaching high particle image densities of ~0.07 to 0.15 ppp by using the Shake-The-Box (STB) method [16, 33].

Ideally, for many relevant flows, time-resolved 3D pressure, temperature and density fields are of high interest as well, but this is quite difficult to achieve directly by LPT methods using, for example, temperature- or pressure sensitive particle tracers or with a separate tomo BOS system [25]. Luckily, at least for incompressible flows, 3D pressure fields can be calculated, for example, directly via integration of the Poisson equation from time-resolved 3D velocity vector fields [10, 29], or implicitly by data assimilation approaches using dense scattered Lagrangian velocity and acceleration vector fields as input [11, 14, 39].

In the following, the preconditions and the methodology of the advanced 3D LPT method STB, some of its recent developments and its application to a large-scale experiment will be described. STB enables reconstructing long time series of a huge number of individual particle trajectories ($N_P \sim$ 105–106) in the investigated flow volume from only a few camera projections. Alternatively, for high speed flows, many short statistically independent and instantaneous (for example, 4, 3 or 2) pulse-tracks, can be reconstructed after having captured particle images in a frame-straddling mode when using multi-pulse and multi-illumination STB techniques [27, 28].

2 Preconditions for 3D LPT Using Shake-The-Box Techniques

For both PTV/LPT and tomo PIV, three-dimensional reconstruction methods of the particle positions, as well as for the estimation of the underlying flow fields, have been developed using various techniques. First approaches used manual tracking of singular particles, dating back to the beginning of the 20th century [24]. 3D LPT techniques operate with particle detection-, triangulation- and tracking algorithms, generating in a first step only 3D position coordinates (x, y, z) and intensities (I) per particle and time step t_i. Classical 3D PTV techniques extract particle tracks with a (each with an individual track ID number) by using a subsequent tracking step [22, 30]. With 3D positions and track IDs (without intensity) we have the sparsest possible representation of the captured flow information, as given by all individually reconstructed particles over time. On the other hand, approaches such as tomo PIV store reconstructed particles as intensity blob distributions in large arrays of voxels

($\sim 10^9$) for each time step, and subsequently operate with local 3D cross-correlation schemes of two (or more) subsequently captured and reconstructed intensity distributions (including ghost particles) to determine a regular 3D grid of (low-pass filtered) three-component (3C) velocity vectors. This procedure has a quite high demand on RAM capacity and computational resources, but has been proven to be robust and reliable when reducing related known bias errors. However, from a perspective of signal processing the use of 3D LPT techniques is advantageous for several aspects, providing that the usable particle concentration is at least comparable to that of tomo PIV. The key advantages are: (a) a faster processing with a sparse representation of particles by using only position, intensity and track ID values, and (b) a maximization of the achievable flow information by tracking each individual particle over time with an improved position, local velocity and acceleration value along the whole track. While the raw position accuracy is already above the one achievable using tomographic methods, the use of a temporal filter function (e.g. 3rd order-B-spline or polynomial) and its derivatives yields further improvements. A third advantage will be demonstrated later when addressing the (advanced) IPR (Iterative Particle Reconstruction) technique (see Sect. 3), which allows individual single-time-step reconstructions at even higher particle image densities than tomo PIV, while producing significantly higher particle position accuracies [16, 52].

Although the 3D LPT method STB relies on particle- position- and tracking-based reconstruction techniques, which in a first view seems very similar to classical PTV approaches, important preconditions for STB have been developed in the process of improving the tomographic PIV technique. The multiplicative nature of the MART (Multiplicative Algebraic Reconstruction Technique) reconstruction process used in tomo PIV [1, 32] requires subpixel calibration accuracies in order to reduce the ghost particle level: Particle image intensities are projected into a voxel space according to the camera calibration along each pixel's lines-of-sight (LOS). This actually acts as a logical "AND" operation, meaning that, in order to produce finally a 3D intensity blob representing a (possible) particle position at the intersection of all camera LOS in the object space, the pixels corresponding to the intersecting LOS on all cameras must exhibit non-zero grey-value intensities. A slight de-calibration of the LOS of a single camera by only one pixel would already create a completely different intensity blob distribution inside the volume, with many more ghost particles and less true reconstructed particles due to the underlying multiplicative nature of MART.

Therefore, an important step for improving the reconstruction quality was the Volume Self Calibration (VSC) technique [50, 51], which corrects the error of the initial target-based 3D camera calibration (typically about one pixel) to errors of ~ 0.1 pixel. VSC uses particle triangulation procedures with sparse particle images (low ppp-values) and calculates their back-projected displacement fields for each sub-volume within the individual cameras in order to correct the initial 2D-3D mapping functions. An accurate 3D camera calibration system is mandatory for a successful MART-based particle field reconstruction, which with PTV had not been recognized as being elementally critical, mainly because PTV tackles much lower particle image densities at < 0.01 ppp (instead of ~ 0.05 ppp for Tomo PIV), so that the epi-polar line intersection error could be treated in a relatively flexible and simple manner. Although the triangulation idea of nearly intersecting epi-polar lines is already an intrinsic part of the PTV technique, the particle based VSC technique had been invented within the framework of voxel-based tomo PIV developments.

Furthermore, building upon VSC, a calibration procedure for the optical transfer function (OTF) [34] of the particles in each sub-volume was developed, accounting for the often varying imaging conditions across the measurement volume between the different cameras and within a single camera frame, thus further reducing the number of reconstructed ghost particles due to more precise projection of the particle image related intensity distribution into the voxel space, followed by the respective subtraction of the found true particle after reprojection into the original images. Note that VSC and OTF are both particle-based in nature and directly use particle imaging properties. This not only makes the gained information ready for use in particle position-based reconstruction schemes such as IPR and STB, but also the necessary algorithmic structures (such as peak detection and triangulation) can be directly used within the IPR algorithm, thereby reducing the developmental effort.

Potentially, the higher the concentration of the seeding, the better the spatial resolution of the reconstructed flow field. While high particle numbers are desirable, the reconstruction process gets increasingly difficult due to increasing particle image densities on the cameras. A growing number of non-existing particles (ghost particles) arises in the reconstruction due to the under-determined nature of the problem. This process heavily depends on the number of imaged particles (N_P), the diameter of the imaged particles (or, more generally, the image source density), the number of cameras (N_C) and the resolution of each camera [9].

In following section, the IPR technique and its recent advances will be presented as a major step towards the dense 3D LPT method STB. The advanced IPR technique [16] has been assessed by reconstructions of 3D particle distributions from synthetic single snapshot particle images at up to 0.16 ppp using four camera projections without typical imaging noise. Furthermore, results and comparisons of various LPT codes, reconstructing 2-pulse, 4-pulse and time-resolved imaging cases of synthetic flows at various ppp-levels, as presented in the 1st International Challenge on Lagrangian Particle Tracking and Data Assimilation, can be found in [48] (which includes IPR and STB code developments at DLR, as presented here, and also the STB code available at LaVision GmbH).

3 Advanced Iterative Particle Reconstruction

The method of Iterative Particle Reconstruction (IPR), introduced by Wieneke in 2012 [52], constitutes a major step forward in Lagrangian Particle Tracking in densely seeded flows [33, 37]. Novel approaches in several key aspects of the algorithm are presented, which, in combination, triple the working range of IPR in terms of particle image densities. The updated method is proven to be fast, accurate and robust against image noise and other imaging artifacts. Most of the proposed changes to the original processing are easy to implement and come at low computational cost (for more details see [16]).

IPR introduces an iterative approach to the triangulation procedure, with an intermediate position optimization and a corresponding removal of apparent ghost particles. Compared to a MART reconstruction, these measures enabled a reconstruction of particle point clouds from images initially with $N_I \leq 0.05$ ppp and with reduced occurrence of ghost particles and higher position accuracies. The position optimization (here also termed 'shaking' due to the process of adjusting the particles positions in 3D space) is performed for each particle individually using image matching techniques; for this reason, the demands on exact calibration with respect to both the triangulation accuracy and the fidelity of the optical transfer function, are very high. As a consequence, an accurate VSC and OTF calibration is mandatory and essential for IPR.

The underlying problem IPR aims to solve is to find a 3D particle distribution $(x, y, z, I)_i$ for $i = 1 \ldots N_P$, whose projection matches a set of measured particle images, simultaneously recorded from multiple cameras from different directions. Here, x, y and z are the three spatial coordinates, I is the particle intensity and N_P the number of particles within the volume. To optimally solve this problem, IPR relies on precise knowledge of the viewing directions of all cameras and the imaging properties of the particles on each camera (Optical Transfer Function).

Given a state vector $(x, y, z, I)_i$, a camera mapping function $(u_i, v_i) = cam((x, y, z)_i)$ and a calibrated 2D particle image shape dependent on the 3D coordinate $shape_i = otf((x, y, z)_i)$, the re-projected camera image can be calculated by adding up the images of all particles:

$$rep_u v = \sum_i I_i shape_i(u - u_i, v - v_i) \tag{1}$$

with (u, v) being the 2D image coordinates. Due to the finite spatial expansion of the particle image shapes, it is sufficient to sum only nearby particles for each pixel uv (a typical particle image diameter is 1.5–4.0 pixel). To find $(x, y, z, I)_i$, cam and OTF so that rep_{uv} optimally assembles the various measured images img_{uv}, the following underlying concepts are applied.

The IPR method is an iterative procedure for the accurate reconstruction of particle positions. At the center of the IPR is the state vector $(x, y, z, I)_i$, containing the interim estimation for the position x, y, z and intensity I of each of the N_P presumed particles. During the reconstruction this state vector gets iteratively replenished with potential missing particles, purged from potential ghost particles and optimized in terms of position and intensity to best match the observed measurement.

The IPR workflow is depicted in Fig. 1 and the individual steps are described below:

1. All peaks present on the current images (being the original measurements in the first iteration and residuals in all following iterations) are detected using a 2D peak detection scheme.

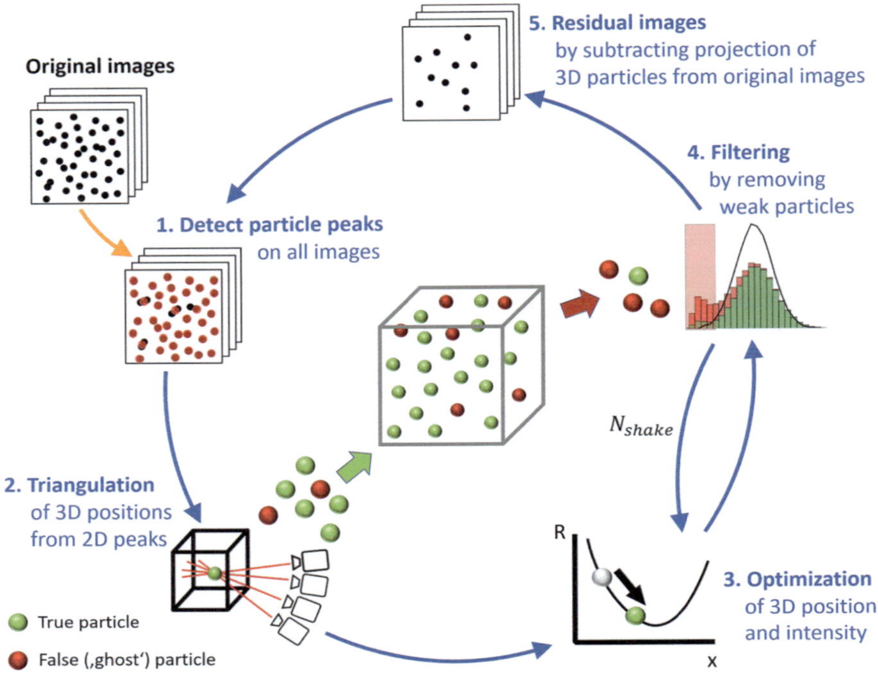

Fig. 1 Scheme of the Iterative Particle Reconstruction (IPR) algorithm. Outer iteration indicated by large arrows. Inner iterations of position/intensity optimization (shaking) and filtering indicated by small arrows (step 3 and 4). Reproduced from [41] (CC BY-NC 4.0)

2. Potential 3D particle positions are triangulated from the found 2D peaks and added to the state vector. Due to ambiguities, inaccuracies of the peak-detection, noisy images and overlapping particle images, the newly added particles might exhibit a significant fraction of ghost particles and suffer from positional inaccuracies.
3. The position and intensity of each particle are optimized individually to best match the measured images on all cameras simultaneously using image matching methods.
4. The state vector is filtered for potential ghost particles, mainly based on the particle intensity (by deleting particles that fall below an intensity threshold).
5. The current interim solution is rendered by projecting all particles onto virtual camera images, as given by the calibration. These projected images I_P are subtracted from the recorded measurement I_{rec}, yielding the residual images I_R.

The process starts anew with peak detection in I_R. Steps 1–5 are repeated until convergence is reached or for a predefined number of outer iterations N_{IPR}. For every outer iteration, N_{shake} iterations of shaking are performed, each of them optionally followed by a filtering step the combination of shaking and filtering we call an inner iteration. At the beginning of every inner iteration the residual images res_{uv}

are determined anew to reflect the changes in position and intensity from the last iteration.

Several key updates to the IPR process have been introduced recently [16] as 'advanced IPR':

1. For the image matching procedure (shaking), the numerical estimation of the derivatives is substituted by analytical derivations of a cost function (being the local residual image). This step enables a direct determination of the gradient (and the Hessian), thereby limiting the number of required evaluations of the cost function.
2. The completeness of the triangulation process benefits from using several permutations of the camera order, as the triangulation result strongly depends on the first two cameras and their relative angle.
3. Instead of using a fixed triangulation error epsilon for all iterations of the IPR procedure, a linearly increasing value of epsilon is applied, which aids in suppressing the generation of ghost particles.
4. Ghost particles can be very effectively suppressed by excluding the camera(s) in which a particle appears the brightest for the optimization of its intensity.
5. In addition to the particle-wise optimization procedure (shaking each single particle independently), an ensemble approach is presented, which optimizes the position of all particles at once ('global shake'). Additionally, this procedure can be used to optimize the camera positions, thus representing a means of performing an in-situ single image calibration correction.

For implementation details, see [16]. In the following, the performance of advanced IPR on a synthetic test data set is shown.

3.1 IPR Performance on Synthetic Test-Cases

A synthetic test case was created, which replicates the one used in [52]: Four virtual pin-hole cameras of 1,300 × 1,300 pixels each are oriented in a pyramidal viewing configuration, where they image randomly created particles in a domain of 50 × 50 × 16 length units. All cameras view the full volume. The particle shape is Gaussian with a diameter of 2 pixel. No variations of the OTF were included. The pixel values are always integers.

First, the performance of the algorithm will be assessed on images free of any defects (case I), and, secondly, with typical measurement noise (case II) (see Fig. 2). For case I the particles have a constant intensity of 1,000 counts in the volume, which is distributed over all pixels of the particle image in each camera (leading to typical peaks of 300–400 counts). No image pre-processing is applied.

For case II (typical noise), the particles have a mean intensity of 1,000 counts in the volume, with a Gaussian distribution with $\sigma^2 = 250$ counts, leading to image peaks of different brightness in the range of ~200–500 counts. A constant Gaussian noise floor with a mean of 40 counts and rms width of 25 counts is added to the

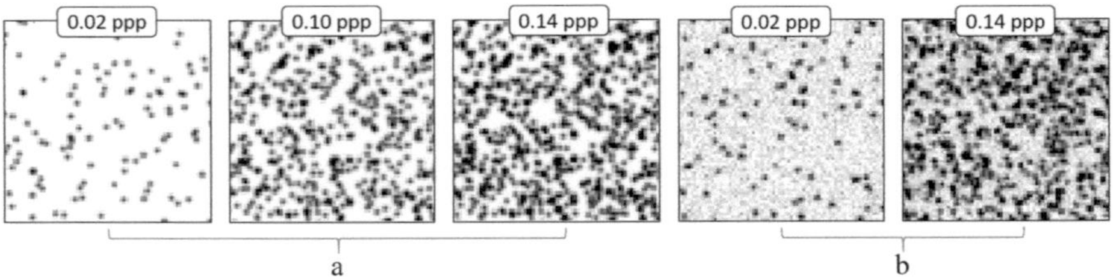

Fig. 2 Excerpts of the synthetic camera images at different particle image densities (ppp-values). **a** without noise and with uniform particle intensities (case I); and **b** with noise (case II). For both cases, effects of particle overlap become increasingly visible at high ppp-values

image. To further simulate photon shot noise, a Poisson process is applied for every pixel, thereby adding noise proportional to the intensity of individual pixels.

Figure 3 shows the convergence behavior of advanced IPR for varying particle image densities, when applied to the described noise-free (left) and the typical moderate image noise (right) synthetic dataset. Shown is the ratio of correctly identified particles R_f (reconstructed within a search radius of 1 pixel around the true particles and normalized to the number of true particles, abbreviated as 'founds'), the ratio of ghost particles R_g (reconstructed particles, for which no true particle could be found within a radius of 1 pixel; also normalized to the number of true particles, abbreviated as 'ghosts') and the average accuracy of the particle positioning of the correctly found particles Δ_f (calculated as the average distance of the re-projected points to the original peak position).

For particle image densities $N_I \leq 0.12$ ppp, the method converges to a stable value of $R_f \approx 1$ in less than 12 to 16 iterations for both the clean (Fig. 3, left) and noisy cases (Scheme of the Shake-The-Box (STB) algorithm. Reproduced from [2] (CC BY-NC 4.0) (Fig. 3, right). Even at $N_I = 0.14$ ppp, the reconstruction converges after 15 iterations for the clean case, where only 8 particles out of 136,477 are not found. For the highest seeding density, eventual convergence is indicated, but has not been achieved within the 20 iterations used.

For low values of N_I the number of ghost particles quickly diminishes after a few iterations. At higher particle image densities, though, a rise in ghost numbers occurs within the first five iterations, which can go up to $R_g = 1$. This behavior can be understood as follows: in these cases many particles could not be identified at this early state and hence the uncertainty of the found ones is high. Therefore, the residual images will show relevant intensities, even in spots where the particles have already been triangulated. These residual intensities will be picked up by the peak finder and new false particles will arise.

However, with the identification of more and more true particles, the ghost level is gradually reduced. As soon as all real particles have been found, the ghosts will then have disappeared. Indeed, for the clean case for $N_I = 0.16$ ppp, ghosts are still present even after 20 iterations (since this case has not yet converged). Similarly for the noisy case (Fig. 3, right), convergence still seems to be possible for $N_I = 0.14$

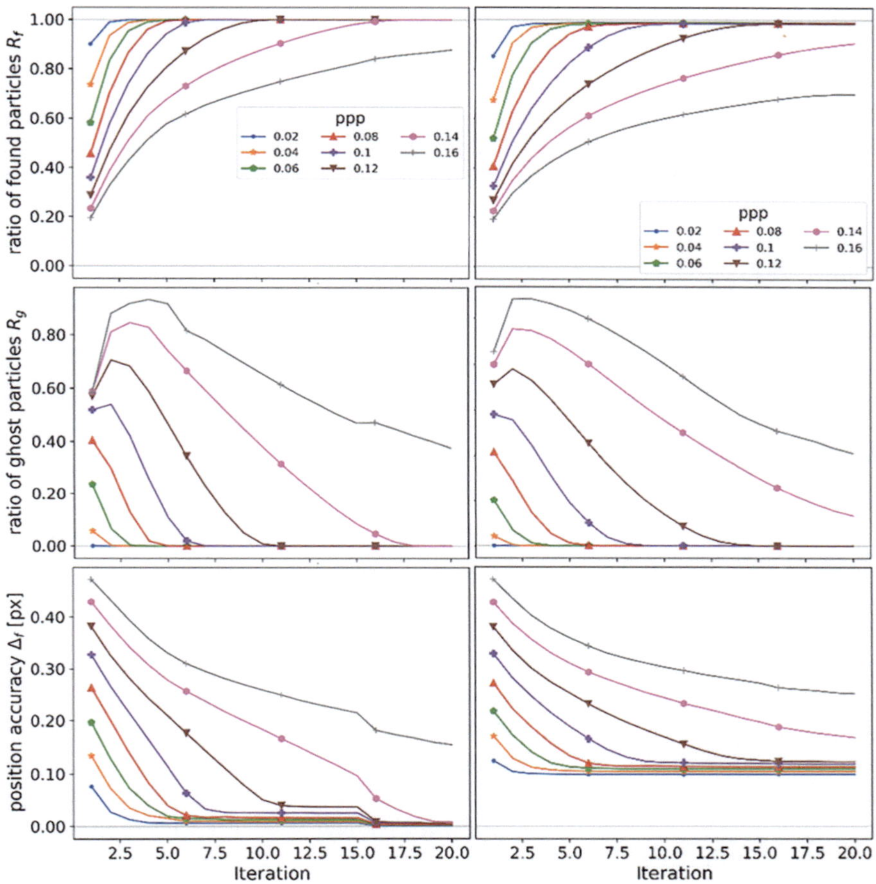

Fig. 3 Convergence of single-image IPR reconstructions for various ppp-values for clean (no noise) (left) and typical image noise (right). Shown are the number of correctly identified particles R_f (top), the ratio of ghost particles R_g (center) and the position accuracy of the true particles Δ_f (bottom) as a function of the iteration number. Reproduced from [16] (CC BY-NC 4.0)

ppp and for all lower ppp values, where the ghost level seems to have been brought down to ~0. However, due to the influence of the image noise, which deteriorates the position optimization scheme by "shaking", a slight decrease of the position accuracy with ppp can be noticed, Δ_f grows slowly to values slightly larger than 0.1 px for the converged phases. As the particle image noise for case II mimics closely that related to typical background optimized experimental conditions, a value of 0.1 px to 0.15 px can be assumed for individual time step reconstruction qualities, when the vast majority of particles have been found and the ghost level is negligible. It will be shown later that optimal temporal filtering after STB track evaluation leads to a further reduction of Δ_f by a factor of ~2.

The position accuracy is very much dependent on the presence of ghost particles and the detection of true particles. As soon as these two have converged, high levels of accuracy have been reached. The reduction of the maximum step width d_{max} at iteration 15 is very visible. For example, for the 0.12 ppp case, the position accuracy dips from a plateau at 0.04 px to 0.007 px. The second reduction of d_{max} at iteration 18 yields a further reduction of Δ_f to 0.005 px at this seeding density. At 0.16 ppp it becomes evident that a reduction of d_{max} leads to a reduction in the speed of convergence, when looking at R_f and R_g. Therefore, this measure should only be applied after convergence is reached.

Compared to the results of Wieneke [52], the method has been greatly extended. Wieneke reported a usable particle image density of $N_I = 0.05$ ppp and a sharp decrease of reconstruction quality for larger ppp, when using 16 outer and 6 inner iterations. Here we find that images at $N_I = 0.14$ ppp produce results completely void of ghost particles and with very few missing real particles (which is far below Wieneke's values at 0.05 ppp). The position accuracy of the advanced IPR is much higher, with values $\Delta_f \leq 0.01$ px for all converged clean cases.

4 3D LPT Using (Time-Resolved) Shake-the-Box

Both previous methods, tomo PIV and 3D PTV, first create a 3D particle reconstruction for each individual time step *separately*.[1] The classical triangulation approach of PTV and the MART-based tomo PIV reconstruction schemes lead to an increasing number of ghost particles with increasing ppp-values of single-particle images. Then, in a second step, the 3D particle distributions of subsequent time steps or along a whole time-series are connected for estimating particle velocities. PTV applies tracking schemes in the temporal domain of the 3D particle position distributions (e.g. [6, 22]). Hereby, high ghost particle levels at high ppp-values produce large ambiguities for the track-connection schemes. Therefore, 3D-PTV has been limited to ppp-values typically below 0.01 ppp, even if various predictive steps and decision-tree strategies for the track building have been implemented (see e.g. [30, 54]). For tomo PIV the second step is to apply a local cross-correlation scheme on the intensity blob distributions for estimating the velocity vector field from two sub-sequent or even several time-steps [31]. Iterative cross-correlation with volume deformation is robust against noise and provides quite reasonable results even at relatively high ghost levels, but only low-pass filtered velocity vectors can be gained (see [19]) and correlated ghosts produce significant bias errors stemming from projections of particles in the bulk flow [9].

Different to that, the key idea of Shake-The-Box [33] is to exploit the *temporal coherence* of the relatively sparse 3D particle trajectories following the flow, which

[1] One exception is the MTE-MART approach to tomographic reconstruction, which iteratively involves two subsequent time-steps in order to reduce the ghost particle problem [26]. S-MTE extended this idea to a time-marching scheme, akin to the STB principle [20].

is already encoded in the time series of particle images. Although it remains an ill-posed reconstruction problem (2D → 3D), we can still use properties of the imaging (e.g. OTF) and the flow physics: *sparsity* of particles in the 3D distribution (mean particle distances are larger in 3D than in projected 2D images), *temporal resolution* of particle image shifts (i.e. sampling rate with respective time-step width below the Kolmogorov time scale) and *smooth 3D trajectories* are given (particle jerk, which is the derivative of the acceleration D^2U/Dt^2, and is small for the vast majority of tracks). Additionally, particles cannot disappear over time inside a 3D volume, so that in STB (after a short four-time-step initialization phase and for each further time step) a *time-marching* approach is applied: *First*, a prediction of the 3D particle positions towards time step t_{n+1} by extrapolation of temporal fitting functions along all found tracks in previous time-steps (e.g. t_{n-3} to t_n) is applied, and then, *secondly*, a correction of the estimated 3D positions is carried out by simultaneous re-projection and image-matching ("shaking") on all cameras for the given time step t_{n+1}. The correction is typically below 1.2 px (less than half the particle image diameter), even for relatively large particle image shifts > 10 px, because of the limited jerk for the vast majority of particle tracks, while local position correction by "shaking" is very accurate (~0.1 px see Chap. 3.1). Ghost tracks which are not coherent over several time steps cannot find respective intensities at the projected image positions at t_{n+1} and are erased when they fall below an intensity threshold. In the following, all already found and extended true particle tracks from previous time steps can be subtracted from the original particle images at time step t_{n+1}, before a *first* 3D particle reconstruction with IPR needs to be done on the less (and in time lesser) populated residual images. In the initialization (the first few time-steps), as well as during the further runtime, new tracks are added by identifying physically reasonable sequences of reconstructed particles (using IPR) in successive frames (typically four). These are then extended to further time steps. This procedure enables a massive reduction of the reconstruction complexity for each further time-step until convergence of the self-optimizing process. The first implementation of these ideas into an STB algorithm had been presented at the ISPIV 2013 [37] and delivered the first 3D LPT results at high ppp-values of a turbulent flow over periodic hills [45].

Once again, STB basically *reverses* the typical process of evaluation: after the initialization step (meant to reconstruct a subset of tracks over the first four recordings), the tracking process *precedes* the reconstruction process. Instead of first determining particle distributions, followed by a deduction of the velocity (by means of correlation or pair-identification), STB uses the available velocity and acceleration information along found tracks to create an (estimated) particle distribution at the next time step. The errors introduced by the estimation are small enough to be easily corrected using image matching; no further partner search is required for known particle tracks, so that the track (with given ID) has already been correctly elongated to the current time-step after the prediction and correction procedure. In this way, the flow physics itself supports the reconstruction process of particle trajectories.

Therefore, STB finally transforms the ill-posed global particle reconstruction problem (present for individual time-steps) into a local simultaneous particle position- and tracking optimization scheme. The prediction-step enables a severe reduction of

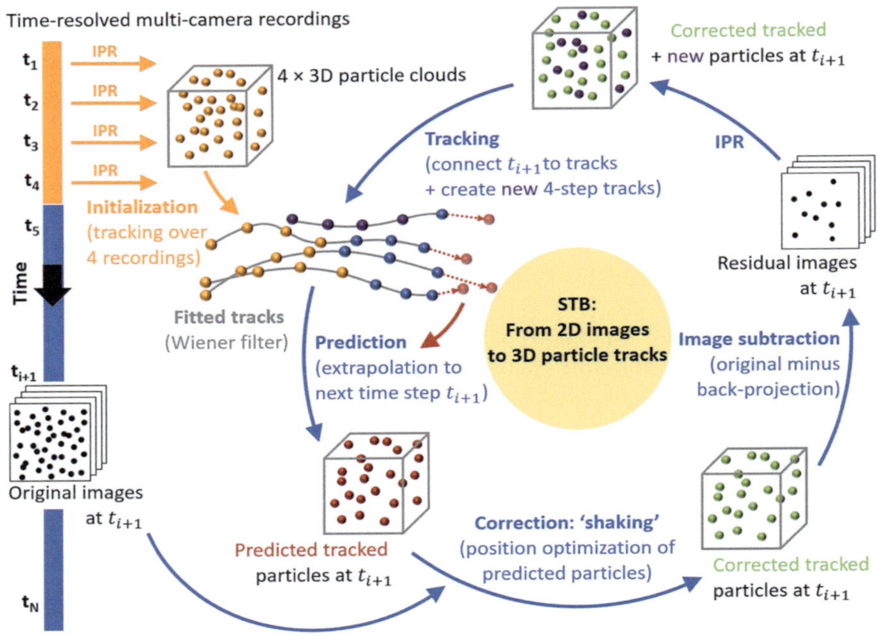

Fig. 4 Scheme of the Shake-The-Box (STB) algorithm. Reproduced from [41] (CC BY-NC 4.0)

iterations and therefore of processing time, as well as significant gains in accuracy due to an efficient suppression of ghost particles (which are rarely temporally coherent). As soon as a sufficient percentage of tracks can be identified, the process is self-stabilizing; one enhancement in a reconstruction property (position accuracy, ghost particle ratio, percentage of found tracks) induces immediate improvements in the others. The STB scheme is depicted in Fig. 4 and described below.

Assuming that the trajectories of (nearly) all particles within the system are known for a certain number of time-steps t_n, the STB-method scheme for the single time-step t_{n+1} is as follows:

1. Perform a temporal fit to the last k positions of tracked particles using an optimal Wiener filter based on an analysis of the track position spectrum [11].
2. Predict the position of the particle in t_{n+1} by evaluating the Wiener filter coefficients with a distribution of weights according to an optimized prediction accuracy.
3. Shake the particles to their correct position and intensity, eliminating the prediction error. This is achieved using an image matching scheme.
4. Find new particles entering the measurement domain on the residual images by IPR.
5. Shake all particles again to correct for residual errors.

6. Remove particles if they leave the volume or if their intensity falls below a certain threshold.
7. Iterate steps 4, 5 and 6, if necessary.
8. Add new tracks for all new particles identified within four consecutive time-steps.

After such processing of a single time-step within the image series, the known particle tracks have been accurately extended to the current time-step and new particle tracks have been added, capturing particles entering the volume. The entirety of these tracks can now be predicted for t_{n+2} and the process starts anew. In this way, STB can work its way through an entire time-series, consisting of possibly thousands of images. The effort needed for every single time-step is low, as the system has been largely pre-solved after the prediction-step, whereafter only minor deviations have to be corrected and a few newly entering particles have to be triangulated.

4.1 Initialization Phase

However, as the knowledge of a vast majority of particle tracks is not initially given (at the beginning the method has to start from scratch), the evaluation of a dataset has to converge to such a stable solution. The progress of the algorithm can be described in three main phases: *Initialization* (trying to find as many particle tracks as possible within a few time-steps provided by IPR); *Convergence* (the complexity of the reconstruction problem is gradually reduced by identifying more and more true tracks); and the *Converged State* (all true tracks are known, the number of newly found particles is balanced by the number of particles leaving the volume). For the sake of simplicity of the method description, the converged phase will be considered first here; this is typically reached after a few time steps by building upon a high quality of the advanced IPR reconstruction and the selective nature of temporally coherent true tracks as against incoherent ones arising from arbitrarily generated ghost particles and tracks. For more details see [33].

4.2 Converged Phase

The algorithm needs only a few time-steps to converge to a stable state, where the number of tracked particles does not change significantly. In this stage, the vast majority of the particles is known and currently tracked. From there on, most tracks end only when the corresponding particles leave the measurement volume and new tracks are found when particles enter the interrogation volume. The general processing remains the same as in the convergence phase.

Figure 5 illustrates the main steps of Shake-The-Box for a single time-step in the converged state and their impact on the residual image of a selected camera. At the beginning of the processing of this time-step, the residual image is the same as the

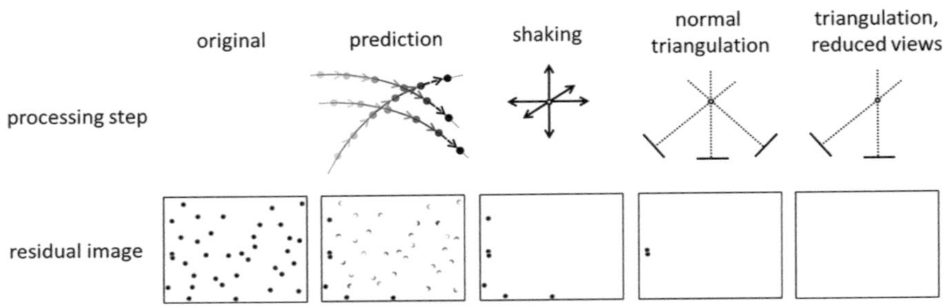

Fig. 5 Schematic description of the Shake-The-Box procedure for one time-step in the converged state by illustrating the effects of the different computational steps on the residual image of one single camera (out of several). Image reproduced from [33] (CC BY-NC 4.0)

recorded camera image. After predicting the positions of the tracked particles, residuals reflecting the errors of the prediction become visible. New particles, entering the measurement domain from the left and from below appear unaltered in the residual images. After performing some iterations of shaking, the residuals of the tracked particles (nearly) completely vanish—only the new particles remain. These are then tackled by the triangulation/IPR process. Due to the low effective particle image density at this point, only particles with overlapping images remain undetected by the triangulation. Most of these situations can be resolved by successively leaving out single cameras during the triangulation. The end result is a (nearly) completely blank residual image.

Following the example given in Fig. 5, it can be seen how much the prediction step—with its inclusion of temporal information—simplifies the problem of particle position detection. At the point where the first triangulations are performed, the perceived particle image density has drastically decreased, enabling a fast and reliable determination of the previously undetected particles. In the converged state, these are mostly new particles that have entered the measurement domain within the last four time-steps. However, the track of a particle can also be lost. Such events are usually caused by overlapping particle images in more than one camera image. The involved particles can be pulled to wrong locations during the shaking process. In this case the prediction for the next time-step will be compromised and the particle will likely get deleted due to its low intensity. Such particles will show up on the residual images of the next time-steps so that the track of the particle will have to be picked up again.

A complete elimination of residuals will only be seen when using synthetic data. With experimental data, the intensities of the images of a single particle on the different cameras will not be as balanced as in a synthetic case. Even a thorough calibration of the OTF will not be able to fully compensate for such effects, as particles do not behave similarly. Polydisperse particles have different scattering properties, which will lead to varying intensities in the different cameras; the same is true for 'potato'-shaped particles (e.g. polyamide seeding particles), which rotate and, depending on orientation, scatter differently. However, even if the residuals of such particles do not vanish completely in all camera images, they will do so for at least some of them (this is because the particle intensity is an average over all the image intensities), therefore preventing the particle from being picked up again by the next triangulation.

4.3 Multi-pass Processing

Even though the tracking process of STB is quite robust compared to traditional 3D-PTV methods, interrupted or incomplete tracks may still occur. Especially in the first time-steps (relative to the Initialization and Convergence phases) where not all particle tracks have yet been identified, there will be many missing particles. One easy method to improve on this situation is to perform a second pass of STB, running temporally backwards through the dataset. In this way, track fragments might be connected and particle tracks can be extended backwards to the time-step when the particle first occurred within the measurement volume. To this end, the tracks identified by the first pass are temporally filtered. For each subsequent time-step, the filtered particles from the previous pass are taken as a predefined particle distribution. In addition, new particle candidates are triangulated, enabling the search for new tracks. If a track ends (i.e. its starting point in the first pass is reached), the track is extended backwards in time by the usual prediction scheme as long as the particle stays within the measurement domain or is lost due to either the intensity-threshold or the other validations discussed above; more passes can be performed, if necessary.

4.4 Variable-Time-Step STB (VT-STB)

A recent development of STB is the variable-time-step (VT-) STB method [36] which solves the problem of ghosts which appear in flows with a very high dynamic velocity range. When tracking particles in flows with almost quiescent (or negligible velocity gradient) regions (e.g. around jets or in recirculation bubbles of separated flows, etc.), overlapping particle images along several camera LOS can be maintained over many time-steps so that the respective ambiguities of the triangulation-, tracking- and correction-process of STB could lead to ghost particle tracks. VT-STB provides a solution for this specific problem by iteratively reducing the time separation of the STB scheme for the first track particles having a larger time-separation (e.g. by using every 15th image) within a time series, which thereby guarantees a sufficiently large particle image shift to accurately distinguish between individual particles. The tracks of the first iteration are stored and used as input for the next STB iteration with a shorter time separation (e.g. by using every 5th image) and so on down to the full temporal resolution, enabling a reconstruction of the fastest tracks in the measurement volume.

The following section will show the application of these developed techniques to experimental data gained from a multi-camera investigation of a quadcopter drone in several flight conditions.

5 Large-Scale Experimental Application

STB has been applied to a large variety of flows and measurement domains, both in air and in water. These include: turbulent boundary layer flows which have been investigated on large scales [5], in the near-wall regions [40, 43] and by wall-normal profiles [46]; water flow around a submerged cube for turbulent and laminar boundary layers [47]; Rayleigh-Bénard convection at high spatial and temporal resolution in air [3, 12] and in water [49]. The developments with Helium-filled soap bubbles (HFSB) as air-flow tracers [2] combined with recent advances in LED illumination allowed the upscaling of the measurement technology to larger scales: The wake of a model helicopter blade could be studied, giving first experimental evidence of secondary vortices [53]. In thermal flows, volumes up to 12 m^3 have been reached by now [42].

Building upon the work in [53], here presented is a time-resolved experimental investigation of a quadcopter wake in volumes of up to 1.85 m^3, thus comprising the full volume around the used models, which are a DJI Mavic 2 pro quadcopter drone (of 56 cm diameter with a blade radius of 11 cm, see Fig. 6c) and a model helicopter with 45 cm blade length (not shown here). To this end, a large test room of 6 × 4

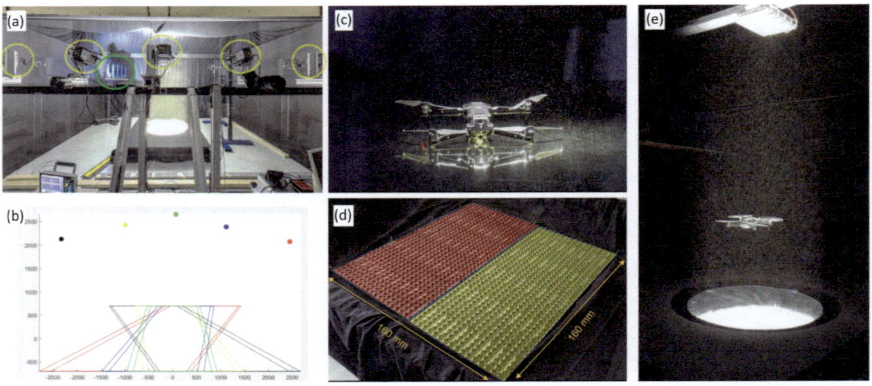

Fig. 6 Experimental setup. **a** 82 m^3 test room with circular floor LED array, HFSB seeding rake (green) and five cameras (yellow) installed behind a polycarbonate glass wall; **b** top view of calibrated camera system, indicating common illuminated volume of ~1.85 m^3; **c** photograph of the DJI Mavic 2 pro quadcopter; **d** the floor LED array without circular mask, with front array in red, back array in yellow; **e** the drone in hovering flight, illuminated by floor and ceiling LEDs and surrounded by HFSB bubbles

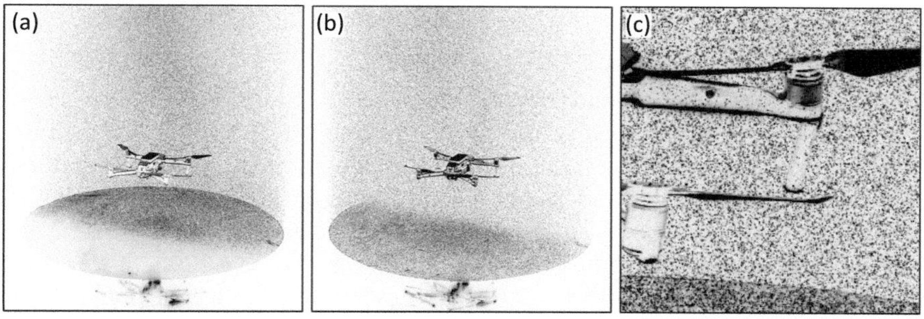

Fig. 7 An example of b/w-inverted images from a selected camera. **a** 'Back' part illuminated by triggering the far half of the LED array in relation to the camera system; **b** 'front' part illuminated by triggering the close half of the LED array in relation to the camera system (recorded with a phase shift relative to **a**) of $\Delta_t/2 = 1/6{,}000$ s); **c** detail view from **a**

$\times$ 3.4 = 82 m^3 was prepared with blackened walls on three sides and a tent-like ceiling (~4 m crest height), constructed from black cloth (see Fig. 6a). In this room, a quadcopter drone was operated in several flight conditions (hovering and forward flight in varying heights above the ground, as well as take-off and landing scenarios). The room was filled with HFSB of approx. 450 μm diameter prior to a recording, which were illuminated by a large array of white Luminous LEDs on the floor (1.6 $\times$ 1.6 m^2 with a round aperture of 1.25 m diameter) and a smaller one on the ceiling to illuminate the region in the shadow of the flight object when illuminated from below (see Fig. 6d, e). The LEDs were operated in pulsed mode with a pulse width of 50 μs.

The camera system consisted of four Vision Research Phantom *v2640* and one *v1840* cameras, capable of recording 16,870 consecutive images per run at 4 MPix resolution. The three central cameras were equipped with 50 mm Zeiss Planar lenses, while the two outer cameras used f = 60 mm Zeiss Makro-Planar lenses. To enable a two-plane scanning, the triggering of the LEDs was separated in a way that the front and back part of the array (with respect to the camera) were each triggered at 3 kHz, offset by a delay of 1/6,000 s (~0.166 ms) between the two LED sub-arrays. The camera system is operated at a recording frequency of 6 kHz, therefore recording alternating images with illuminations of the front and the back part of the volume. By extracting every even or odd image, the time-series of each sub-volume can be isolated.

Figure 7 shows two consecutive images captured with one of the measurement cameras for the drone in hovering flight in ground effect. Seeding densities of more than 0.1 ppp per subvolume could be achieved (see detail view in Fig. 7c). In this way the camera system was able to capture both time-series for 8,435 consecutive images at 3 kHz within the time-series of 16,870 images recorded at 6 kHz.

5.1 Data Evaluation

The camera system was calibrated by placing a large calibration target ($2.5 \times 1.25\,\text{m}^2$) in the center of the illuminated volume and traversing it $\pm 500\,\text{mm}$ within the main viewing direction of the cameras (here always the y-coordinate). From the recordings of the target at different volume depths, a two-plane calibration model (Wei and Ma 1991) could be determined for each of the cameras. The geometrical calibration was further refined using Volume-Self-Calibration (VSC [51]) and a 2D-field correction of the residual calibration errors that cannot be described by the second order camera model [38]. The particle imaging properties are characterized by a calibration of the OTF.

Preprocessing of the images was optimized to remove as much of the imaged quadcopter as possible, while still retaining peaks from particles between it and the camera. The first step is to subtract the minimum image of each run. Secondly, a mask is created based on the intensity of a Gaussian smoothing of the image. Within the mask, the intensity of the Gaussian smoothed image is subtracted from the original image. This procedure restricts the region of influence of the filter to connected bright regions (only the reflecting parts of the drone), while retaining the peaks visible on top of the bright areas by subtracting a local average intensity. Figure 8 demonstrates the effects of image preprocessing for a single camera.

The preprocessed images were then further processed using the DLR implementation of the Shake-The-Box (STB) method. In the current state, the two time-series from front and back illumination were reconstructed independent of each other. For each of the time-series, a separate tracking was performed using the advanced Variable-Timestep STB processing. The evaluation consists of four passes with three different time-separations. The first pass uses a time-separation of $\Delta_S = 15$ time-steps and tracks the particles in the bulk. The second pass, going backwards in time uses $\Delta_S = 8$ time-steps and adds some faster particles to the tracking system. The third pass, going again forwards in time, uses $\Delta_S = 1$ time-step (therefore using all available images) and adds the missing fast particles. In the end a fourth pass with $\Delta_S = 1$—but walking backwards in time—is performed to retrace all tracked particles to their first appearance in the measurement domain.

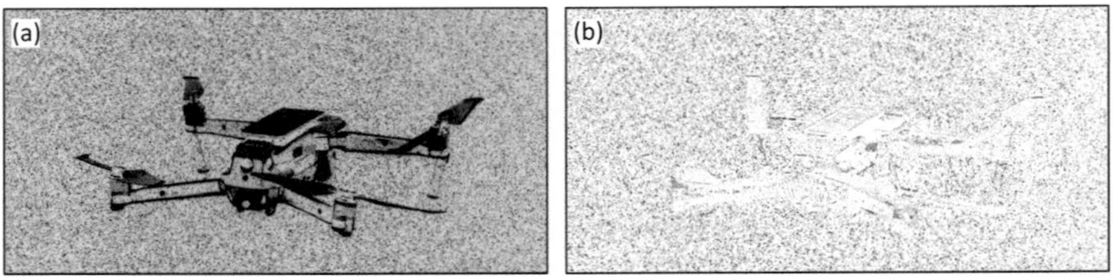

Fig. 8 Image preprocessing for removal of background and the bright regions of the imaged quadcopter: **a** original camera image **b** preprocessed image

5.2 Results and Discussion

Evaluations of the case with the drone in hovering flight close to the ground shown in Fig. 7 yield approx. 354,000 instantaneously tracked particles for the 'back' time-series and approx. 338,000 for the 'front' time-series. As can be seen from Fig. 7, the full resolution of 4 MPx could not be used in this case, since an image area of only approximately 2.8 MPx is actually filled with particle images. Each of the time-series therefore operates with an effective particle image density of around 0.125 ppp. Figure 9a, b show the velocity vectors for the two time-series of a singular time-step. It can be seen that due to the use of LEDs that emit light with an opening angle of around 7°, the two sub-volumes overlap to a certain extent and the particles in the middle of the volume are reconstructed in both time-series. The perceived, or effective, seeding density from the combined sub-volumes is estimated to be around 0.2 ppp. The use of LEDs with a narrower opening angle or of a scanning laser light

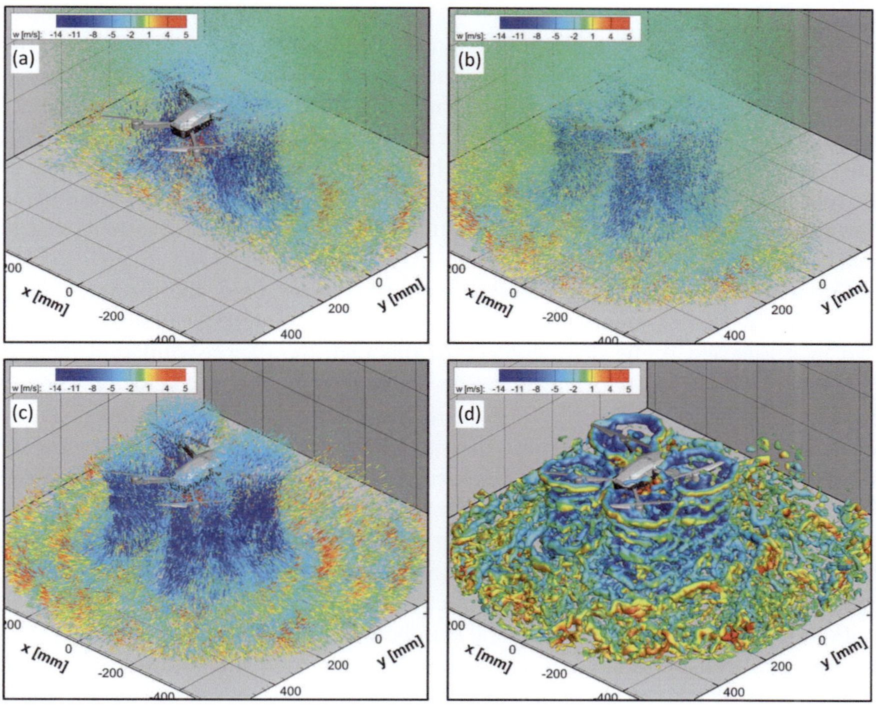

Fig. 9 Experimental results of the scanning LPT experiment for hovering flight with ground effect. **a** Velocity vectors of a single time-step for the 'back' part of the volume; **b** the same for the 'front' part; **c** combination of the two parts for the same time-step. Particles with a velocity lower than 3 m/s have been blanked out; **d** FlowFit data assimilation result of the same time step, showing iso-surfaces of Q-criterion ($Q = 20{,}000/s^2$). Note that the phase of the rotors of the rendered Quadcopter is not synchronized to the measurement

sheet would allow for less overlap, thereby increasing the effective seeding density and with it the value of particles per volume (ppv).

The overlap region in this experiment coincides with the most interesting region of the quadcopter wake, where strong accelerations due to the interaction of the four wakes occur. By modifying the processing to operate in a way that integrates the two time-series to a single evaluation, the particles in the overlap region can be tracked with a temporal resolution of 6 kHz for an enhanced capture of high-acceleration events. To this end, the STB scheme will have to be modified in such a way that it can cope with the majority of particles being visible in only every second image. Currently, tracks will be terminated if the intensity of the tracked particle falls below a certain threshold. The algorithm has to be informed about which particles are visible at what time. To this end, the volumetric light intensity of each sub-volume will be determined and used as a weighting factor. Based on this factor, the algorithm can decide whether a particle is visible for the current time-step or not. If it is not regarded as visible, its (predicted) position will not be altered by shaking, nor will it be deleted due to the intensity threshold. As a result of these modifications, the Lagrangian tracks of the HFSB will extend over both sub-volumes at double the recording frequency of each time-series (here 6 kHz). Once implemented, the procedure can easily be adapted to cope with more than two scanned (sub-) volumes.

Figure 9c visually combines the two individual time-series by showing velocity vectors from both reconstructions for three consecutive time-steps. For better clarity, only particles with a velocity magnitude larger than 3 m/s are shown. The wake of the four rotors is clearly evident: The fast flow impacts onto the ground plate and the flow is diverted and spreads circularly from the central point. Vortical structures can also be identified in the particle field.

For a better understanding of the instantaneous flow field, the discrete values of velocity and acceleration at the particle locations are interpolated to a Cartesian grid in a Eulerian frame of reference. To this end, the data assimilation method FlowFit [11] is applied, which reconstructs continuous functions of velocity, velocity gradients and pressure of single time-steps. It applies physical constraints given by the governing equations to increase the spatial resolution beyond the sampling by the particles. Figure 9d shows the result of applying FlowFit to the particles from Fig. 9a–c. The individual wakes of the four rotors emerge as spirals of vortices, being transported downwards towards to the floor, where they decompose into less coherent structures. Beneath the center of the quadcopter, a region with upward flow is visible, documenting a so-called fountaining effect.

For these initial results, the temporal shift between the two sub-volumes was simply ignored. In the overlap region the particles from both sub-volumes were used jointly, resulting in a double representation of the particle tracks in this region. However, no apparent effects or artifacts through using these measures are visible with the instantaneous FlowFit results. Nevertheless, the planned integrated evaluation (see previous paragraph) will ensure that each particle is represented only once at a harmonized timing.

Figure 10 displays results from an experiment of forward flight in ground effect (height of the quadcopter above ground 0.55 m, which corresponds to approximately

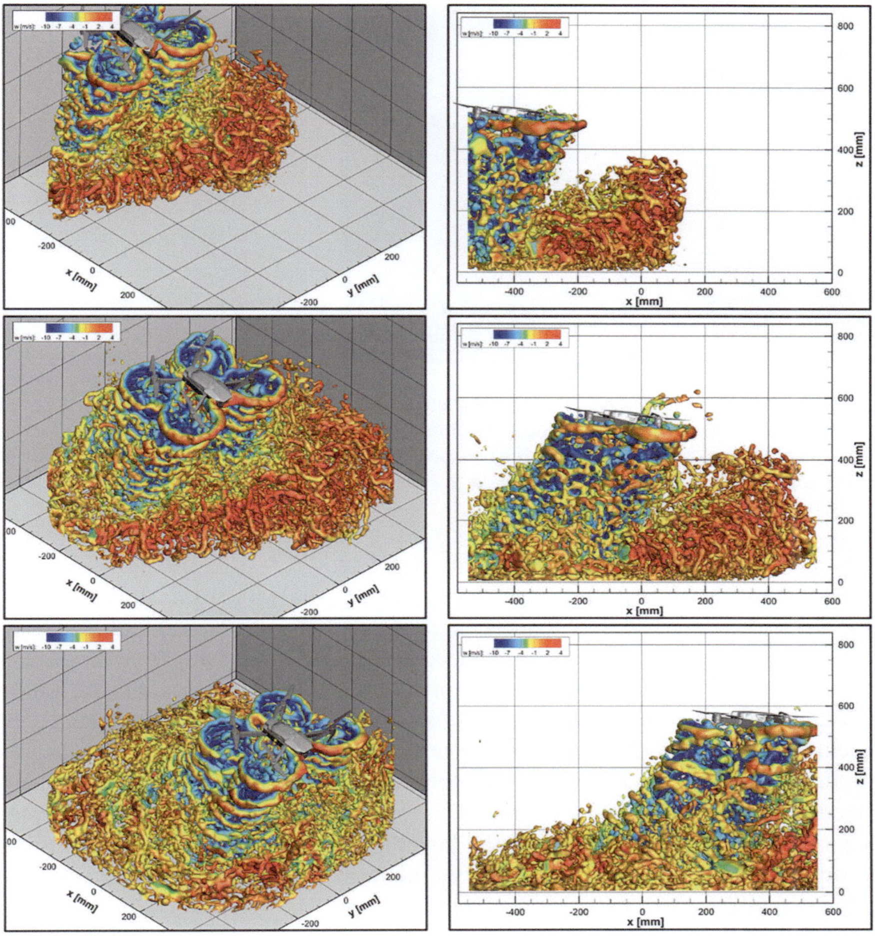

Fig. 10 Isosurfaces of Q-criterion ($Q = 5{,}000/s^2$) of a forward flight maneuver in ground effect at three subsequent times viewed from an inclined and a side view (left and right column). Time separation between the frames: 500 time-steps (166.6 ms). Results gained by FlowFit interpolation of the tracks from scanning STB-evaluation (~300,000 tracked particles per subvolume). The drone has been rendered for visualization purposes only. Neither position nor phase have been adapted to the data

one drone diameter). Three different points in time, separated by 166.6 ms, are shown to provide a better understanding of the temporal dynamics. When entering the measurement domain, the quadcopter pushes a strong frontal bow wave of small structures, while the main wake of the rotors is inclined in flight direction. The bow wake shows an asymmetry which might be caused by history effects. Effects of brown-out (situations where the vision of a pilot is blocked by dust or dirt that is stirred up by the wake of an aircraft) are visible (see middle row), where vortical

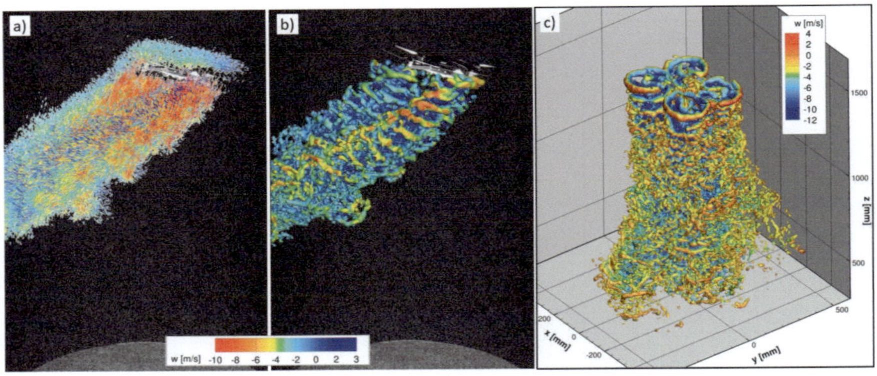

Fig. 11 Snapshots from evaluations of forward and hovering flight out of ground effect. **a** Forward flight: Particles tracked by STB (velocity magnitudes < 3 m/s have been blanked out), overlaid on a camera image of the quadcopter; **b** Isosurfaces of Q-criterion (Q = 15,000/s^2) from FlowFit processing of the same time-step of image **a**; **c** Hovering flight: Isosurfaces of Q-criterion (Q = 10,000/s^2)

structures of the bow wave pile up and are eventually drawn towards the rotors. The rear wake of the quadcopter remains comparatively flat and attached to the floor.

Examples for test cases with the quadcopter located further away from the ground are provided in Fig. 11 for hovering and forward flight. In these cases, a much longer development of the rotor wakes can be observed. While the wake structures are quickly interacting in the central parts beneath the quadcopter, long-lived structures can be identified in the outer regions, especially for the forward flight case, where coherent structures created by the rotors can still be seen more than one quadcopter diameter away from the drone. In the hovering case these structures seem to be decomposed more quickly into secondary structures, likely due to the more stationary turbulent field.

6 Fields of Application for Lagrangian Particle Tracking

The example described in Sect. 5 demonstrates the abilities of STB in a large-scale experiment. It has been shown in a series of measurements [4, 13, 15, 42, 53] that the combination of HFSB and LED illumination offers great potential for particle tracking experiments, because the LED illumination with its low coherent length (as opposed to laser light) avoids any artifacts such as speckle or interference which arise from this coherence. This feature, combined with the quasi transparent nature of the bubbles, leads to a nearly perfect summation of overlapping particle images, resembling the one seen in synthetic experiments. When using laser-based illumination, the light scattered from two or more particles located along closely neighboring lines-of-sight of a camera would interfere with each other. The resulting multi-peak will not be the sum of two Gaussian blobs, but it will be deformed according to the interference effects, which is detrimental to the precise positioning using the shaking-procedure, which assumes a particle image according to the OTF calibration. Despite these shortcomings when using illumination with coherent light, STB can be effectively used in such scenarios, though the seeding density is typically more limited ($\leq$0.1 ppp). Examples of dense tracking of oil-droplets in air include the near-wall region of turbulent boundary layers [43, 45] and subsonic jets using Multi-Pulse STB [23]. Examples for laser-based experiments in water are the flow around a surface-mounted cube [47] or within homogenous turbulence [44]. However, LED illumination is not just limited to HFSB seeding. Due to the comparable density of water to many plastics (e.g. polyamide), relatively large seeding particles (20–100 μm) can be applied to water experiments, which scatter enough light when illuminated by LEDs. Examples are the measurement of a jet impinging on a cylinder [18] or the simultaneous measurement of wing deformation and the surrounding 3D flow field [35]. These experiments also exhibit an image quality that is well suited for particle tracking, which can be even further enhanced by using fluorescent particles. The use of suitable filters on the cameras eliminates background reflections, as shown, for example, in measuring within bubble plumes [17] or Rayleigh-Bénard convection cells [49].

It can be stated that at the current level of development, LPT can be applied to a wide range of flow scenarios in water and air in volumes ranging from few cubic centimeters to several cubic meters. Care has to be taken with the choice of the seeding material, which should both accurately follow the flow (low Stokes number) and scatter enough light to yield a sufficiently high intensity on the cameras. At the same time, the illumination has to be sufficiently strong to allow the use of a small aperture for the lenses (greater depth of field), which guarantees a sharp imaging of the particles over the whole measurement domain. This requirement can be difficult to fulfill for large volume depths. It is therefore recommended to position the cameras so they look as perpendicularly as possible to the leanest side of the illuminated volume. Restricting the volume depth, preferably by using a sharp edge, avoids illumination of out-of-focus particles.

The repetition rate of the system should be chosen in a way that the highest expected particle accelerations do not lead to mispredictions of more than the particle image diameter. In this case, the position of the predicted particles can typically be corrected by the shaking procedure and the track can be continued. As an easy rule of thumb, if the fastest particles shift by around 8–10 pixels, a reliable tracking of the vast majority of particles is likely. Installing the cameras with a large opening angle between the outermost cameras ($\sim 90°$) will guarantee equal position accuracy in all directions in space. A small opening angle will significantly increase the noise in the depth direction.

After the preceding results and discussion, the reader should now have become aware that a key to successful particle tracking experiments is a careful tuning of the many setup aspects. Optimizing camera positions, the particle imaging and the illumination will facilitate the tracking process itself, as well as the time spent in fine tuning the post-processing.

Abbreviations

BOS	Background oriented schlieren
HFSB	Helium filled soap bubbles
IPR	Iterative particle reconstruction
LED	Light emitting diode
LOS	Line of sight
LPT	Lagrangian particle tracking
MART	Multiplicative algebraic reconstruction technique
OTF	Optical transfer function
PIV	Particle imaging velocimetry
ppp	Particles per pixel
ppv	Particles per volume
PTV	Particle tracking velocimetry
px	Pixel
STB	Shake-The-Box
VSC	Volume self calibration
VT-STB	Variable timestep Shake-The-Box

Acknowledgements Parts of this work received funding from the Clean Sky 2 Joint Undertaking under the European Union's Horizon 2020 research and innovation program under grant agreement N° CS2-LPA-GAM-2020-2023-01.

References

1. Atkinson C, Soria J (2009) An efficient simultaneous reconstruction technique for tomographic particle image velocimetry. Exp Fluids 47:553–568. https://doi.org/10.1007/s00348-009-0728-0
2. Bosbach J, Kühn M, Wagner C (2009) Large scale particle image velocimetry with helium filled soap bubbles. Exp Fluids 46:539–547. https://doi.org/10.1007/s00348-008-0579-0
3. Bosbach J, Schanz D, Godbersen P, Schröder A (2019) Dense Lagrangian particle tracking of turbulent Rayleigh Bénard convection in a cylindrical sample using Shake-The-Box. In: 17th European turbulence conference (ETC 2019). Torino
4. Bosbach J, Schanz D, Godbersen P, Schröder A (2021) Spatially and temporally resolved measurements of turbulent Rayleigh-Bénard convection by Lagrangian particle tracking of long-lived helium-filled soap bubbles. In: 14th International symposium on particle image velocimetry, vol 1. https://doi.org/10.18409/ispiv.v1i1.208
5. Bross M, Schanz D, Novara M, Eich F, Schröder A, Kähler C (2023) Turbulent superstructure statistics in a turbulent boundary layer with pressure gradients. Eur J Mech B/Fluids 101:209–218. https://doi.org/10.1016/j.euromechflu.2023.05.010
6. Dabiri D, Pecora C (2019) Particle tracking velocimetry. IOP Publishing, pp 2053–2563. https://doi.org/10.1088/978-0-7503-2203-4

7. Discetti S, Coletti F (2018) Volumetric velocimetry for fluid flows. Meas Sci Technol 29. https://doi.org/10.1088/1361-6501/aaa571
8. Elsinga G, Scarano F, Wieneke B, Oudheusden B (2006) Tomographic particle image velocimetry. Exp Fluids 41:933–947. https://doi.org/10.1007/s00348-006-0212-z
9. Elsinga G, Westerweel J, Scarano F, Novara M (2011) On the velocity of ghost particles and the bias errors in Tomographic-PIV. Exp Fluids 50:825–838. https://doi.org/10.1007/s00348-010-0930-0
10. van Gent P, Michaelis D, Oudheusden B, Weiss PE, Kat R, Laskari A, Jeon Y, David L, Schanz D, Huhn F, Gesemann S, Novara M, McPhaden C, Neeteson N, Rival D, Schneiders J, Schrijer F (2017) Comparative assessment of pressure field reconstructions from particle image velocimetry measurements and Lagrangian particle tracking. Exp Fluids 58:33. https://doi.org/10.1007/s00348-017-2324-z
11. Gesemann S, Huhn F, Schanz D, Schröder A (2016) From noisy particle tracks to velocity, acceleration and pressure fields using B-splines and penalties. In: 18th International symposium on applications of laser and imaging techniques to fluid mechanics. Lisbon
12. Godbersen P, Bosbach J, Schanz D, Schröder A (2021) Beauty of turbulent convection: a particle tracking endeavor. Phys Rev Fluids 6:110509. https://doi.org/10.1103/PhysRevFluids.6.110509
13. Huhn F, Schanz D, Gesemann S, Dierksheide U, van de Meerendonk R, Schröder A (2017) Large-scale volumetric flow measurement in a pure thermal plume by dense tracking of helium-filled soap bubbles. Exp Fluids 58:116. https://doi.org/10.1007/s00348-017-2390-2
14. Huhn F, Schanz D, Gesemann S, Schröder A (2016) FFT integration of instantaneous 3D pressure gradient fields measured by Lagrangian particle tracking in turbulent flows. Exp Fluids 57. https://doi.org/10.1007/s00348-016-2236-3
15. Huhn F, Schanz D, Manovski P, Gesemann S, Schröder A (2018) Time-resolved large-scale volumetric pressure fields of an impinging jet from dense Lagrangian particle tracking. Exp Fluids 59:81. https://doi.org/10.1007/s00348-018-2533-0
16. Jahn T, Schanz D, Schröder A (2021) Advanced iterative particle reconstruction for Lagrangian particle tracking. Exp Fluids 62:179. https://doi.org/10.1007/s00348-021-03276-7
17. Kim D, Schanz D, Novara M, Seo H, Kim Y, Schröder A, Kim KC (2022) Experimental study of turbulent bubbly jet. Part 1. Simultaneous measurement of three-dimensional velocity fields of bubbles and water. J Fluid Mech 941:A42. https://doi.org/10.1017/jfm.2022.301
18. Kim M, Schanz D, Novara M, Godbersen P, Yeom E, Schröder A (2023) Experimental study on flow and turbulence characteristics of jet impinging on cylinder using three-dimensional Lagrangian particle tracking velocimetry. Sci Rep 13:10929. https://doi.org/10.1038/s41598-023-37970-9
19. Kähler C, Cierpka C, Scharnowski S (2012) On the uncertainty of digital PIV and PTV near walls. Exp Fluids 52:1641–1656. https://doi.org/10.1007/s00348-012-1307-3
20. Lynch KP, Scarano F (2015) An efficient and accurate approach to MTE-MART for time-resolved tomographic PIV. Exp Fluids 56:66
21. Maas HG, Gruen A, Papantoniou DA (1993) Particle tracking velocimetry in three-dimensional flows. Exp Fluids 15:133–146. https://doi.org/10.1007/BF00190953
22. Malik NA, Dracos T, Papantoniou DA (1993) Particle tracking velocimetry in three-dimensional flows. Exp Fluids 15:279–294. https://doi.org/10.1007/BF00223406
23. Manovski P, Novara M, Depuru Mohan NK, Geisler R, Schanz D, Agocs J, Godbersen P, Schröder A (2021) 3D Lagrangian particle tracking of a subsonic jet using multi-pulse Shake-The-Box. Exp Therm Fluid Sci 123:110346
24. Nayler JL, Frazer BA (1917–1918) Preliminary report upon an experimental method of investigating, by the aid of kinematographic photography, the history of eddying flow past a model immersed in water. Technical report of the Advisory Committee for Aeronautics
25. Nicolas F, Donjat D, Léon O, Le Besnerais G, Champagnat F, Micheli F (2017) 3D reconstruction of a compressible flow by synchronized multi-camera BOS. Exp Fluids 58. https://doi.org/10.1007/s00348-017-2325-y

26. Novara M, Batenburg KJ, Scarano F (2010) Motion tracking-enhanced MART for tomographic PIV. Meas Sci Technol 21:035401
27. Novara M, Schanz D, Geisler R, Gesemann S, Voss C, Schröder A (2019) Multi-exposed recordings for 3D Lagrangian particle tracking with multi-pulse Shake-The-Box. Exp Fluids 60:44. https://doi.org/10.1007/s00348-019-2692-7
28. Novara M, Schanz D, Reuther N, Kähler C, Schröder A (2016) Lagrangian 3D particle tracking in high-speed flows: Shake-The-Box for multi-pulse systems. Exp Fluids 57:128. https://doi.org/10.1007/s00348-016-2216-7
29. Oudheusden BWV (2013) PIV-based pressure measurement. Meas Sci Technol 24:032001. https://doi.org/10.1088/0957-0233/24/3/032001
30. Ouellette NT, Xu H, Bodenschatz E (2006) A quantitative study of three-dimensional Lagrangian particle tracking algorithms. Exp Fluids 40:301–313
31. Raffel M, Kähler CJ, Willert CE, Wereley ST, Scarano F, Kompenhans J (2018) Particle image velocimetry: a practical guide, 3rd edn. Springer. https://doi.org/10.1007/978-3-319-68852-7
32. Scarano F (2012) Tomographic PIV: Principles and practice. Meas Sci Technol 24:012001. https://doi.org/10.1088/0957-0233/24/1/012001
33. Schanz D, Gesemann S, Schröder A (2016) Shake-The-Box: Lagrangian particle tracking at high particle image densities. Exp Fluids 57:70. https://doi.org/10.1007/s00348-016-2157-1
34. Schanz D, Gesemann S, Schröder A, Wieneke B, Novara M (2012) Non-uniform optical transfer functions in particle imaging: calibration and application to tomographic reconstruction. Meas Scie Technol 24:024009. https://doi.org/10.1088/0957-0233/24/2/024009
35. Schanz D, Konrath R, Geisler R, Erdogdu A, Agocs J, Schröder A (2022) Full-volume investigation of moving rigid and flexible wings using Lagrangian particle tracking of surface markers and tracer particles. In: Proceedings of the 20th international symposium on the application of laser and imaging techniques to fluid mechanics. Lisbon
36. Schanz D, Novara M, Schröder A (2021) Shake-The-Box particle tracking with variable time-steps in flows with high velocity range (VT-STB). In: Christensen KT (ed) Proceedings of 14th international symposium on particle image velocimetry, vol 1. ILLINOIS Tech/Paul V. Galvin Library, Chicago, USA, pp 1 – 11
37. Schanz D, Schröder A, Gesemann S, Michaelis D, Wieneke B (2013) Shake The Box: a highly efficient and accurate Tomographic Particle Tracking Velocimetry (TOMO-PTV) method using prediction of particle positions. In: Proceedings of 10th international symposium on particle image velocimetry. Delft
38. Schanz D, Schröder A, Novara M, Geisler R, Agocs J, Kähler C (2019) Large-scale volumetric characterization of a turbulent boundary layer flow. In: Proceedings of the 13th international symposium on particle image velocimetry. Munich
39. Schneiders JFG, Scarano F (2016) Dense velocity reconstruction from tomographic PTV with material derivatives. Exp Fluids 57:151. https://doi.org/10.1007/s00348-016-2225-6
40. Schröder A, Schanz D, Geisler R, Novara M, Christian W (2015) Near-wall turbulence characterization using 4D-PTV "Shake-The-Box". In: 11th International symposium on particle image velocimetry—PIV2015
41. Schröder A, Schanz D (2023) 3D Lagrangian particle tracking in fluid mechanics. Annu Rev Fluid Mech 55:511–540. https://doi.org/10.1146/annurev-fluid-031822-041721
42. Schröder A, Schanz D, Bosbach J, Novara M, Geisler R, Agocs J, Kohl A (2022) Large-scale volumetric flow studies on transport of aerosol particles using a breathing human model with and without face protections. Phys Fluids 34:035133. https://doi.org/10.1063/5.0086383
43. Schröder A, Schanz D, Geisler R, Godbersen P, Agocs J, Simhan A (2024) Near-wall flow features in ZPG-TBL at various Reynolds numbers using dense 3D Lagrangian Particle Tracking. In: 21st International symposium on applied laser techniques to fluid mechanics. Lisbon
44. Schröder A, Schanz D, Gesemann S, Huhn F, Paz DG, Perez-Munuzuri V, Bodenschatz E (2019) Measuring the full velocity gradient and dissipation rate tensor in homogeneous turbulence using Shake-The-Box and FlowFit. In: 17th European turbulence conference 2019. Turin
45. Schröder A, Schanz D, Michaelis D, Cierpka C, Scharnowski S, Kähler C (2015) Advances of PIV and 4D-PTV "'Shake-The-Box" for turbulent flow analysis—the flow over periodic hills. Flow Turbul Combust 95:193–209. https://doi.org/10.1007/s10494-015-9616-2

46. Schröder A, Schanz D, Novara M, Philipp F, Geisler R, Knopp T, Schroll M, Willert C (2018) Investigation of a high Reynolds number turbulent boundary layer flow with adverse pressure gradients using PIV and 2D- and 3D-Shake-The-Box. In: 19th International symposium on the application of laser and imaging techniques to fluid mechanics. Lisbon
47. Schröder A, Willert C, Schanz D, Geisler R, Jahn T, Gallas Q, Leclaire B (2020) The flow around a surface mounted cube: a characterization by time-resolved PIV, 3D Shake-The-Box and LBM simulation. Exp Fluids 61. https://doi.org/10.1007/s00348-020-03014-5
48. Sciacchitano A, Leclaire B, Schröder A (2021) Main results of the first Lagrangian particle tracking challenge. In: 14th International symposium on particle image velocimetry, vol 1. https://doi.org/10.18409/ispiv.v1i1.197
49. Weiss S, Schanz D, Erdogdu AO, Schröder A, Bosbach J (2023) Investigation of turbulent superstructures in Rayleigh–Bénard convection by Lagrangian particle tracking of fluorescent microspheres. Exp Fluids 64:82. https://doi.org/10.1007/s00348-023-03624-9
50. Wieneke B (2001) Improvements for volume self-calibration. Meas Sci Technol 29:084002
51. Wieneke B (2008) Volume self-calibration for 3D particle image velocimetry. Exp Fluids 45:549–556
52. Wieneke B (2012) Iterative reconstruction of volumetric particle distribution. Meas Sci Technol 24:024008. https://doi.org/10.1088/0957-0233/24/2/024008
53. Wolf CC, Schwarz C, Kaufmann K, Gardner A, Michaelis D, Bosbach J, Schanz D, Schröder A (2019) Experimental study of secondary vortex structures in a rotor wake. Exp Fluids 60:175. https://doi.org/10.1007/s00348-019-2807-1
54. Xu H (2008) Tracking Lagrangian trajectories in position–velocity space. Meas Sci Technol 19(7):075105. https://doi.org/10.1088/0957-0233/19/7/075105

Index

F. Boden (ed.), *Numerical Methods and Experimental Means for the Integration of Future Propulsion Systems*, Springer Aerospace Technology,
https://doi.org/10.1007/978-3-031-97725-1

MIX
Papier aus verantwortungsvollen Quellen
Paper from responsible sources
FSC® C105338

If you have any concerns about our products,
you can contact us on
ProductSafety@springernature.com

In case Publisher is established outside the EU,
the EU authorized representative is:
Springer Nature Customer Service Center GmbH
Europaplatz 3, 69115 Heidelberg, Germany

Printed by Libri Plureos GmbH
in Hamburg, Germany